W9-BAA-014

Concise Encyclopedia of Geography

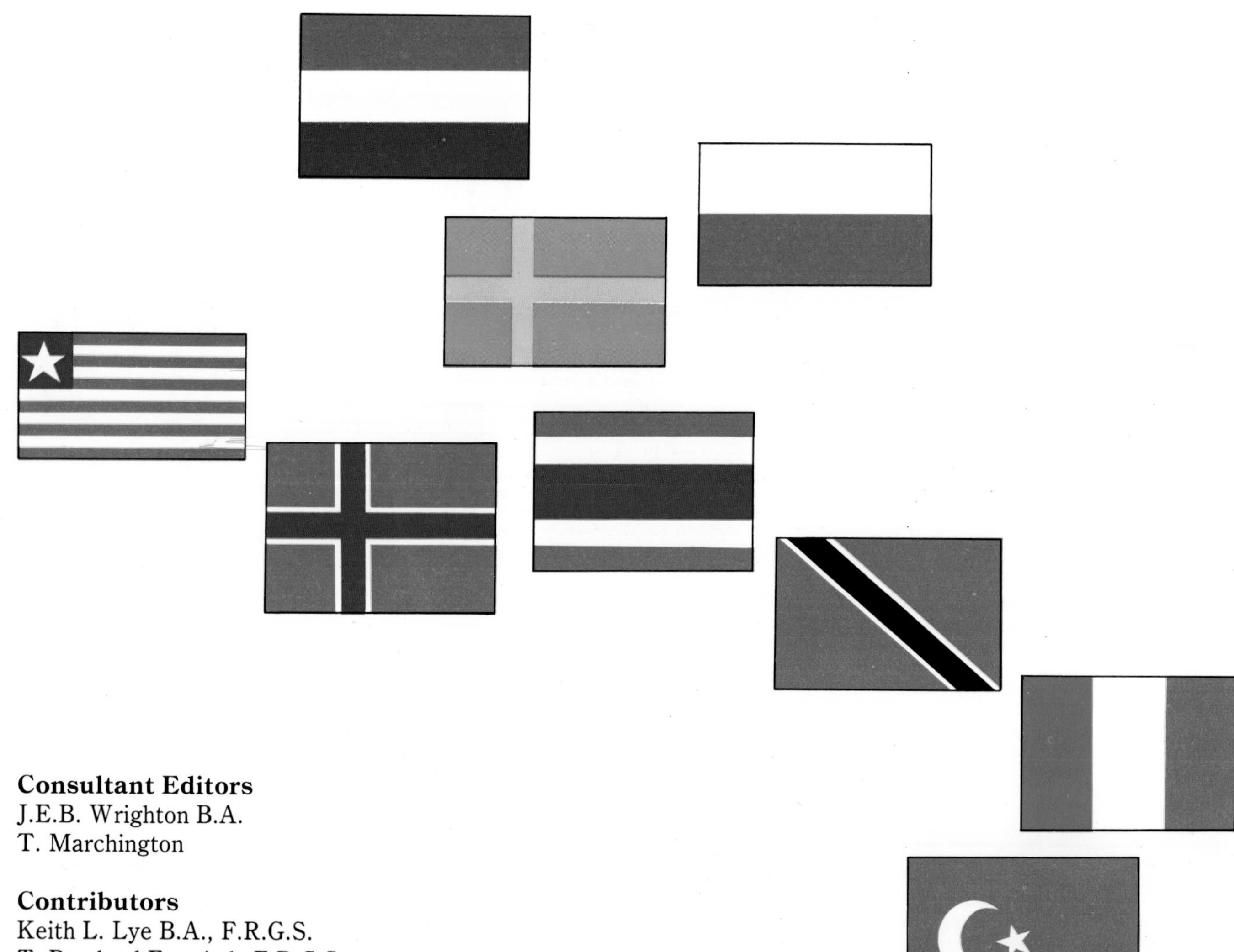

Consultant Editors
J.E.B. Wrighton B.A.
T. Marchington

Contributors
Keith L. Lye B.A., F.R.G.S.
T. Rowland-Entwistle F.R.G.S.
A.S. Butterfield
Diane James
Lionel Grigson B.A.
Marion Bloch Ph.D.

ISBN 0 361 06489 6

Published 1976 by Purnell Books,
Paulton, Bristol BS18 5LQ,
a member of the BPCC group.

Printed in Hong Kong

Reprinted 1978, 1979, 1980
New edition 1984

Concise Encyclopedia of Geography

C.J. Tunney M.A.

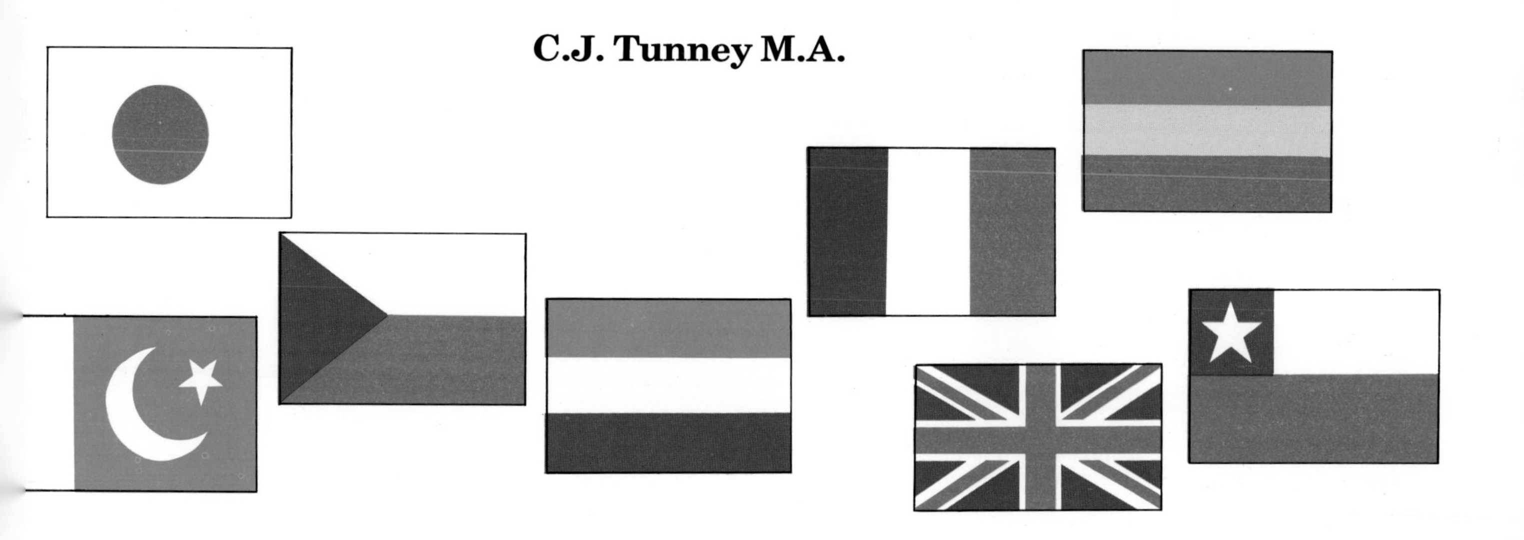

Contents

Our Varied World

A traveller from Outer Space studying our planet might be forgiven if he called it by his word for 'Ocean' or 'Water'. The large volume of water on Earth makes our planet unique in the Solar System. Some 70.8 per cent. of the Earth's surface is covered by five great, interconnected oceans—the Pacific, the Atlantic, the Indian, the Antarctic, and the Arctic. The oceans are important to life on Earth because their currents act like a thermostat, cooling some warm areas and warming some cold areas.

Land covers only 29.2 per cent. of the Earth's surface. A space traveller would be impressed by the variety of habitats on our planet. He would be struck by the fact that, while some plant life and some animal life are adapted to live almost everywhere, the distribution of people is very uneven. Most of the world's people live on about 10 per cent. of the Earth's land surface.

The continent of Antarctica is a vast ice sheet which is so inhospitable to people that they can survive there only if they import everything they need. In the Northern Hemisphere, around the polar region, there are vast tracts of almost uninhabited treeless plains, called *tundra*. To the south of the tundra lie cold, coniferous forest regions.

Other parts of the Earth are too hot and too dry for much human settlement. The world's largest desert, the Sahara, is populated by a small number of nomads. Its only farmers cluster around a few scattered oases.

However, in regions where it is not too hot, too cold, or too dry, Man has created great civilizations. His ingenuity has made it possible for farming and manufacturing industries to support the needs of the world's enormous and fast-growing population.

This success has resulted in many problems and a great challenge. The world's population growth will have to be matched by increased food production if the people are all to be adequately fed. Even now more than a third of the population does not enjoy a full, balanced diet. There are great contrasts in the standards of living of different peoples. Careless farming practice has resulted in the loss of invaluable top soil. The success of industry has resulted in large proportions of the Earth's minerals being almost exhausted and pollution in some areas is a serious problem. The human race has the technical ability to solve all of its problems. Will it continue to overcome them as in the past?

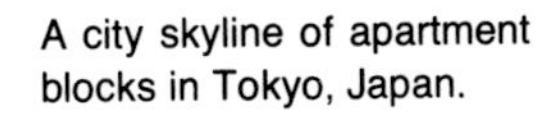
A city skyline of apartment blocks in Tokyo, Japan.

Old houses line a canal in Ghent, in Belgium.

A prosperous farm on its hillside in France.

A lonely mountain road in Peru, flanked by inhospitable snow-clad peaks.

A fishing village in the Philippines, built on stilts. Men have established settlements wherever they can manage to live.

The Earth in Space

Right: Polaris, the North Star, appears to us to be always in the same position because it is almost directly in line with the Earth's axis. But the Earth 'wobbles' as it rotates and its axis line describes a circle in the sky—taking 25,800 years to do so. Over thousands of years, various stars in turn become the North Star.

Below: The Solar System. Each of the nine planets moves round the Sun in an elliptical orbit. The Earth's orbit is completed in just over 365 days.

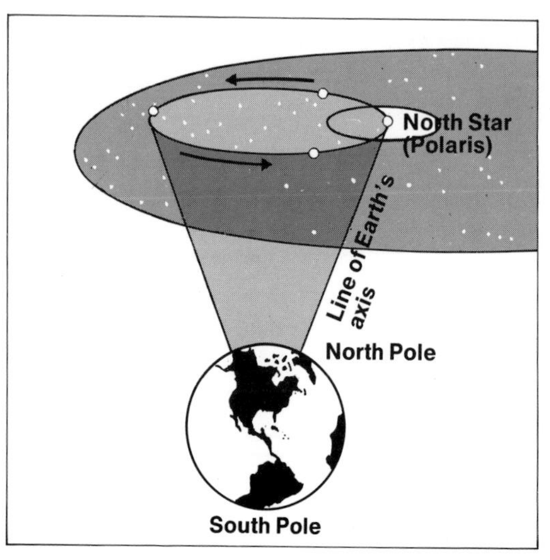

The nine planets of the Solar System are bound to the Sun by the force of gravity. The time taken for a planet to orbit once around the Sun is called the planet's 'year'. The orbits are elliptical in shape and lie roughly in the plane of the Sun's equator. Each of the planets also spins on its own axis and the time it takes to spin once is called the planet's 'day'. All the axes of the planets are tilted from the vertical.

In the Solar System, the Earth is the third planet from the Sun and it is the fifth largest of the planets. However, by comparison with the Sun, the Earth is very small. The Sun's mass is 330,000 times as great as that of the Earth.

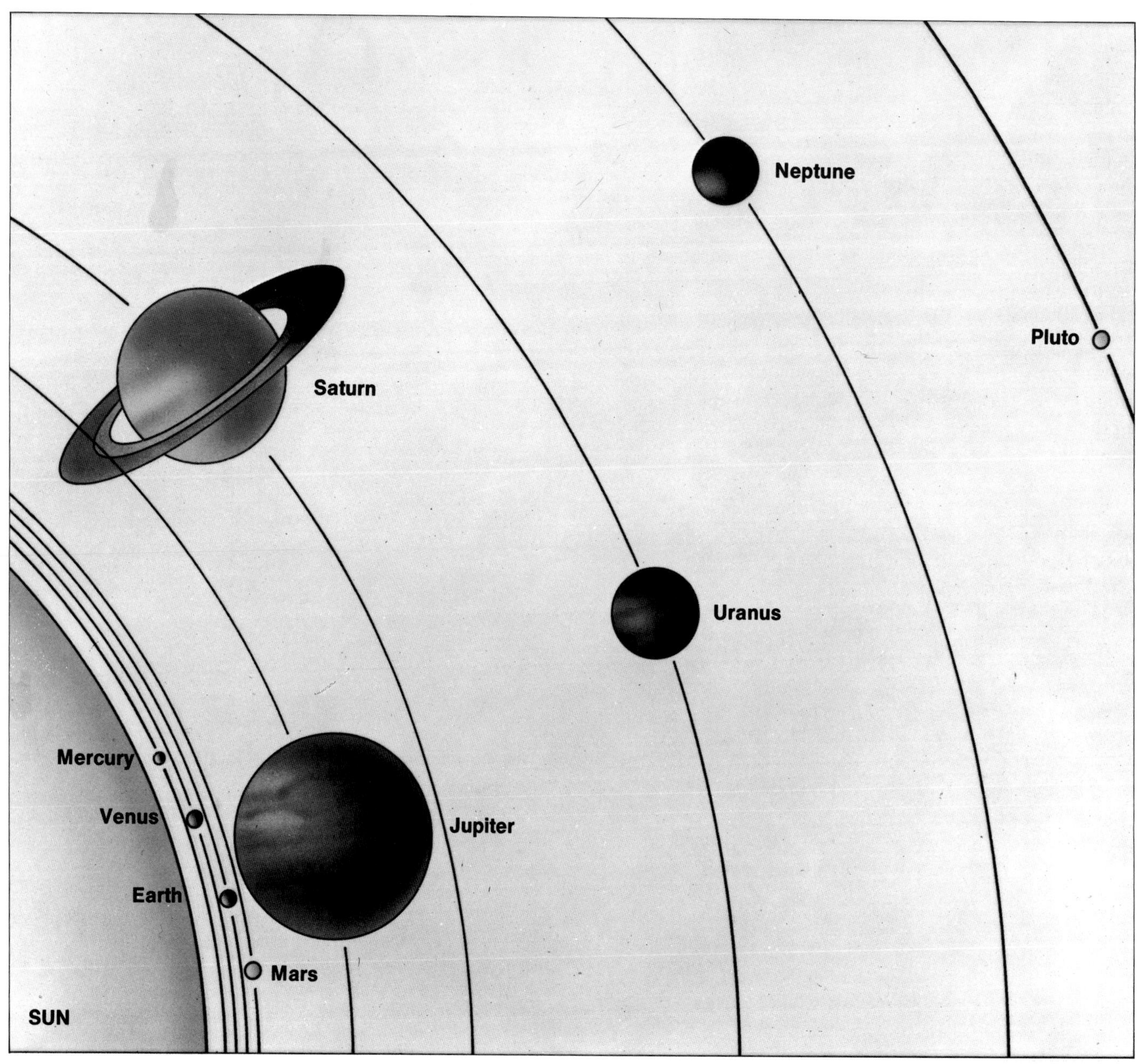

The Earth moves in three main ways. It rotates around the Sun once every 365 days, 5 hours, 48 minutes, and 46 seconds at a speed of 66,000 mph (106,200 kph). Second, the Earth spins on its axis at a speed of 1,038 mph (1,670 kph) at the equator. This speed decreases towards the poles. Like other planets, the Earth's axis is tilted. The tilt of the Earth's axis is 23½° from a line perpendicular to its path around the Sun. The axes of all planets are tilted at less than 30° from the vertical, apart from Uranus, whose axis is tilted at slightly over 90°.

Together with the rest of the Solar System, the Earth moves in a third way. It is moving around the Milky Way at a speed of 43,000 mph (69,200 kph). The Milky Way is a galaxy (star system) which contains about 100,000 million stars. One of them is the Sun in our Solar System. The Solar System takes about 200 million years to rotate once around the centre of the galaxy. Millions of galaxies form the Universe, but no one knows just how big the Universe is.

Like several other planets, the Earth has a satellite, the Moon. Recent studies of Moon rocks, obtained by Russian and American missions, have shown that the Moon is of the same age as the Earth (about 4.5 million years old). The Moon's diameter is 2,160 miles (3,476 km)—about one-fourth the size of the Earth. The Moon orbits the Earth at an average distance of 238,887 miles (384,400 km). The Moon rotates once on its axis and also completes one revolution around the Earth every 27⅓ days. As a result, from the Earth we always see the same hemisphere of the Moon.

Most scientists think that, about 5,000 million years ago, a large cloud of dust and gas collected. Part of this dust and gas concentrated into a dense core to form the Sun. The rest of the material rotated around the core. Smaller bodies formed in this rotating dust and gas and these bodies swept up more and more material. By 4,500 million years ago, these bodies had grown into the planets which form the Solar System.

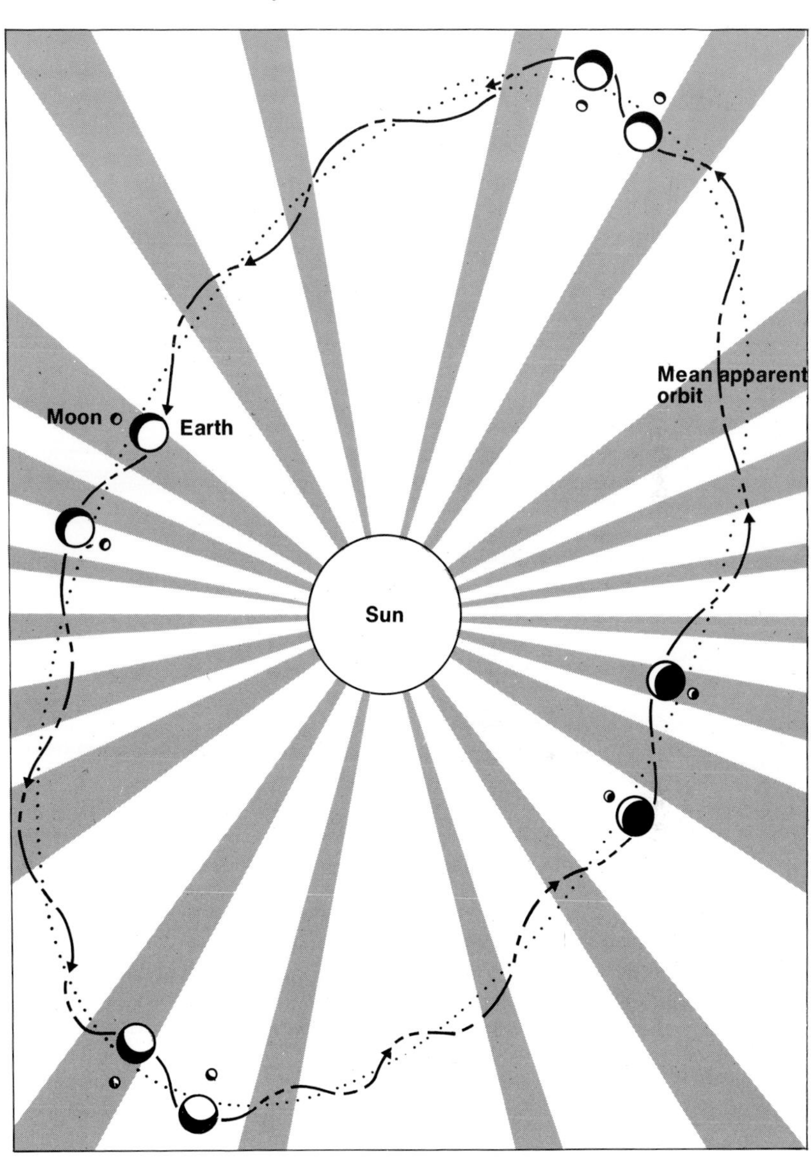

Earth and Moon Orbits. As the Earth and the Moon orbit the Sun, the Moon also revolves around the Earth, weaving in and out of the Earth's orbit. Only one side of the Moon is ever completely visible from the Earth because the Moon rotates once on its axis in the same time that it takes to revolve once around the Earth.

THE EARTH AND THE SOLAR SYSTEM

Heavenly body	*Diameter in miles (and kilometres)*		*Average distance from Sun in millions of miles (and kilometres)*		*Period of orbital revolution*
Mercury	3,100	(4,989)	36.0	(57.9)	88 days
Venus	7,700	(12,392)	67.2	(108.1)	225 days
Earth	7,917	(12,741)	92.9	(149.5)	365 days
Mars	4,200	(6,759)	141.5	(227.7)	687 days
Jupiter	88,700	(142,749)	483.3	(777.8)	11.9 years
Saturn	71,500	(115,068)	886.1	(1,426.0)	29.5 years
Uranus	30,900	(49,729)	1,783.0	(2,869.4)	84.0 years
Neptune	28,000	(45,062)	2,793.0	(4,494.9)	164.8 years
Pluto	3,100–3,600	(4,989–5,794)	3,666.0	(5.899.9)	249.0 years

The Changing Year

The Other Side of the World. Above: January in Europe—clearing a road in France. Centre: January in the South Pacific, with summer sunshine.

Below: The Seasons. The Earth is tilted on its axis as it orbits the Sun. Each Hemisphere receives most sunlight when tilted towards the Sun.

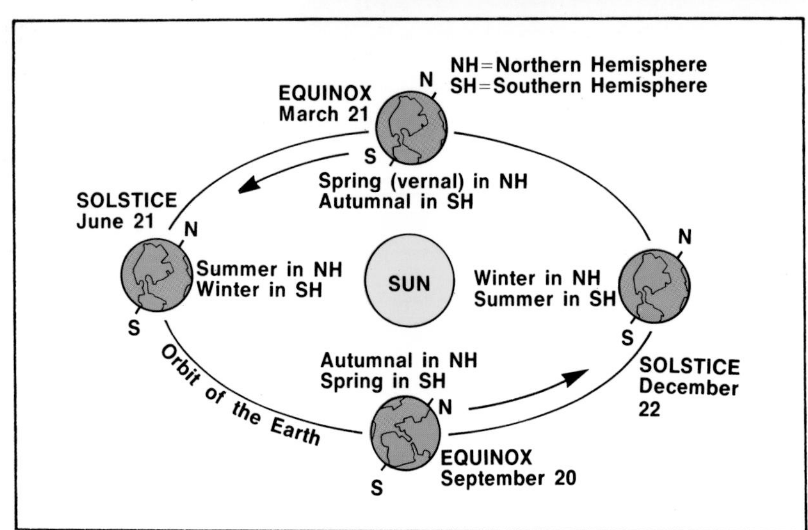

The Earth takes one *solar year* (365 days, 5 hours, 48 minutes, and 46 seconds) to complete one revolution round the Sun. Some early peoples, however, measured time in relation to the Moon. The month (or Moon-th) was the period taken by the Moon to go from one new Moon to the next. This period is 29½ days. This method of measuring caused much confusion because no number of 'Moon-ths' could be made to equal a solar year. Twelve months of 29½ days is 354 days and 13 months totals 383½ days.

In 1582, however, the Gregorian calendar was adopted, named after Pope Gregory XIII. This calendar is based on the idea of a seven-day week. A year of 52 such weeks totals 364 days, which is 1 day, 5 hours, 48 minutes, and 46 seconds less than a solar year. To eliminate the odd day, each year in the Gregorian calendar begins one day later in the week than the previous year. Leap years are fixed for every year that can be divided by four, such as 1976 and 1980. As a result, six hours are added on average to every year. This leaves a difference between the calendar and the solar year of 11 minutes and 14 seconds. Most of this error is eliminated by making century years that cannot be divided by 400—such as 1700, 1800, and 1900—*non*-leap years. But centuries, such as 2000, that can be divided by 400 are leap years. In this way, the difference between the Gregorian calendar and the solar year is reduced to 26.3 seconds. However, another error occurs because the solar year is slowly getting slightly shorter. This error amounts to 0.53 seconds every 100 years.

While the Earth rotates on its axis, half of the Earth is facing the Sun while the other half is facing away from the Sun. As a result, we experience day and night. The Earth takes 23 hours, 56 minutes and 4.1 seconds to spin round once on its axis. This period of time, measured in relation to the stars, is called the *sidereal day*.

Because the Earth is divided into 360° of longitude and it completes one spin on its axis in a day (24 hours), we can see that time alters by one hour for every 15° of longitude (360 ÷ 24). Each place on Earth has its own *local time*. This can be measured forwards or backwards from the moment when the Sun reaches its highest point in the sky, which is noon or midday.

The Earth is divided into *time zones*. The

simplest time zones would be based on dividing the globe into equal segments of 15° and changing the time by one hour from one zone (segment) to the next. Time zones have irregular boundaries that make allowance for political frontiers.

The prime meridian, or 0° longitude, from which the time zones are measured, passes through Greenwich, England. A person travelling west of Greenwich for 180° of longitude would have to put his clock back by 12 hours. If another traveller were going east of Greenwich for 180°, he would put his clock on by 12 hours. When they met on the *International Date Line,* there would be one day's difference between them. As a result, travellers crossing the International Date Line from east to west lose a day. Travellers crossing over the Date Line from west to east gain a day.

Hours of daylight vary because the Earth's axis is tilted. The North Pole is tilted towards the Sun in the northern summer and so it enjoys continuous daylight. At the same time, the South Pole tilts away from the Sun, and is in continuous darkness. The length of the day varies between the poles and the equator. The longest day in the Northern Hemisphere is usually June 21. The shortest is December 22.

The tilt of the Earth's axis is also responsible for seasons. Spring in the Northern Hemisphere begins on March 21, when the Sun is overhead at the equator. March 21 is called the *vernal equinox.* After March 21, the overhead Sun moves northwards until, on June 21, the *summer solstice,* it reaches its northernmost point. The overhead Sun then moves south back towards the equator, where it is again overhead on about September 20, the *autumnal equinox.* On December 22, the *winter solstice,* the overhead Sun reaches its southernmost point in the sky.

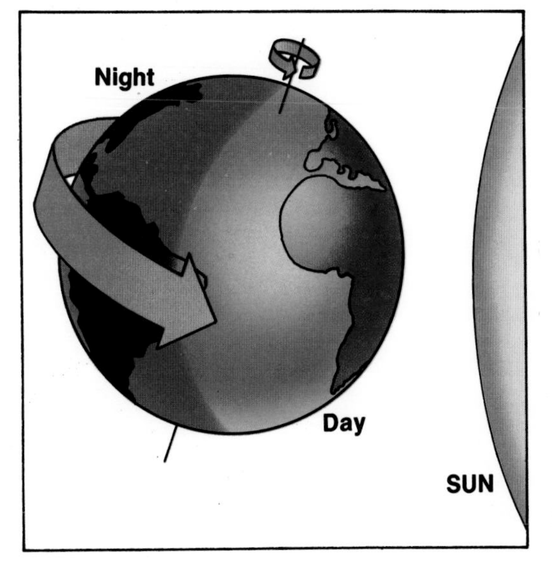

Left: Day and Night. While the Earth orbits the Sun, it also rotates on its axis. This rotation, towards the east, is responsible for day and night. Each point on the Earth's surface has 'day' when the Sun's rays can reach it, 'night' when it has spun 'round the back' out of reach.

Below: Standard Time Zones of the world showing how they differ from (fast or slow of) Greenwich Mean Time. The time in any particular zone is one hour earlier than that in the zone to its east. But at the International Date Line there is a variation of 24 hours.

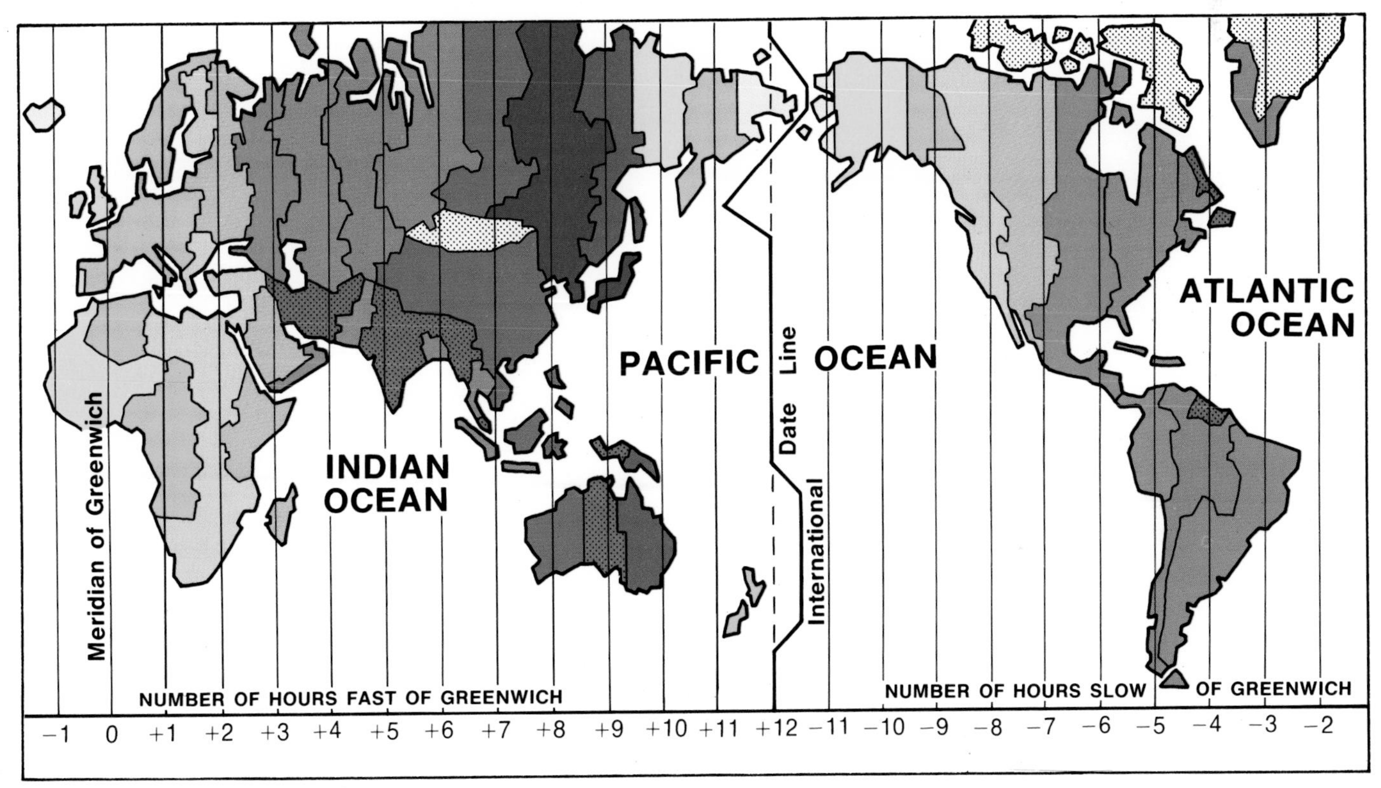

Weather and Climate

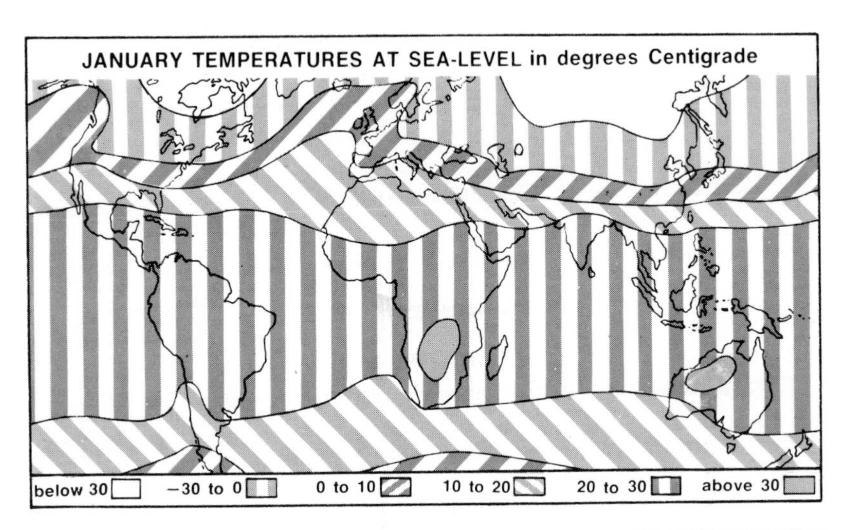

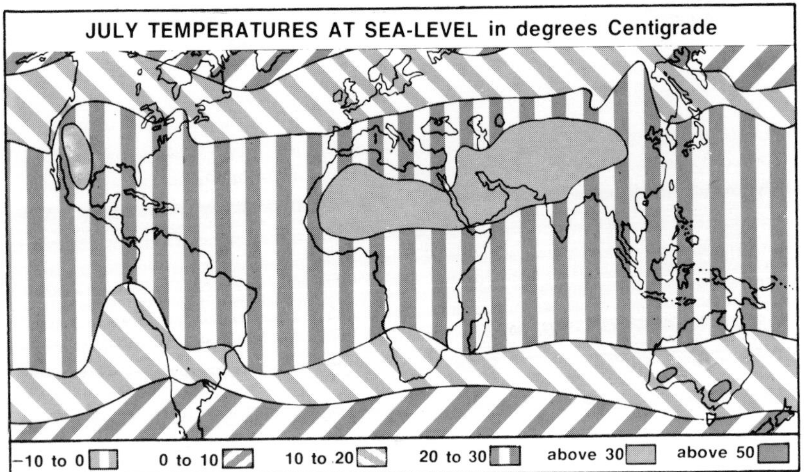

The climate of a country or other region is its average or typical weather over a period of time. Climate strongly affects our way of life. It influences, for example, the way we dress, the kinds of houses we live in, and the foods that farmers can grow. We can control the effects of climate in some ways. For example, central heating and air-conditioning make buildings more comfortable to live and work in. And scientists have succeeded in experiments to produce rain in dry regions by seeding clouds with certain chemicals.

Climate is determined by several factors, one of which is latitude. The Sun's heat is concentrated at the equator. But at the poles much of the heat is absorbed by the atmosphere, and heat that does reach the ground is spread over a large area.

Mount Kilimanjaro, Africa's highest mountain, lies almost on the equator in Tanzania. This mountain is capped by snow and ice throughout the year. The reason is that, as one ascends a hill or mountain, the temperature falls by about 3.5°F for every 1,000 ft (6°C per 1,000 m). Mountains also affect rainfall. Onshore winds, bearing much moisture evaporated from the sea, are chilled as they rise over coastal mountain ranges. As air cools, it loses its capacity to hold moisture, which falls as rain or snow.

Oceans and seas influence climate. For

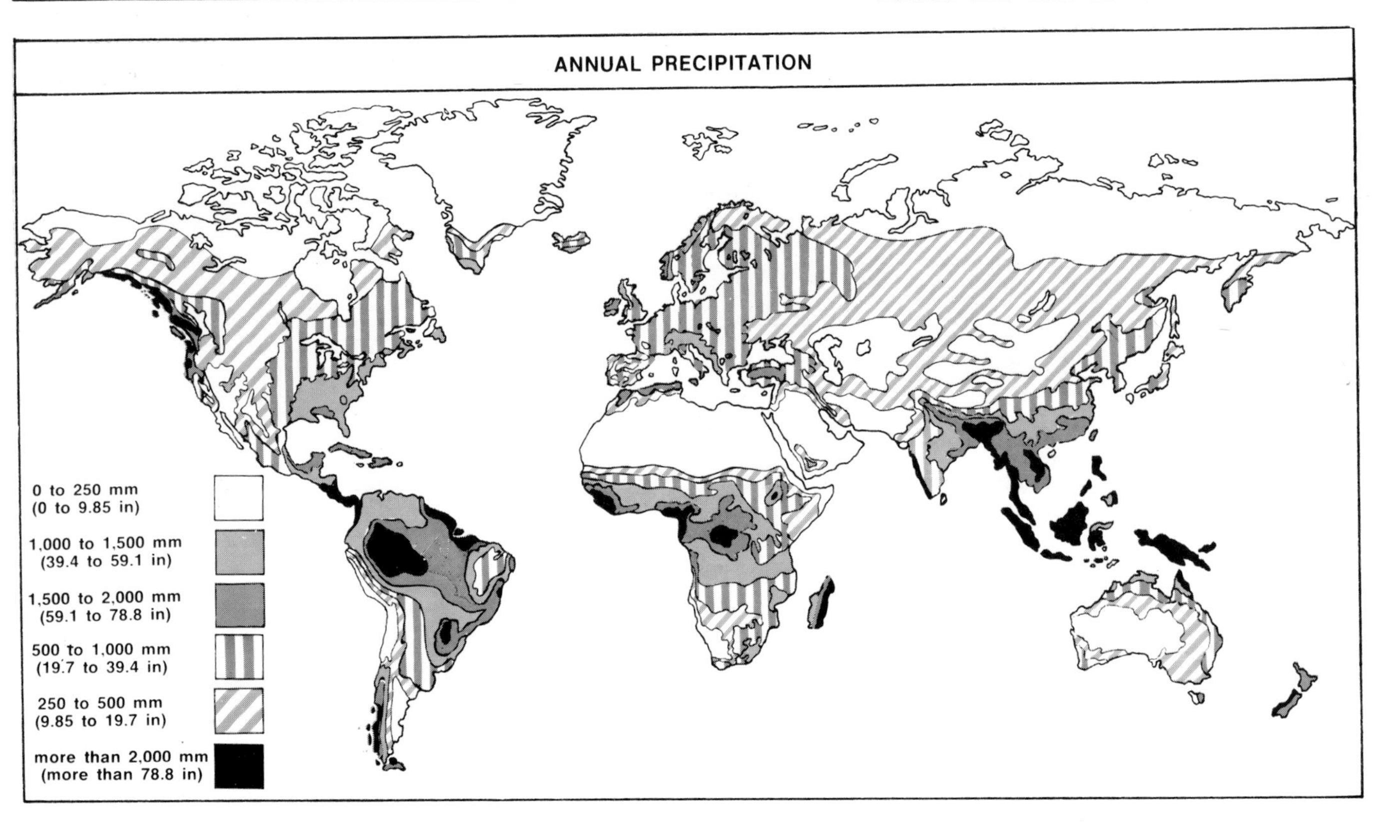

example, the Gulf Stream is a warm ocean current which flows from the Gulf of Mexico across the Atlantic. One branch of this current flows past Norway. Onshore winds are warmed by the current and bring mild weather to Norway. In fact, the average January temperatures are 28°F (10°C), higher than is usual for Norway's latitude. Inland, beyond the moderating influence of the sea, conditions become more severe.

The definition of climatic regions is complicated by these and other factors. One simple classification that is widely used was made by a Russian, Vladimir Köppen. He divided the world into five main regions: tropical rainy climates with no cool season; dry climates; middle latitude rainy climates with severe winters; tundra climates; and polar climates. Each of these regions is subdivided according to rainfall and temperature, and also according to the distribution of rainfall season by season.

Weather is the day-to-day state of the atmosphere. The weather is determined by the temperature and pressure of the atmosphere, by the way the atmosphere moves, and by the amount of moisture it contains. Moisture may be present in the atmosphere as an invisible gas—water vapour—or it may be observable as clouds, fog, or dew. Clouds are the source of rain as well as of hail, sleet, and snow.

At weather stations on land and at sea, readings are taken of atmospheric conditions, usually every six hours. The readings from weather stations are sent by radio or teleprinter to forecast centres. At these centres, the reports are analysed. Weather maps, called *synoptic charts*, are drawn. These charts show temperatures, wind speeds and direction, and cloud amounts for each station. They are marked with lines, called *isobars*, that join points of equal atmospheric pressure.

Finally, cold and warm fronts are added to the charts. Fronts are features of depressions—complex air systems which form in temperate latitudes where the cold, polar air mass meets warm, sub-tropical air. The edges, or fronts, of the cold and warm air are zones where cloud and rain occur.

Weather forecasters examine the synoptic charts of the last 24 hours or so, and study information about conditions in the upper atmosphere. Then they prepare a forecast helped by computers.

Weather and climate profoundly affect Man's life. They largely determine where he can live, what shelter and clothes he needs, and what food is available to him. The chief elements of weather, apart from air pressure, are temperature, moisture and wind.

Temperature is the most important factor in weather. The Earth's heat comes from the Sun. Because heat is trapped in the atmosphere, most parts of the Earth's surface have some warmth, even when on the side away from the Sun.

Moisture. About three-quarters of the Earth's surface is water. The heat of the Sun causes water to evaporate from the seas, from lakes and rivers, and from land and vegetation, forming water vapour in the atmosphere. Some of it condenses as clouds. More falls back to Earth as precipitation—rain, hail, snow and dew.

Wind is air in movement. Some winds follow defined patterns across the surface of the Earth—the monsoons, for example, and the trade winds. Others are part of local weather conditions. Some winds are gentle, some destructive.

The Structure of Earth

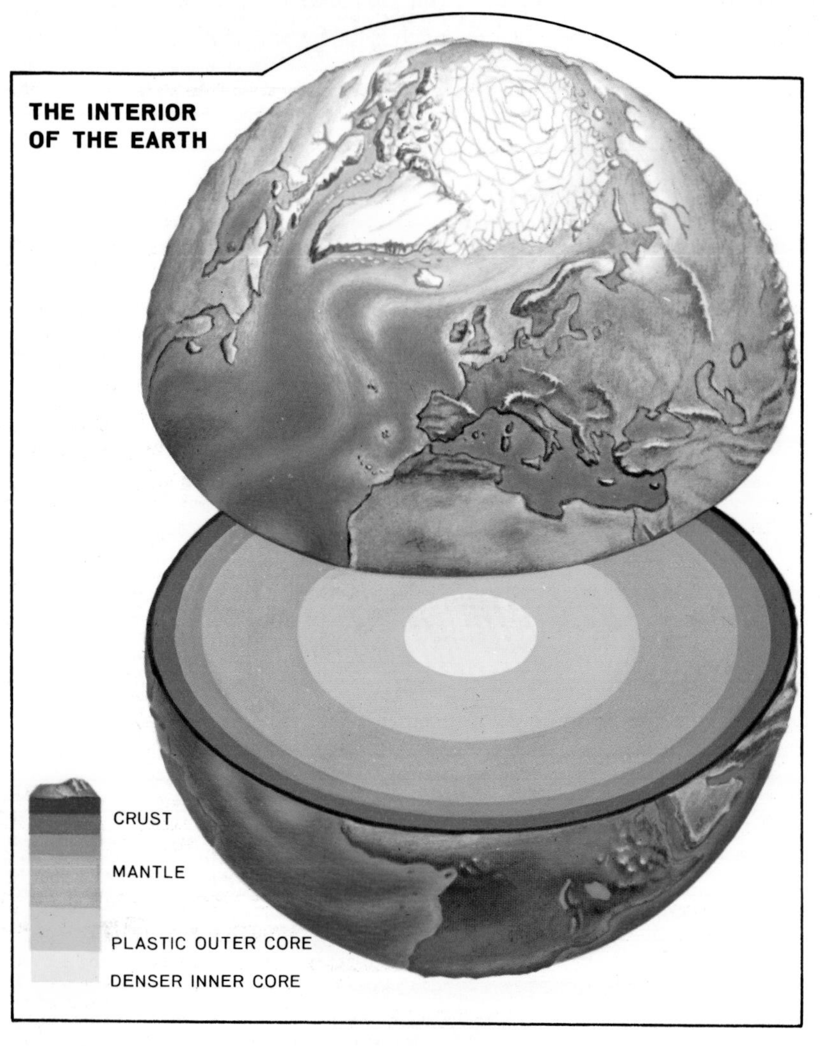

The Earth's crust averages around 20 miles (32 km) in thickness beneath the continents and about six miles (9.7 km) under the oceans. The crust and the first land masses were formed in the earliest era of Earth history. But many changes have taken place since that time.

About 200 million years ago, the land masses were all grouped in one large continent, which geologists call *Pangaea.* This continent split into two parts, *Laurasia* and *Gondwanaland,* which drifted northwards. These two smaller continents later broke up again, and moved slowly to their present positions.

This movement, called *continental drift,* is still continuing today. The Earth's crust is cracked and split into a number of large, rigid *plates.* These plates are moved by powerful forces in the Earth's mantle. An example of movement occurs in the mountain ranges called *mid-oceanic ridges* that have been found in the oceans.

These ridges are centres of volcanic activity and earthquakes. New crustal rock is being added to them by *magma* (molten rock) which wells up from the top of the mantle. This new rock is widening the oceans.

But the Earth is not growing in size. To balance the addition of new crustal rock, the

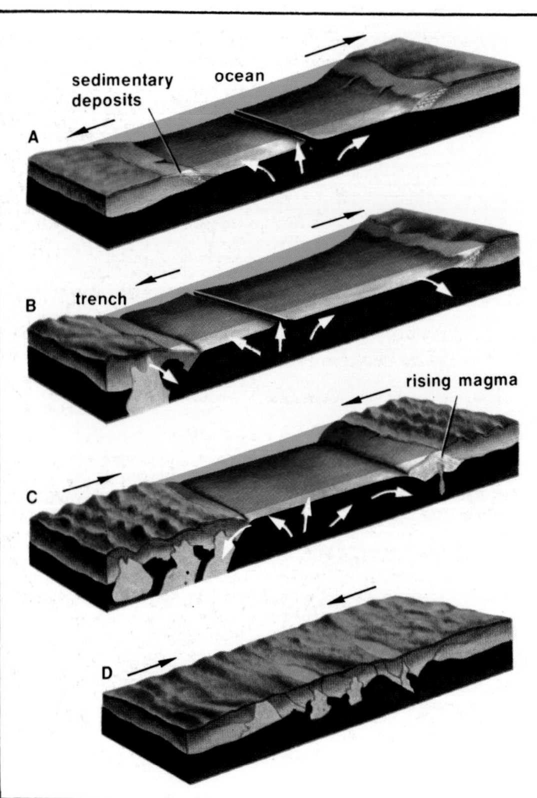

FORMATION OF HIGHLANDS AND MOUNTAINS

Rocks found in highland and mountain areas are partly sedimentary in origin, partly igneous, and partly metamorphic. The sedimentary rocks originated as deposits in seas and lakes, and the igneous—such as basalt and granite—as molten magma rising from deep in the Earth. Metamorphic rocks were formed from sedimentary and igneous types by heat and pressure transformation. Some sedimentary rocks contain fossils, usually of sea animals, showing that the rocks were once part of the sea bed.

Several theories exist to explain these facts. One is that, over millions of years, sediments—sand and mud—piled up on the sea bed in thick layers. The weight of these sediments may have forced the crust down until it melted and became weak, allowing the sediments to become squeezed up into mountains. The new theory of plate tectonics allows a different explanation. The crust is formed of plates on which the continents ride. In mid-ocean these are spreading ridges where the plates are moving apart, allowing magma to rise from below to form new crust (A). At the margin of the ocean a trench may form and the ocean crust may slide beneath the continent (B). Deep down it melts but the magma rises into the crusted rocks (C). Finally the continents may move together to close the ocean and thrust both sediments and volcanic rock upwards to form mountains (D).

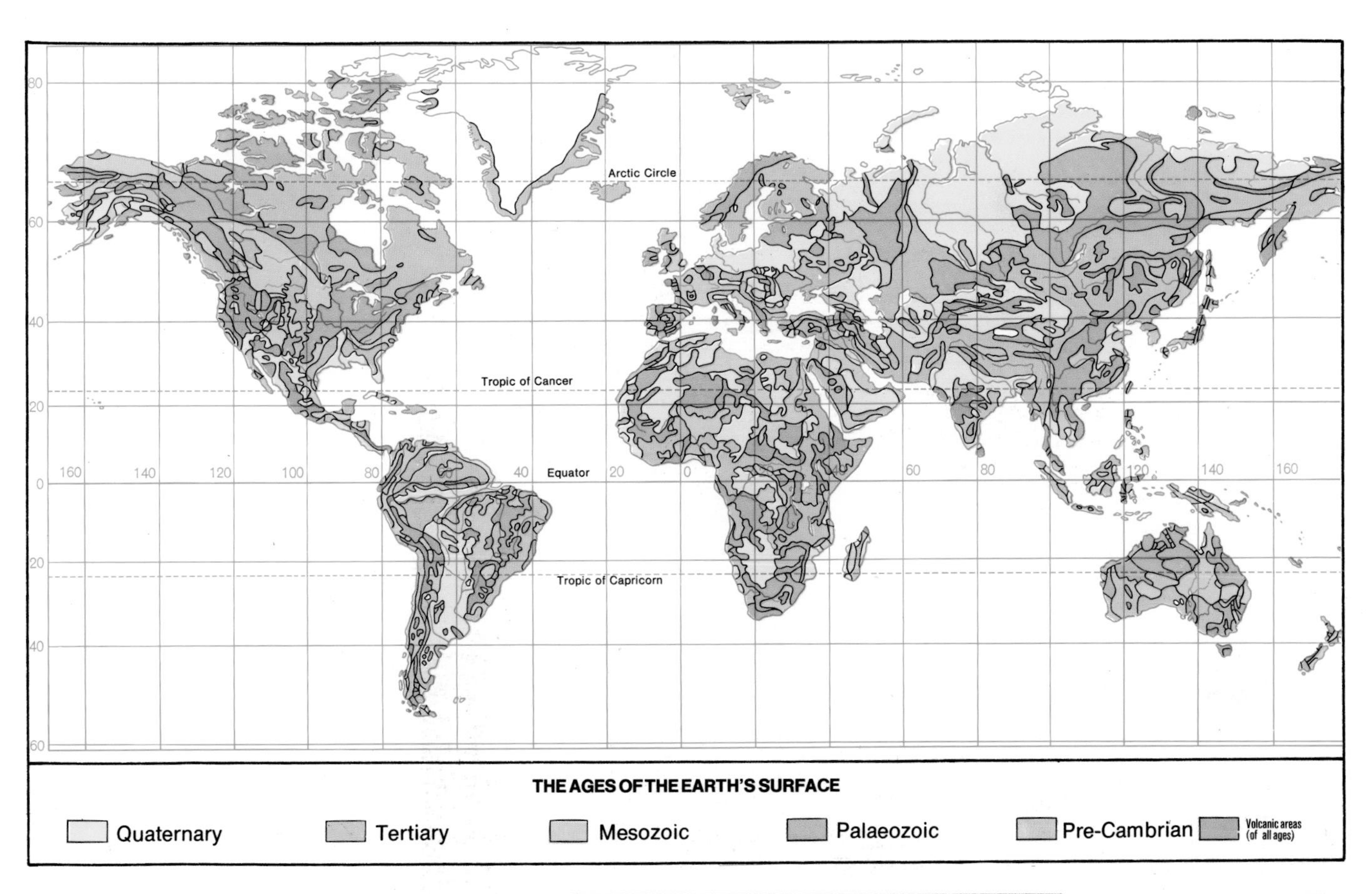

crust in other places is being destroyed. This occurs when two plates are forced against each other. The edge of one plate is pushed beneath the other. As it descends, the crustal rock melts. Some of the melted rock returns to the surface through volcanoes and some returns to the Earth's interior. Sometimes, when two plates collide, sediments on the ocean floor between the plates are squeezed up into mountain ranges. For example, when the Indian sub-continent drifted against the Asian mainland, the ocean sediments between were forced up to form the Himalayas.

THE EARTH: FACTS AND FIGURES

Diameter: 7,926 miles (12,755 km) at the equator; 7,899 miles (12,712 km) at the poles
Circumference: 24,901 miles (40,074 km) at the equator; 24,857 miles (40,003 km) at the poles
Area: 196,940,400 sq. miles (510,073,270 sq km), of which 70.8 per cent. is covered by water
Highest point: Mount Everest 29,028 ft (8,848 m) above sea-level.
Lowest point: Challenger Deep (Pacific Ocean) 36,198 ft (11,033m) below sea-level

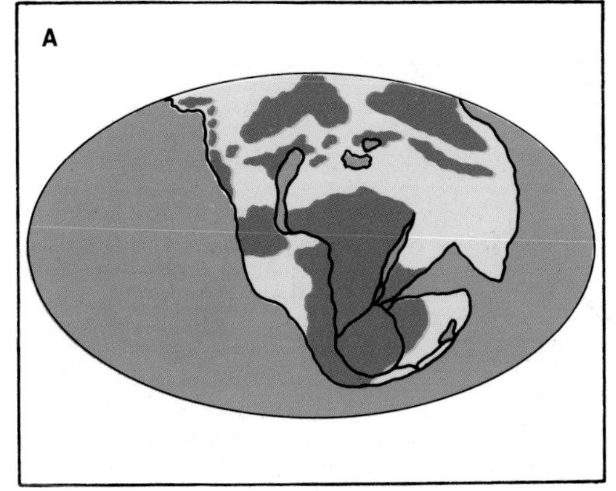

Dark blue: deep sea. Light blue: shallow sea. Brown: land.

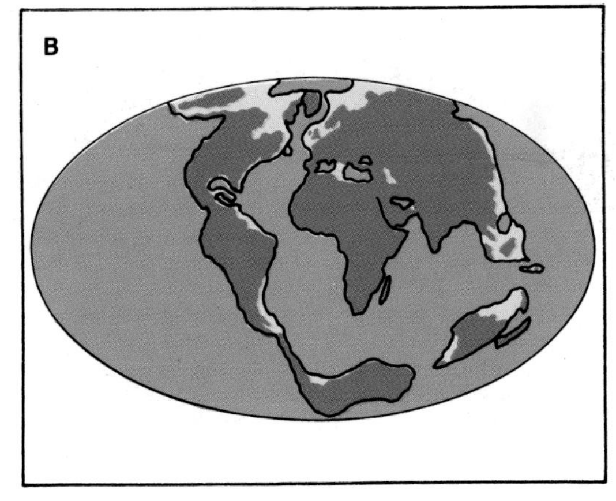

CONTINENTAL DRIFT

The idea that the continents of today are parts of one or two vast land masses that existed millions of years ago is an old one. It was given scientific form in 1910 by the German geologist and meteorologist Alfred Wegener. His maps, left, show (A) the continental masses in the late Carboniferous Period and (B) in the early Pleistocene Epoch.

Some of the evidence for continental drift relates to the shape of the continents: the eastern bulge of South America, for example, would fit fairly well into the concave western coast of Africa. But there is also more compelling evidence based on studies of geological structure, the past distribution of plants and animals, and past climatic changes. Modern plate tectonic theory and ocean studies tend to bear out the idea of continental drift.

A broad outline of the Earth's history has been traced by *geologists* —scientists who study the rocks and other materials that make up the Earth's crust—and by *palaeontologists*—scientists who study the remains of ancient plants and animals preserved in rock. A fossil may consist of all or part of an animal's body; or it may be merely a rocky mould or cast showing the shape of a plant or animal that was pressed into the rock before it hardened, millions of years ago. Fossils give a clue to the age of rocks. The oldest rocks contain fossils of the simplest types of living things. More recent rocks have fossils of evolving plants and animals.

The History of Earth

A time-scale of the Earth's history has six major divisions, called *Eras*, each of them representing many millions of years. They are, from the most distant to the most recent, the Azoic, Archaeozoic, Proterozoic, Palaeozoic, Mesozoic, and Cenozoic Eras. Often, the first three are grouped together as the Pre-Cambrian Era. Very little is known about the Earth in the Pre-Cambrian years. Eras are divided into *periods*, and periods further divided into *epochs*.

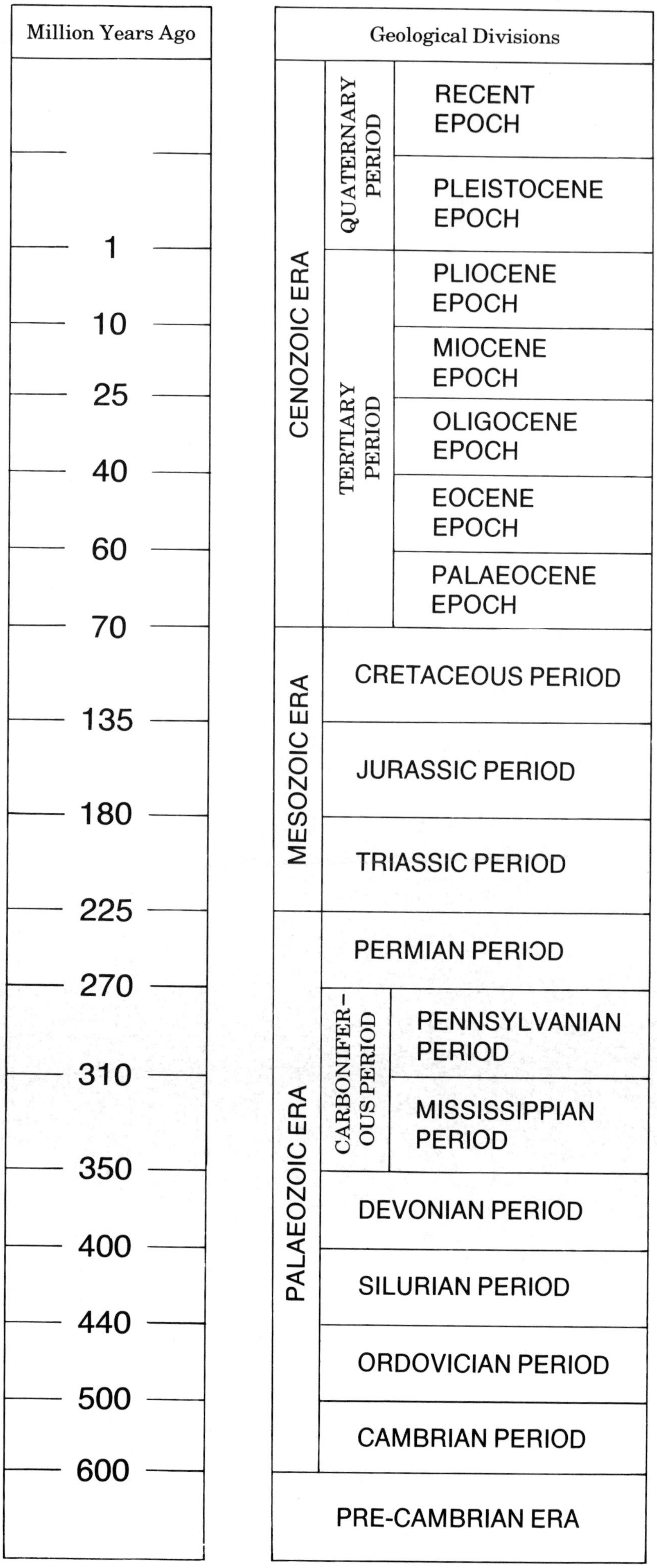

Million Years Ago	Geological Divisions		
	CENOZOIC ERA	QUATERNARY PERIOD	RECENT EPOCH
1			PLEISTOCENE EPOCH
10		TERTIARY PERIOD	PLIOCENE EPOCH
25			MIOCENE EPOCH
40			OLIGOCENE EPOCH
60			EOCENE EPOCH
70			PALAEOCENE EPOCH
135	MESOZOIC ERA	CRETACEOUS PERIOD	
180		JURASSIC PERIOD	
225		TRIASSIC PERIOD	
270	PALAEOZOIC ERA	PERMIAN PERIOD	
310		CARBONIFEROUS PERIOD	PENNSYLVANIAN PERIOD
350			MISSISSIPPIAN PERIOD
400		DEVONIAN PERIOD	
440		SILURIAN PERIOD	
500		ORDOVICIAN PERIOD	
600		CAMBRIAN PERIOD	
	PRE-CAMBRIAN ERA		

Conditions on Earth	Animal and Vegetable Life
Ice sheets decreased in size; glaciers melted. Sea levels rose. Land was eroded, and coastlines and landscapes were formed as we know them.	Forests spread. Marine and land animals developed as we know them. Man tamed other animals, learnt to grow food, and used various natural resources.
A time of great climatic changes. Ice sheets and glaciers covering much of Europe and North America melted four times. Land masses rose.	Many plants died out. An ape-like creature, identified as Man, appeared and discovered how to make stone implements. Mammoths and woolly rhinos died out.
Decrease in mountain building. The Earth's features took on the shapes we know. Climates cooler.	Animal life increasingly took on the forms we know. Man-like apes developed and became common.
The seas continued to retreat. Much mountain building and volcanic eruption.	Flowering plants developed much as today. Grazing mammals spread. Apes evolved further and spread.
The seas retreated from the land. There was extensive mountain building and volcanic eruption.	Cats, dogs, rodents, and bears evolved. Plant-eating animals became numerous. A primitive ape appeared.
Land subsidence caused the oceans to inundate coasts. The Atlantic Ocean was shaped much as today.	Grains and grasses appeared. Many fish, mammals, and insects began to take the form we know.
New mountain ranges continued to grow. In some parts of the world deep, fertile soil formed.	Flowering plants were dominant in vegetation. Dinosaurs had become extinct.
A major period of mountain building. Chalks formed in many regions. The seas continued to advance over the land. Coal swamps formed.	Flowering plants developed. Sea and land were still dominated by reptiles, many of them now giant and armoured. Mammals continued to evolve.
Much erosion of mountains, and much land inundated by the sea. The continents were cut up by swamps. Limestones as well as sandstones were formed.	Reptiles were dominant in the sea and on land. *Archaeopteryx*, the first bird-like creature, appeared. Dinosaurs grew to enormous sizes.
Much of the land was still arid. There was some volcanic activity, and marl and sandstone deposits formed. Copper and uranium deposits formed.	Conifers, ferns, and cycads were common. The first mammals appeared. Turtles and crocodiles appeared, as did dinosaurs—as yet only a few inches long.
A time of much Earth movement and mountain building. Melting glaciers left vast sedimentary deposits.	Deciduous plants appeared. Trilobites and eurypterids died out. Reptiles and insects were dominant.
The second period of the Coal Age. Land areas sank, producing vast swamps or regions of shallow sea. In other places, sea-beds rose and became dry land.	Huge evergreen trees and ferns grew. Fish were common, and large amphibians developed. Large insects evolved. Reptiles appeared on land.
The first period of the Coal Age: rotted vegetation became peat and, later, coal. Other minerals formed, including lead and zinc.	Mosses appeared. Amphibious creatures continued to develop. Large coral reefs began to form. Trilobites became less common.
The area of dry land increased. Much mountain building. Shales, slates, and sandstone formed.	Forests and ferns grew on land. Many fish evolved, including sharks. Amphibians and insects appeared.
Sea levels rose and fell. Mountain building in Europe. Limestones and sandstones were formed.	Leafless plants appeared on land. Eurypterids, animals that breathed air, developed in the sea.
Some mountain building. Boundaries between land and sea changed constantly. Great deposits of silt.	Still no life on land. Invertebrates were still common in the waters. Jawless fishes developed.
Seas covered much of the Earth's surface. Climates became milder. Some volcanic activity.	Invertebrate animals, such as trilobites, were common. The first fishes appeared late in the period.
The Earth's crust formed. Vapour surrounding the Earth condensed to form water masses. Probably, much of the Earth's surface was desert.	No life on land. Algae and bacteria lived in the sea. Later, invertebrates—such as jellyfish, sponges, and worms—lived in the warmer waters.

Exploring the Earth

NORTH AMERICA: INDIAN TRIBAL AREAS AND LANGUAGES AND EARLY EUROPEAN EXPLORATION

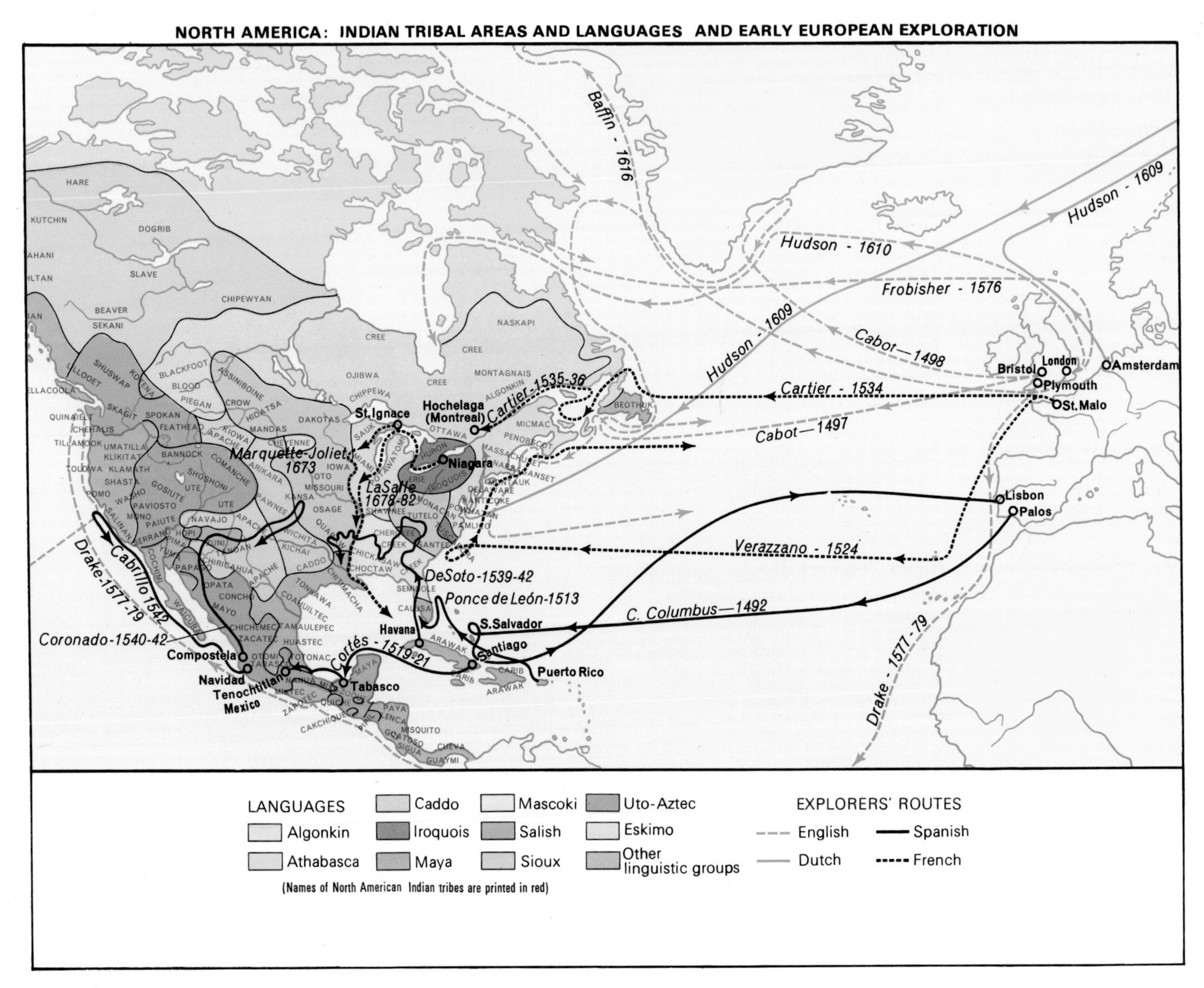

New Worlds to Explore. An American space rocket on the launch pad at Cape Canaveral. By the 1900s, men had explored most corners of the Earth's surface. Explorers of a new type—scientists rather than adventurers—were turning their thoughts towards the other planets. In 1969, two American astronauts landed on the Moon, the first human beings to set foot on non-terrestrial dry land.

Throughout history there have been periods when Man's desire for knowledge seems to have been particularly intense.

The ancient Greeks lived through one of these periods of great intellectual activity. For example, as early as the 4th century B.C., Greek philosophers had learnt enough about the world to conclude that the Earth was a sphere. Greek knowledge of geography was summarized by Ptolemy, a scholar in Alexandria in the A.D. 100s. The lands known to the Greeks included Europe, northern Africa, and part of Asia. Their knowledge of Asia owed much to Alexander the Great's march to the Indus Valley in the 300s B.C.

Discovery was not confined to the Europeans. While the Greeks, and later the Romans, explored the West, a great civilization was flourishing in China. Chinese traders extended their knowledge of geography over most of Asia.

After the Greek and Roman empires declined, Europe ceased for a time to be a centre of learning. Ptolemy's work was temporarily forgotten there and most people reverted to the idea that the Earth was flat. But, fortunately, the Arabs knew the works of Ptolemy and preserved Greek knowledge. In fact, during the Dark Ages of Europe, the Arab world became a centre of science and exploration. We owe our knowledge of the medieval empires of Africa to Arab traders and geographers who crossed the Sahara and recorded what they saw about the Negro empires in the south.

In the 800s, Norsemen began to sail the seas in search of plunder and lands to colonize. A Norseman, Leif Ericson, was probably the first European to land in North America. By the 1200s, the traders of Venice began to extend their horizons. Their search for trade culminated in the travels of Marco Polo through Asia (1271–95).

In the 1400s, another brilliant period of exploration began in Europe. It was linked with a development of science and art, called the Renaissance. Greek ideas were revived (often from Arab sources) and developed. In 1487, Bartholomeu Dias, a Portuguese captain, sailed as far south as the Cape of Good Hope at the tip of Africa.

Christopher Columbus questioned the belief that the Earth was flat and interested the Spanish king and queen in the notion that he could reach Asia by sailing west. On October 12, 1492, Columbus reached the Bahama Islands, but he believed that he had reached Asia. Two other Italians, John and Sebastian Cabot, reached Canada in 1497, also thinking that they were in Asia.

But the Portuguese still wanted to reach Asia by sea to obtain gold and spices, so they sent Vasco Da Gama on a journey around Africa to India in 1497–98. This historic voyage was followed by another Portuguese achievement in 1519–21, when Ferdinand Magellan led an expedition that sailed around the world.

The discovery of the New World led quickly to further expeditions. Between 1519 and 1521, the Spaniard Hernando Cortes conquered Mexico with a tiny force. He owed his success to the belief of the Aztec Indians that he was an invincible god. Another Spaniard, Francisco Pizarro, crushed the Incas in Peru in the 1530s. Also in the 1530s, a Frenchman, Jacques Cartier, explored the St. Lawrence River. In the 1540s, Spanish adventurers were exploring the southern parts of North America.

Rivalries developed in North America between various European powers, but settlement continued through the 1600s and

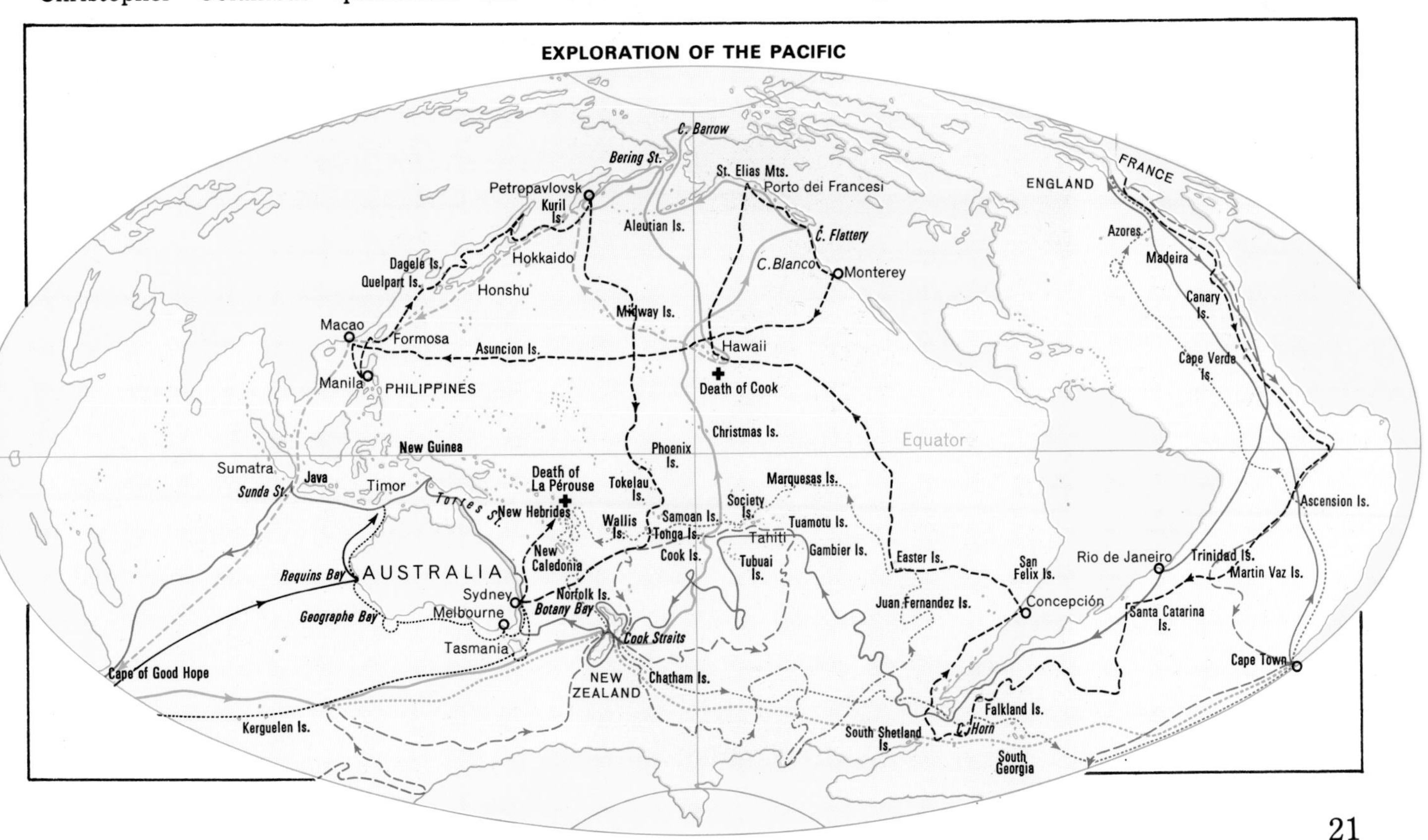

Above: The ruins of Machu Picchu in Peru. Some of the more inaccessible spots on the Earth's surface are 'discovered' more than once. At Machu Picchu, 6,750 ft (2,000 m) high in the Andes Mountains, the ancient Incas built a fortress city that they later abandoned. For hundreds of years, the ruins remained untouched and forgotten, until the American explorer Hiram Bingham happened on them in 1911.

Centre: The Australian Alps, the highest mountains in Australia, rise at the southern end of the Great Dividing Range. For many years, the Range was an obstacle that prevented settlers on Australia's eastern coast discovering the great plains to the west. In 1813 the Range was penetrated by William Wentworth and other explorers.

Below: Mount Kenya, a huge, extinct volcano, is Africa's second highest mountain. Its existence was first recorded by a German missionary in 1849. Africa was for long called the 'Dark Continent' because so little was known about its interior. Much of the continent was not explored until the 1800s.

1700s. Between 1804 and 1806, two explorers, Meriwether Lewis and William Clark, finally made the long journey from the Missouri River to the Pacific coast.

Since the days of the ancient Greeks, there had been talk of a continent in the southern hemisphere. The first certain European landfall in Australia was made by a Dutch ship in 1606. However, Abel Tasman's voyage of 1642–43 was the peak of the Dutch exploration. He discovered Tasmania and New Zealand and, in 1644, he explored the north Australian coast. From 1768, the British Captain James Cook explored the Pacific region. He proved that Australia was a continent and explored Pacific islands before he was killed in Hawaii in 1779. By this time, the outline of the world was known and charted on maps.

By 1800, navigators and explorers had travelled along the coastlines of all the continents, except for that of Antarctica. The mainland of Antarctica was discovered in 1820, although the first landing was not made until 1821. In the 1800s, explorers travelled into the interiors of the unmapped continents. They were often dogged by disease and they suffered great hardships. In Africa, Mungo Park, a Scotsman, explored the River Niger in 1796 and in 1805–06. He was killed on this last expedition. Richard Burton and John Hanning Speke suffered many illnesses and mishaps before reaching Lake Tanganyika in 1858. The greatest European explorer in Africa was the missionary David Livingstone. He explored the River Zambezi and found the Victoria Falls. He later joined the search for the source of the Nile, and was still on his travels when he died in 1873. Henry Morton Stanley explored the Congo region (now Zaire) and enabled the king of Belgium to establish a colony there. The exploration of the desert interior of Australia also created many problems for explorers. Robert O'Hara Burke led the first north–south crossing of Australia, but on the return journey he and his deputy William Wills died of starvation.

In the late 1800s and early 1900s, serious exploration began in the polar regions. Robert E. Peary of the U.S. Navy first reached the North Pole, on April 6, 1909.

The South Pole lies at the heart of the bleak continent of Antarctica. On December 14, 1911, the Norwegian Roald Amundsen reached the Pole. A month later, a British expedition under Captain Robert Falcon Scott also reached the South Pole, but the entire party perished on the return journey.

The development of aircraft made exploration far easier. The American Richard E. Byrd flew over the North Pole in 1926 and over the South Pole in 1929.

Even more dramatic than the arrival of the air age was the start of the space age in 1957, when Russia launched the first artificial satellite to orbit the Earth. Russia achieved another first in 1961, when cosmonaut Yuri Gagarin was sent into an orbit of the Earth. The United States Apollo project began in 1966 and reached its peak when Neil Armstrong stepped from his spacecraft onto the surface of the Moon on July 20, 1969. The Apollo project came to an end in 1975, when an Apollo craft linked up with a Russian Soyuz craft in orbit.

The space age has added to our knowledge of the Earth. Satellites circling the Earth send back photographs and other information which helps us in many ways. Satellites help us with weather forecasting, and aid navigators in fixing their position. Earth resources satellites help geologists to identify sources of valuable minerals, and reveal areas that are in danger of pollution. Also, the study of the Moon has contributed to our knowledge of the history of our planet.

Mount Everest in the Himalayas, the world's highest mountain. Many lives were lost in attempts to reach its summit. It was climbed in 1953.

FAMOUS EXPLORATIONS

Achievement	Explorer	Nationality
Possibly the first European landfall in North America (1003)	Ericson, Leif (?———?)	Norse
Travels in central Asia and China (1271-1295)	Polo, Marco (1254?-1324?)	Italian
Voyage around the Cape of Good Hope (1487)	Dias, Bartholomeu (1450-1500)	Portuguese
Exploration of the Caribbean (1492-1504)	Columbus, Christopher (1451-1506)	Italian
Landfall in Canada,(1497)	Cabot, John (1450-1498)	Italian
First to sail around Africa to India (1497-1498)	Da Gama, Vasco (1469-1524)	Portuguese
First sea journey around the world (1519-1521)	Magellan, Ferdinand (1480-1521)	Portuguese
Conquest and exploration of Mexico (1519-1521)	Cortes, Hernando (1485-1547)	Spanish
Conquest and exploration of Peru (1530-1538)	Pizarro, Francisco (1478-1541)	Spanish
Discovery of Tasmania and New Zealand (1642)	Tasman, Abel Janszoon (1603-1659)	Dutch
Exploration of the South Pacific (1768-1779)	Cook, James (1728-1779)	English
Exploration of southern and central Africa (1841-1873)	Livingstone, David (1813-1873)	Scottish
First north-south crossing of Australia (1860-1861)	Burke, Robert O'Hara (1821-1861)	Irish
Exploration of central Africa (1874-1889)	Stanley, Henry Morton (1841-1904)	Welsh
First man to reach the North Pole (1909)	Peary, Robert E. (1856-1920)	American
First man to reach the South Pole (1911)	Amundsen, Roald (1872-1928)	Norwegian
First man to orbit the Earth (April, 1961)	Gagarin, Yuri (1934-1968)	Russian
First man to step onto the Moon (July, 1969)	Armstrong, Neil (1930-)	American

The World's Features

Above: Rich farmland in Minnesota, in the United States. Some plains are deserts; others are the most fertile parts of the Earth's surface.

Centre: The Himalayas, the most magnificent and forbidding of the world's mountain ranges. Mountains occupy about one-fifth of the Earth's land surface.

Below: The Iguaçu Falls on the Iguaçu River, on the border between Brazil and Argentina. Rivers cut valleys, irrigate the land, and, by spreading silt, make land fertile.

Surface Features

Water covers 70.8 per cent. of the Earth's surface. The largest ocean, the Pacific, has an area of 63,800,000 sq miles (165,241,000 sq km). The world's greatest ocean depth of 36,198 ft (11,033 m) is in the Pacific. The Pacific contains many volcanic islands. In a sense, some of these islands are the highest mountains on Earth. For example Mauna Loa in Hawaii rises more than 30,000 ft (9,150 m) from the sea floor, although it is only 13,680 ft (4,170 m) above sea-level. Earthquakes and volcanoes are common, especially around the rim of the Pacific.

The Atlantic Ocean is the second largest ocean, with an area of 31,530,000 sq miles (81,660,000 sq km). Volcanic activity and earthquakes are common along the Atlantic mid-oceanic ridge, an underwater mountain range. Iceland is part of this ridge that juts above sea-level. The Atlantic is connected to the Mediterranean (and the Black Sea) by the narrow Strait of Gibraltar. The Baltic Sea is linked to the Atlantic by the Skagerrak. In the west, the Hudson Strait connects Hudson Bay with the Atlantic. In the tropics, the Caribbean Sea and the Gulf of Mexico are extensions of the Atlantic.

The third largest ocean is the Indian Ocean which covers 28,357,000 sq miles (73,444,000 sq km). It is linked to the Red Sea, the saltiest part of the world's oceans. The Arctic ocean covers 5,440,000 sq miles (14,090,000 sq km). The Antarctic Ocean surrounds the continent of Antarctica. Some geographers consider the Antarctic Ocean as part of the Pacific, Indian, and Atlantic oceans.

The average salt content of the oceans is 35 per mille—35 parts of salts per 1,000 parts of water. In the Red Sea, salinity is increased by high evaporation to 41 parts per mille. Some inland seas have even higher salinities. The Dead Sea's salinity is 192 parts per mille, and the Great Salt Lake of Utah, in the United States, has a salinity of 203. There, the water is so dense that a swimmer cannot sink. Supplies of fresh water—from rain, melting ice, and rivers—reduce salinity. The world's largest inland body of water, the Caspian Sea, receives a lot of fresh water from rivers. Its salinity is 13 parts per mille.

Land areas occupy 29.2 per cent. of the Earth's surface. The nature of the land is largely determined by height. Three main

GREAT MOUNTAINS

Mountain	*Height in ft. (metres)*
Everest	29,028 (8,848)
Godwin-Austen (K2)	28,250 (8,610)
Kanchenjunga	28,168 (8,586)
Mount Communism	24,590 (7,495)
Aconcagua	22,834 (6,960)
McKinley	20,320 (6,194)
Cotopaxi	19,344 (5,896)
Kilimanjaro	19,340 (5,895)
Mont Blanc	15,781 (4,810)

LARGEST ISLANDS

Island	*Area in sq miles (sq km)*
Greenland	840,000 (2,176,000)
New Guinea	316,856 (820,653)
Borneo	286,967 (743,241)
Madagascar	230,000 (596,000)

LONGEST RIVERS

River	*Length in miles (km)*	*Location*
Nile	4,160 (6,700)	North-east Africa
Amazon	3,900 (6,300)	Brazil
Yangtze	3,100 (5,000)	China
Zaire	2,718 (4,374)	Zaire
Missouri	2,714 (4,368)	North America
Amur	2,700 (4,345)	China
Hwang Ho	2,700 (4,345)	China
Lena	2,645 (4,257)	Siberia

height zones occur: plains near coasts; plateaux; and mountains.

Plains are good environments for people. Gentle slopes make farming and travel easy, and the well-watered flood plains and deltas of tropical and temperate regions are among the most fertile places on Earth. The early civilizations of Mesopotamia, Egypt, and China grew up in river valleys with rich alluvial soil and plenty of water. But flooding sometimes causes tragedy. The Hwang Ho in China has burst its banks many times, drowning hundreds of thousands of people.

Some plateaux in tropical regions are attractive places for settlement, because they are cooler than the hot lowlands of the same latitude. But high plateaux of mountains are often hostile environments.

Above: The island of Morea, one of the Society Islands in the South Pacific.

Below: Inhospitable Land. Life within the Arctic Circle, in Alaska.

Right: Sea Power. Cape Reinga, at the northern tip of the North Island of New Zealand. The action of the waves continuously wears away the rocks, while at the same time building up the sandy beach.

Below: River of Ice. A glacier flows slowly to the sea in Magdalena Fiord, in Svalbard. Glaciers tend to deepen the valleys through which they pass, laying bare the rocks.

Forces of Change

The Earth's surface is constantly changing. The most dramatic and sudden changes are caused by forces originating within the Earth. These forces are linked with continental drift (*see* pages 16–17). They are responsible for volcanic activity, earthquakes, and mountain building. Other changes are caused by the forces of erosion which act on the Earth's surface.

Volcanic activity has been responsible for creating most of the rocks in the Earth's crust. Volcanoes are supplied with molten rock, or *magma*, in two main ways. First, heat sources in the Earth's mantle, which are probably radioactive, melt rock. The magma then rises through a conduit (channel) to the Earth's surface. Second, magma is created when one plate in the Earth's crust is forced beneath another. The descending rock is melted by friction and pressure.

Volcanic eruptions are of three main kinds: explosive, quiet, and intermediate. *Explosive eruptions* usually occur when explosive gases are trapped in the molten magma. When an explosive volcano erupts, the gases shatter the magma, which shoots out of the volcano as ash, as cinders, or sometimes in lumps called *volcanic bombs*. The greatest explosive eruption in history occurred in August 1883 at Krakatoa, a volcanic island between Java and Sumatra, in Indonesia. The eruption caused the 2,600-ft (792-m) high volcano to explode. Most of the island disappeared. The explosion also generated a destructive wave called a *tsunami*, which battered the coasts of Java and Sumatra, killing 36,000 people. Tsunamis are also caused by earthquakes.

In *quiet volcanoes*, such as Mauna Loa on Hawaii, the magma contains little gas, or at any rate, the gas can escape easily. As a result, great explosions do not occur. Instead, the magma flows through the vent of the volcano or through fissures (openings) in the ground as lava. *Intermediate volcanoes*, such as Vesuvius in Italy, combine features of explosive and quiet volcanoes.

Earthquakes occur around the edges of the plates in the Earth's crust. They are caused by movements along faults (cracks) in rocks. The major breaks in the Earth's crust are the plate edges themselves. For example, the San Andreas Fault in California is 600 miles (960 km) long. It is the boundary between two plates that are moving in opposite directions. In 1906, a great earthquake destroyed the city of San Francisco. This earthquake occurred when the two plates along the San Andreas Fault jammed together, creating tension. The tension was released in a sudden jerk, making the ground shake. Near San Francisco, the shift along the fault was 15 ft (4.5 m).

Other earthquakes are caused when two plates are pushed against each other. The edge of one plate is forced down beneath the other. This is not a gradual movement, but one that proceeds in a series of jerks. Fold mountains are formed by a lateral pressure created when continental drift makes two plates collide. Sediments between the plates are compressed and squeezed up in large ripples called *folds*. The Alps, for example, are compressed sediments that accumulated on the sea bed. The formation of fold mountains is very slow. The Alps were formed as one plate, bearing the continent of Africa, drifted northwards and pressed against Europe. Geologists think that the Alps may still be rising. But this is difficult to prove, because as new mountains rise, the forces of erosion wear them down.

One erosive force acting on mountain slopes is called *weathering*. An example of this process is water freezing in a crack; it expands and splits the rock. Sudden changes in temperature in hot regions also shatter rocks. Even the roots of plants that grow in rocky terrain can split rocks. Fragments of rock tumble down mountain slopes and pile up in heaps known as *scree*.

In dry regions, the wind is an important force. Strong winds erode rock in a process rather like sand-blasting. They pick up the sand from dunes where it has piled up.

Water is a major agent of erosion. Some rainwater seeps into the ground; and rocks, such as limestone, that are soluble in water are dissolved. In this way, large caves are formed. Some rain, too, flows over the surface into streams and rivers. The rivers erode valleys, and carry eroded material to the sea. The Mississippi River in the United States carries 440 million tons of eroded material into the Gulf of Mexico in an average year.

Snow in mountain regions is compacted into ice and forms glaciers and ice sheets. The ice is pulled downhill by gravity and acts like a giant file. Bits of rock frozen in the base of the ice scrape the ground, wearing out deep U-shaped valleys. At the base of glaciers, eroded material is dumped in *moraines*.

Natural erosion is a slow process and may hardly be noticeable in one person's lifetime. But Man sometimes speeds up erosion by misusing the land. When grasslands are ploughed and forests cut down, the soil is exposed to the wind and rain. Intensive farming robs the soil of its fertility and the soil becomes unstable.

Left: The volcano of Osorno in southern Chile. It is in the Andes Mountains. Many of the high Andean peaks are volcanoes, some of them active.

Below: The Dolomites, in the eastern Alps. Erosive forces have shaped the mountains into jagged peaks of spectacular beauty.

Vegetation

In general, belts of vegetation run E–W and are related to latitude. But there are many complications. For example, the height of the land in any particular place affects the climate and, therefore, also affects the vegetation.

Vegetation changes as one travels up a high mountain, because the temperature falls by 3.5°F for every 1,000 feet (6°C per 1,000 metres). In the Himalayas, the lower mountain slopes are covered by tropical forest. Gradually, the tropical forest gives way to deciduous forest—to trees that shed their leaves during one season of the year. Higher still, there are coniferous forests. Trees become more and more widely spaced and stunted as one nears the *tree line*—the upper limit of tree growth. Rich, grassy meadows which are located around the tree line finally give way to tundra, a region of grasses, lichens, and mosses. The tundra ends around 17,000 ft (5,000 m) above sea-level, where snow and ice begin.

This sequence of vegetation is broadly similar to the vegetation belts that encircle the Earth. Around the icy wastes of the North Pole is a region of treeless tundra, covering the far N of Europe, Canada, and Alaska. The soil just under the surface (the subsoil) is often frozen, and even in summer the temperature does not rise above 50°F (10°C). As one travels S, the temperatures gradually rise until coniferous trees appear. These trees become increasingly frequent until they form the boreal (northern) coniferous forest belt. Conifers, such as firs and pines, are specially adapted to cold conditions.

To the S of the coniferous forests are the deciduous forests of the temperate zone. Deciduous trees include oaks, elms, and maples. Such forests need at least six months annually when the temperature exceeds the minimum necessary for tree growth, which stops entirely in winter.

In some regions there are vast areas of grassland. These are called *prairies* in North America, *steppes* in Russia, *pampas* in Argentina, and *veld* or *velt* in Africa. The chief economic activity is livestock rearing, but large areas have been cultivated with such crops as cereals and cattle fodder.

Another forest belt occurs in warm Mediterranean regions. Mediterranean trees include evergreen oaks, corks, and such

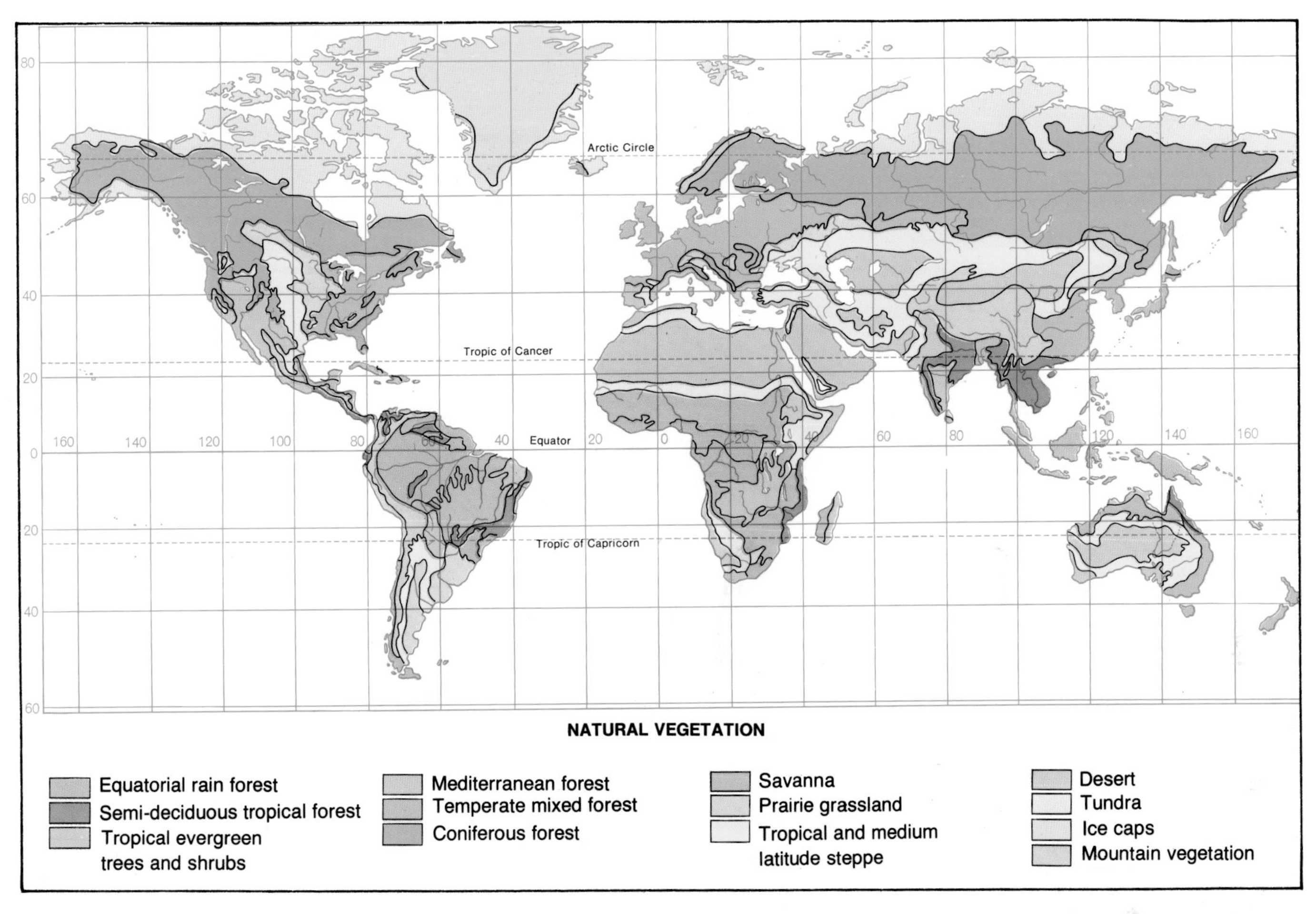

conifers as pines, firs, cypresses, and cedars. Such trees are adapted to summer drought and cool, moist winters.

The world's greatest desert, the Sahara, lies S of the Mediterranean zone. It covers about 3 million sq miles (7.8 million sq km). Other African deserts include the Namib and the Kalahari, which is really a semi-desert. The Middle East is largely desert, and desert extends into central Asia. The heart of Australia is desert, and deserts also occur in North and South America. The world's highest shade temperatures have been recorded in deserts, such as 136.4°F (58°C) at Azizia, in Libya, and 134°F (56.7°C) at Death Valley, California. Sand dunes build up in deserts, but large areas are bare, rocky wastes. Some plants survive these conditions. Cacti, for example, have spines instead of leaves to check water loss.

Bordering the deserts of Africa are vast tracts of savanna, which is tropical grassland with scattered trees. Savanna finally gives way to the hot, wet tropical forests around the equator. These forests cover Zaire and parts of West Africa, much of the Amazon and Orinoco basins of South America, and much of south-eastern Asia, including Indonesia. They include tall trees, often more than 100 ft (30 m) high. The foliage at the top may block out light from the forest floor. Trees in tropical forests include mahogany and rubber trees.

Opposite page: The lush vegetation of a tropical rain forest.

Below: The scant vegetation of the Kalahari Desert.

The Influence of Man

Right: Harvesting hemp. Not all crops are grown for food, some—such as hemp, cotton, and flax—are used in making textiles, others have varied uses. Soya-beans, for example, are a valuable food, and can be used also in the manufacture of paints, adhesives, linoleum, and many other products.

Below: Rice cultivation in South-East Asia. Rice is the world's chief food crop. Probably mankind's biggest single step in the process of civilization was learning how to grow food crops. Primitive men had to exist on whatever food they could find; if they could not kill animals or find edible plants within reach of their caves or forest shelters, they had to move elsewhere.

Agriculture

Before early men began to establish farming settlements and to plant crops and rear animals, they lived by hunting and gathering. This is still the way of life of such people as the Bushmen of the Kalahari in Africa. By studying them, we find that they use up most of their energy in the struggle to stay alive.

Farming began about 8,000 years ago when people discovered that some of the plant species that they collected yielded more food than others. By planting the seeds of the best food crops, they increased their food supplies. They also stopped moving around from one place to another, and built permanent homes. The earliest farm settlements were probably in Turkey and Palestine. From these areas, farming spread to the valleys of the Tigris-Euphrates Rivers in Mesopotamia and of the Nile in Egypt. There, people learned how to irrigate the land with water from the rivers. Irrigation was important in lands where the rainfall was light and infrequent. Around 5,000 years ago, the iron plough was in use in Egypt and Mesopotamia. Farming also began at an early date in China and, later, the lands of the Roman Empire became highly developed agriculturally, and crop rotation was introduced to conserve the fertility of the soil.

In farming societies, food supplies were reliable and people were relieved of the constant pressure for survival. As a result, civilizations developed, with advances in science and the arts. This led to improvements in farming techniques, the discovery of new plant hybrids, which yielded more food than the old varieties, and the breeding of better quality livestock.

Farming involves changing the environment. It has led to the destruction of much of the natural vegetation in many areas. For example, most of the original deciduous forests of temperate zones have been cleared to create farmland. This happened gradually in Europe; but in the United States, after European settlement began in the 1500s, it happened quickly. In the last 300 years, most of North America's deciduous forest has disappeared and wild animal populations have dwindled. Fast development has its dangers, such as soil erosion.

Crops of grass, lucerne, and clover are produced as food for animals. Cattle are raised for their milk, and for meat and hides. Huge cattle ranches are located on the great grasslands of North and South America and Australia. Dairy farming requires rich grass. Other important farm animals are pigs, sheep, goats and poultry.

Developments in agriculture have become

Vines growing on a sunny slope below the ancient walls of Monteriggioni, near Siena in Italy. In addition to grapes—for eating and for making wine—luxury crops include tea, coffee, spices, and tobacco.

especially important as the world's population has increased in the 1900s. In 1925, the world's population was less than 2,000 million. By 1975 it was almost 4,000 million, and it is expected to exceed 6,250 million by the year 2000.

The most important recent development in agriculture has been the 'Green Revolution'. This revolution of farming techniques and plant hybrids has been particularly important in the countries where much of the population expansion is taking place. Scientists have produced new hybrids of such important food crops as maize, rice, and wheat—hybrids that yield bigger harvests. Agricultural experts have also been advising farmers on improved techniques and the use of fertilizers. In the 1940s Mexico—to take an example of what can be done—had to import wheat, but by 1964 it was exporting half a million tons a year. The new varieties of rice introduced in the Philippines yield about 15 times more rice than old varieties.

Another important source of food is the sea. Today fishing techniques are rapidly improving. Radar, spotter planes, and even satellites are used to locate fish.

Climate and soils are the factors that determine what the chief crops in any area will be. For example, rice is the world's principal food crop, in terms of the number of people who depend on it. The main producers are China and India; and it flourishes throughout the wet, hot tropical lands of eastern Asia. Wheat, barley, and rye are more important cereals in cooler regions. The world's leading barley producers, for example, are the U.S.S.R., France and the United Kingdom.

A Polynesian islander carries a fish he has harpooned. People living in coastal areas or on islands where there is little or no agricultural land rely heavily on fish as a source of protein. To others, fish may be one item in a varied diet.

Shelter

Food, clothing and shelter are three essentials of life. Numerous archaeological finds have shown that prehistoric men often found shelter in caves. But once men began to farm the land, they built shelters for themselves and their livestock.

Shelter for a Few. Top: A village on a Samoan island, in the Pacific Ocean. Half of the world's people live in small villages. Centre: Aborigines in a desert region of Australia shelter in their rough tent.

Below: Shelter for Millions. The skyline of Manhattan, in New York City. As civilization developed, certain points on the Earth's surface were found to have particular advantages for defence, communications, trade, or industry. In these spots, people grouped their shelters, forming villages, then towns and cities. The biggest cities grew upwards as well as outwards. Today, many people would like to see the size of cities limited.

The types of houses men build vary from place to place, according to the climate and the building materials available. For example, houses in forest regions are usually made predominantly of timber; and stone houses are common where easy-to-work stone is quarried. The Eskimo's igloo consists of blocks of compacted snow; the Mexicans use sun-baked bricks called *adobe*; and many Africans weave grass huts.

Climate affects the design of houses. In hot, dry regions, houses often have thick walls to keep out the heat, as well as flat roofs where the infrequent rainfall can be collected and where people can sleep at night. In cold regions, houses have to be sturdy, too, to keep out the cold, but they have steeply-inclined roofs to carry away heavy rainfall and snow.

In regions where earthquakes are common, people often live in extremely light structures which do little harm to their inhabitants if they cave in. In constructing larger buildings, architects have to take special precautions. They establish extremely secure foundations and use reinforced materials that will sway but not collapse when the ground shakes.

Modern technology has contributed much to comfortable living. Electric lighting, air-conditioning, central heating and refrigeration have all helped to make life comfortable for people living in areas where the weather outside is harsh.

The distribution of population over the Earth's surface is very uneven. More than two-thirds of the world's people live in areas which total less than 8 per cent. of the available land. About one-third of the world's people live in urban settlements—ranging from towns with a few thousand inhabitants to cities of several millions. The proportion of people living in towns as opposed to rural areas varies, of course, from country to country. For example, about 90 per cent. of the people of Britain now live in urban areas, although about 150 years ago 80 per cent. were country-dwellers. In large parts of Africa, less than 10 per cent. of the people live in urban settlements today; and Bangladesh, China, India and Indonesia have enormous rural populations—80 to 85 per cent. In Russia and Japan, just over half the people live in urban areas; this is a low proportion by comparison with most European countries and the United States.

The first cities grew up some time after Man began to practise agriculture. Farming of the land led to a surplus of food and materials, enabling some people to turn their energies to making goods and providing services. Some of the earliest towns were built in Mesopotamia in the 5000s and 4000s B.C. At a later date, the great city of Babylon probably had a population of 80,000.

Early towns may have been primarily trading centres, markets through which materials that were unavailable locally could be imported. Later, towns assumed an administrative role and handicraft industries were established. Writing and record-keeping were essential to carry on town business, and such developments stimulated the rapid growth of civilization.

From Mesopotamia, town development spread to Egypt, the Indus Valley and China, and later to the Mediterranean. Greek and Roman towns were small by modern standards. Ancient Athens reached a peak population of about 150,000. Ancient Rome probably had a population of 200,000, but both Athens and Rome were great administrative capitals and centres of learning.

In the Middle Ages, increasing trade furthered the growth of cities. The Industrial Revolution from the 1700s led to an acceleration of city growth and to a relative decline in rural populations. Canals and railways made it possible for food and raw materials to be carried swiftly into cities, and manufactured products were exported. Towns tended to be divided into distinct zones. Administrative and business areas developed, and industrial zones were often separated from residential areas. Cities differed greatly in character. Some were centres of trade or industry. Others were cultural centres, or centres of government.

Cities today face many problems. Shortage of building space has led to the construction of tall buildings. Many residents of these buildings complain that they have no feeling of living in a genuine community. As cities grow in size, so traffic problems and certain types of crime increase. Some city authorities anticipate that they will soon be facing other severe problems if growth continues, such as water shortages or insufficient electricity supplies. Furthermore, in a nuclear war, the concentration of people in cities makes them extremely vulnerable to attack.

Top: The castle of Almanza, in Spain, stands guard over its clustering village. In the past, such villages were usually self-contained communities, able to meet most of their own needs. Their life revolved around the church and the castle of the local nobleman.

Centre: The Meteora monasteries near Kalabaka, in Greece. Built in the Middle Ages on the top of natural rock pinnacles 1,000 ft (300 m) high, the monasteries were cut off from the world and felt themselves safe from invaders. Access was by means of steps cut in the rock and baskets that were hauled up with ropes.

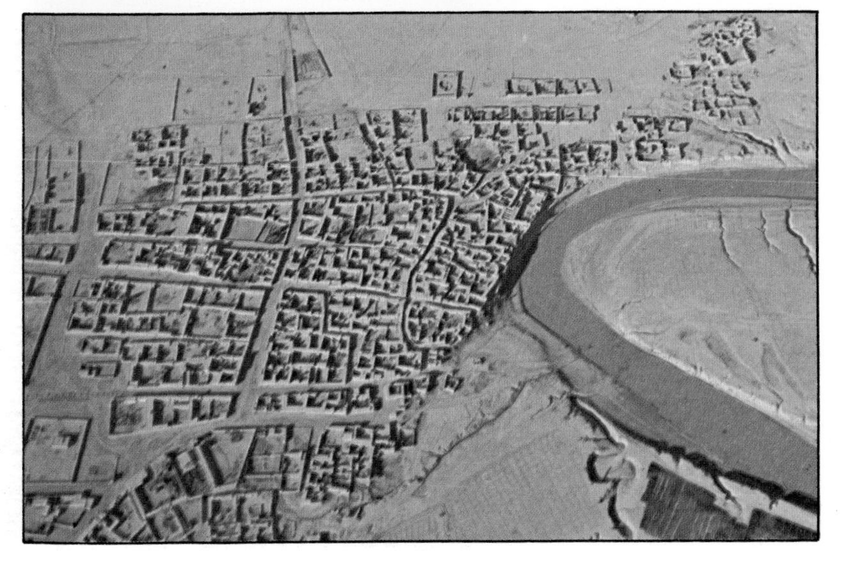

Below: A City in the Sand. The old settlement of Ahwaz, in Iran, was restored to importance in modern times by the discovery of petroleum in its neighbourhood. Cities have developed in many unexpected places: in jungles, in deserts, or cut into cliffs. But, always, there is a reason for their existence. One, for example, may be sited at a point where jungle trails once crossed, another where a caravan route crosses a river.

Above: Approach roads to the George Washington Bridge in the U.S.—a highway system used by thousands of New York commuters daily.

Below: A railway in the Andes Mountains, between Mendoza in Argentina and Santiago in Chile.

Communications

Transport and communications are essential to the existence of all societies except the simplest subsistence economies. The people of early civilizations, such as the Chinese, Incas, and Romans, were mostly great road-builders. Along their roads, animals and human labourers hauled the goods they needed.

Animals are still used for transport today; camels in the Middle East, for example.

Land Transport

Land transport includes roads, railways and pipelines. The construction of motorways has speeded up road transport and it is of vital importance in most of the world's industrial countries. The United States has about one-third of the world total of commercial vehicles, and Japan and western Europe combined account for another third. The United States also has about half of the world's passenger cars.

The U.S.S.R., however, relies more on railways. In tons per kilometre, the U.S.S.R. handles about 40 per cent. of the world's total rail trade. This is partly explained by the long haulages necessary in that vast country. Railways are also important in eastern European countries. Railways developed in the 1800s and they have played a major role in opening up the interiors of Africa, North America, and the U.S.S.R. For example, the Trans-Siberian Railway, which runs 5,800 miles (9,400 km) from Moscow to Vladivostok, has contributed greatly to Siberia's growing prosperity.

Pipelines serve areas where petroleum and natural gas are mined or distributed. Pipelines also transport such essential things as water and gas.

Water Transport

Water transport includes coastal and lake shipping, river and canal transport, and ocean shipping. Coastal shipping is important in many countries where bulky raw materials, such as coal and ores, must be transported from mines to industrial areas. Haulage is slow but costs are low. Lake shipping is well developed in such areas as the Great Lakes of North America and the lakes of Africa. Large rivers, such as the Rhine and the Danube, are major transport arteries, and canals remain important in some areas. But canal usage has declined in many countries because of competition with faster forms of travel.

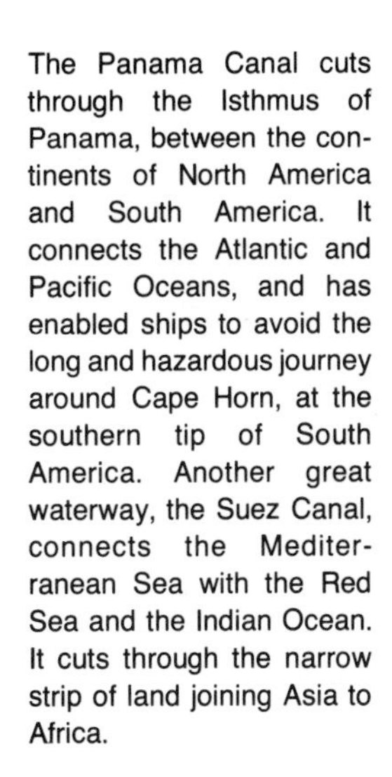

The Panama Canal cuts through the Isthmus of Panama, between the continents of North America and South America. It connects the Atlantic and Pacific Oceans, and has enabled ships to avoid the long and hazardous journey around Cape Horn, at the southern tip of South America. Another great waterway, the Suez Canal, connects the Mediterranean Sea with the Red Sea and the Indian Ocean. It cuts through the narrow strip of land joining Asia to Africa.

Air Transport

Ocean travel is a relaxing means of travel for passengers, but is comparatively slow. Jet aircraft can circle the world in 26 hours, and supersonic jets travel at speeds of 1,400 mph (2,250 kph). Freight traffic by air is, however, expensive compared with sea transport, though the use of helicopters may reduce the cost, and these aircraft may have a bright future in the transport of freight. Modern technological developments suggest that the world will continue to 'shrink' as faster and faster methods of travel are devised.

The great canals, however—chiefly the Panama Canal and the Suez Canal (which was reopened in 1975 after having been closed for eight years)—play a major role in international shipping. These canals are short cuts on ocean routes, saving much time for cargo ships small enough to use them.

Shipping has been important in trade since ancient times, progressing from sails to steam engines. Today most ships are powered by oil-burning engines. Giant oil tankers are especially important. Some of them can carry 300,000 tons of crude oil. A major recent advance has been the introduction of container ships. These vessels are designed to transport large metal boxes, called containers, which are packed by the sender and loaded complete into the ships. This method simplifies the handling of goods and saves time.

The Golden Gate Bridge connects the peninsula of San Francisco to northern California. It crosses a sea channel, but most bridges have been constructed to carry roads and railways across rivers.

The effects of communications technology on the environment cause increasing concern. Motorways speed traffic, but add to the concrete jungle. Aircraft pollute the atmosphere, and the noise they make may often be almost intolerable.

Above: Industrial plant on Delaware Bay, in the United States. Though refineries, factories, and storage depots have added little to the beauty of the world, they have made an immense contribution to mankind's comfort and health.

Trade and Industry

Trade is an essential part of contemporary life and has a major influence on the standards of living of peoples throughout the world. Trading involves the buying and selling—or the exchange—of all kinds of products, foods and other agricultural goods, minerals, and manufactured items.

About half of the world's people work on farms. But in industrialized countries only a small proportion are engaged in agriculture. In Britain, for example, only about four out of every 100 people work on the land; whereas in such countries as Nepal and Tanzania more than 90 per cent. of the population are farm workers. British farming is highly efficient; but it does not produce all the food that the country needs and so food must be imported. However, despite this fact, the British people enjoy a far higher standard of living than either the Nepalese or the Tanzanians. The reason for Britain's superiority in standards of living is a reflection of two factors: trade and industry.

The most valuable items involved in world trade are not farm products, but minerals and the products manufactured from them. As a result, countries where most people work on farms find it very difficult to achieve high

Below: An oil well in Libya. Petroleum is the world's largest single source of energy.

standards of living, even if their agriculture is very efficient and they export farm produce.

Mineral extraction industries employ comparatively few people, but they are very important to the economies of many countries. Minerals fall into two categories: those that are used to supply energy; and those, such as metallic ores, that are used in manufacturing industries.

Minerals used to produce energy are petroleum, natural gas, coal and lignite. In the late 1960s, about 40 per cent. of the world's energy came from petroleum products, 18 per cent. from natural gas, and 38 per cent. from coal and lignite. The remaining four per cent. came from hydroelectricity, nuclear power and other sources. Hydro-electricity is important in regions where mineral fuels are scarce.

Petroleum plays an extremely important part in world trade. Since the 1930s, world production of petroleum has increased by more than seven times. Some countries, such as the United States and the U.S.S.R., use most of their own petroleum resources at home, but many petroleum producers sell most of their production to other countries. About four-tenths of all petroleum is exported. In the 1970s, many petroleum-producing nations decided to raise the price and restrict the supply of this fuel, so as to conserve their resources.

Other minerals, especially such metallic ores as copper and bauxite (aluminium ore), are costly and are, consequently, very important economically. Some developing countries rely on the production of one or two minerals for their export income. For example, the export of copper is vital to Zambia and bauxite exports are extremely important to Surinam and Guyana.

Most industrialized countries have to import minerals. The process of industrialization, which has changed the world, began in Britain in the late 1700s. During the 1800s, it spread to other countries of western Europe, and to the United States, Canada, Japan, Russia and Sweden. In the early stages, industries were based mainly on local supplies of minerals. But as world trade developed, so minerals were extracted in some countries and sold to other, industrialized, countries. The industrialized countries then sold manufactured products, such as machinery, on the world market.

Broadly, the countries of the world today can be divided into two groups: the industrialized nations and the non-industrialized or 'developing' nations. North America, Europe, the U.S.S.R., Australia, and Japan combined have only about 30 per cent. of the world's population, but they produce about 85 per cent. of the world's industrial goods. The rest of the world has some industrial regions (for example, in Brazil, China and South Africa). But despite these pockets of industry, the rest of the world has 70 per cent. of the world's peoples and makes only 15 per cent. of the world's industrial products. In the industrialized world, North America is the chief region. With seven per cent. of the world's population, it manufactures 35 per cent. of the total industrial production.

The chief group of trading nations is western Europe. Its prominence is partly due to the fact that it consists of small nations that trade between themselves. Western Europe and North America combined handle about two-thirds of the world's exports. Japan, the U.S.S.R., and eastern Europe are also important; consequently, it can be seen that the industrialized nations are also the chief trading nations.

Top: Heavy industry at Pittsburgh, in the United States. Centre: An atomic-power plant in France. Below: Mass production in a Japanese factory.

Dictionary

Abrasion The wearing away of rock surfaces or other surfaces by the rubbing action of materials carried by wind, running water, or moving ice. *See* ATTRITION; CORRASION; DENUDATION; EROSION; WEATHERING.

Accretion The gradual building up of water-borne material such as mud and sand in an area that is regularly flooded.

Advection A horizontal movement of air or a liquid. The term is often used in connexion with the movement of warm air from tropical to temperate latitudes or from sea to land. *See* CONVECTION.

Agriculture The science and practice of cultivating the land in order to produce crops. The term is often used in a wider sense to include the rearing of livestock.

Air *See* ATMOSPHERE.

Air mass A mass of air having a uniform temperature and humidity. Its boundaries are called *fronts*. An air mass develops when the air over a broad land or sea surface remains still long enough to adopt the surface temperature and humidity. It retains these qualities when it moves elsewhere. Air masses are described as *tropical* or *polar* (hot or cold); *maritime* or *continental* (moist or dry). The four main types are tropical maritime, tropical continental, polar maritime and polar continental. *See* FRONT.

Air-stream A breeze or wind; any moving current of air.

Alfalfa *See* LUCERNE.

Alluvium Solid material carried by a river and eventually deposited on the river bed, on nearby flooded land, or in lakes or deltas. It may include silt, sand, gravel and organic material.

Alp A mountain pasture, especially in Switzerland, formed of a gentle slope high up in the mountains. It is usually covered with snow in winter, but provides rich grazing for farm animals in summer.

Alpine peaks enclose the Engadine, part of the valley of the River Inn, in Switzerland. The village of St. Moritz in the valley is a popular resort.

Alpine Having to do with the Alps, the largest mountain range in Europe. The word is also used (with small initial *a*) to describe any high mountains, or their climate, plants and wildlife.

Altitude The height of any object, or the vertical distance above mean sea level. The altitude of any feature on the Earth's surface, such as a mountain, is usually measured in feet or metres. *See* MEAN SEA LEVEL.

Anemometer An instrument for measuring the speed and direction of the wind.

Aneroid barometer An instrument that measures the pressure of air without using mercury or any other fluid. It is made up of a hollow metallic box from which almost all the air has been removed. Changes in outside air pressure cause the sides to bulge out or cave in. These movements are registered by a needle on a graded scale. *See* ATMOSPHERIC PRESSURE; BAROGRAPH; BAROMETER.

Antarctic As an adjective, the term describes the South Polar regions. As a noun, it refers to that part of the Earth's surface that lies within the Antarctic Circle (latitude 66° 32′ S).

Anticline An arch-like fold in the layers of rock of the Earth's crust, formed when pressure from each side forced a section of rock upwards in a bulge. *See* FOLD.

Anticyclone A region of high atmospheric pressure, with the highest pressure at its centre. A spiral flow of air moves clockwise from the centre in the Northern Hemisphere; anti-clockwise in the Southern Hemisphere. An anticyclone is usually accompanied by settled weather. *See* CYCLONE.

Antipodes Points that are opposite to each other on a diameter of the Earth (180° of longitude apart). People in Britain sometimes call Australia and New Zealand 'the Antipodes' because they are 'at the opposite end' of the Earth.

Archipelago A group of islands in a sea or other broad expanse of water.

Arctic The region of the North Pole and that near the Pole. Strictly speaking, the Arctic is the region within the Arctic Circle (latitude 66° 32′ N).

Artesian basin A basin-shaped layer of porous rock in the Earth's crust that is trapped between layers of non-porous rock, thus also trapping water. *See* ARTESIAN WELL.

Artesian well A deep, narrow well sunk into an artesian basin and producing a continuous stream of water at the surface. The water is forced upwards because the outlet of the well is below the level of the water's source. *See* ARTESIAN BASIN.

Atmosphere Air; the envelope of mixed gases that surrounds the Earth, up to a height of some 300 miles (about 480 km). It consists of about 78% nitrogen and about 21% oxygen. Argon, carbon dioxide, helium and some other gases make up the remaining 1%. Often, the atmosphere also contains water vapour.

Atmospheric pressure The pressure of the atmosphere on the surface of the Earth as a result of its weight of about 6,000 million million tons. The average atmospheric pressure at sea level is 14.7 pounds per sq inch (1,033 g per sq cm). Atmospheric pressure decreases with altitude. *See* ATMOSPHERE; BAROMETER.

Atoll A circular or horseshoe-shaped coral reef that encloses a lagoon. *See* CORAL REEF; LAGOON.

Attrition The constant wearing down of pieces of rock into ever finer particles as they are carried along by wind, water, or ice. It differs from *abrasion*, the wearing away of rock surfaces. *See* ABRASION; CORRASION; DENUDATION; EROSION; WEATHERING.

Aureole (i) The zone of rocks immediately surrounding a mass of igneous rock and metamorphosed as a result. (ii) A corona or halo round the sun or moon. *See* IGNEOUS ROCK; METAMORPHIC ROCK.

Aurora Spectacular coloured lights seen in the sky at night. Those seen in the Northern Hemisphere are called the *Aurora Borealis* or *Northern Lights*; those in the Southern Hemisphere, the *Aurora Australis* or *Southern Lights*.

Autumn The season of the year between summer and winter. *See* SEASON; SUMMER; WINTER.

Avalanche The fall down a mountainside of masses of snow, ice, rocks, or earth. *See* LANDSLIDE.

Backing An anti-clockwise change in the direction of the wind; for example, from N to NW. *See* VEERING.

Badlands A region where erosion by wind and water has resulted in deep ravines and steep hills. The name was originally given to typical areas of South Dakota and western Nebraska in the United States. *See* EROSION.

Balloon-sonde, Ballon-sonde A balloon carrying instruments for checking the weather.

Bar (i) A ridge of material such as mud, stones, or sand that forms across a bay or the mouth of a river. (ii) A unit of atmospheric pressure equal to one million dynes per sq cm. It is equivalent to 29.53 inches (750 mm) of mercury at 0° C in latitude 45°. Usually the *millibar* is used in measuring atmospheric pressure. *See* MILLIBAR.

Barograph An aneroid barometer that records atmospheric pressure on a moving drum. A pen shows pressure changes hour by hour as a continuous line drawn on graph paper fastened round the drum. *See* ANEROID BAROMETER; ATMOSPHERIC PRESSURE; BAROMETER.

Barometer An instrument for measuring the pressure of the atmosphere. A common type of barometer, the *mercury barometer*, indicates atmospheric pressure by balancing the weight of a column of mercury against the weight of a column of air. *See* ANEROID BAROMETER; ATMOSPHERIC PRESSURE; BAROGRAPH.

Barrier reef A coral reef lying parallel to the shore and separated from it by a wide stretch of deep water. Usually there are many openings in the reef. *See* CORAL REEF.

Barysphere (i) The core of the Earth. (ii) The whole of the Earth's interior, including mantle and core. *See* CORE; EARTH; MANTLE.

Basin (i) A hollow in the Earth's crust; if filled with water it may be a *lake basin* or an *ocean basin*. (ii) The area drained by a river system. *See* ARTESIAN BASIN.

Bay A wide, open, curving arm of the sea, or a lake within the coastline or shoreline.

Beach The shore of a sea, consisting of a strip of pebbles, sand, or mud lying between low and high water marks.

Bearing The direction of one object from another; the horizontal angle between the meridian and an object seen by the observer, measured in degrees clockwise from N. *See* MERIDIAN.

Beaufort Wind Scale A internationally accepted series of numbers devised by Admiral Sir Francis Beaufort in 1805 to indicate different wind strengths.

Beaufort No.	*Description*	*Speed (mph)*
0	Calm	Less than 1
1	Light air	1–3
2	Light breeze	4–7
3	Gentle breeze	8–12
4	Moderate breeze	13–18
5	Fresh breeze	19–24
6	Strong breeze	25–31
7	Near gale	32–38
8	Fresh gale	39–46
9	Strong gale	47–54
10	Whole gale	55–63
11	Storm	64–75
12	Hurricane	above 75

Dwarfed by the huge snow-laden mountains of Antarctica, a research ship sails through the frozen waters of the Lemaire Channel.

Bedrock Solid, unweathered rock lying beneath the soil and subsoil with which most of the Earth is covered.

Belt A zone of vegetation or rock outcrops that has its own definite characteristics, making it distinct from the rest of the region.

Bill A long, narrow cape or headland jutting out to sea. *See* CAPE; HEADLAND; PENINSULA; POINT.

Biosphere The surface zone of the Earth where living things can exist. It extends a short distance into the Earth's crust and upwards into that part of the atmosphere that contains sufficient oxygen to support life.

Blizzard A blinding snowstorm with strong icy winds.

Blow-hole Hole resulting from the partial collapse of the roof of a coastal cave. Spray shoots through the hole at high tide and can be seen above the cliff-top level.

Right: The Great Barrier Reef, off the coast of Queensland, in Australia, stretches for about 1,250 miles (2,000 km).

Below: A tranquil inlet on the Dalmatian coast of Yugoslavia, with the village of Sveti Stefan.

Bluff A steep headland or cliff. Specifically, the term refers to the remains of a spur that has been cut back by the wearing away of a meandering river valley. *See* SPUR.

Bog An area of waterlogged, spongy ground with rotting vegetation lying on the surface. Eventually the vegetation turns into a layer of peat. *See* PEAT; MARSH.

Bore A wall of foaming water that travels at speed up a river. It occurs when a tidal wave enters an estuary and breaks as it proceeds up river. *See* ESTUARY; TIDAL WAVE.

Boreal The *boreal zone* is a climatic zone in which winters are snowy and summers short.

Boulder clay Thick clay with stones of all sizes formed of rock that has been crushed and deposited by glaciers and ice sheets. *See* CLAY.

Bourne A stream that flows along a normally dry valley and that appears only when the water table rises. *See* WATER TABLE.

Breaker A wave that breaks as it reaches shallow water near the shore.

Breakwater A man-made barrier built out from the sea coast to break the force of the waves and so protect the shore.

Bush An area of wild and uncleared vegetation, shrubby or wooded. The term is commonly used in southern Africa and in Australia.

Buttress A mass of rock that juts out of a mountainside.

Canal A man-made inland waterway used for transport. Sometimes a river that is too narrow or too shallow for navigation is 'canalized' to enable it to take boats. Small canals are also constructed to irrigate arid land. *See* IRRIGATION.

Cancer, Tropic of A line of latitude at 23° 32′N of the equator where the sun shines directly overhead on June 21. It marks the northern boundary of the Tropical Zone. *See* CAPRICORN, TROPIC OF.

Canyon A steep-sided, usually deep gorge cut by a river through soft rock in a dry region.

Cape A large, rocky headland jutting out to sea.

Capricorn, Tropic of A line of latitude at 23° 32′ S of the equator where the sun shines directly overhead on December 21. It marks the southern boundary of the Tropical Zone. *See* CANCER, TROPIC OF.

Cardinal points The four main points of the compass: north, south, east and west. *See* COMPASS.

Cartography The art of drawing maps and charts. *See* CHART; MAP; MAP READING.

Catchment area The area from which a river draws its water supply; limited by the watershed—the ridge on the other side of which the water flows away from the area. Also called *drainage area*. *See* DRAINAGE; RIVER; WATERSHED.

Cave, Cavern A hole in the Earth's crust, usually running horizontally or almost horizontally.

Celsius scale A scale used on thermometers, an alternative name for the *centigrade scale*. Anders Celsius (1704–1744), a Swedish astronomer, invented the scale in 1742 by dividing the interval between the freezing and boiling points of water into 100 parts. Freezing point = 0°, boiling point = 100°. (On the original scale these markings were reversed.) To convert Celsius into Fahrenheit, multiply the C reading by 9/5 and add 32. Thus F = 9C/5 + 32. *See* FAHRENHEIT.

Centigrade scale *See* CELSIUS SCALE.

Channel (i) The deep navigable part of a harbour. (ii) The deepest part of a river bed. (iii) A stretch of sea betwen two land masses. (iv) An irrigation ditch. (v) A tapering estuary.

Chart (i) A map of the sea and coastline, drawn up for use by navigators. (ii) A weather map. (iii) Sometimes, loosely, any map. *See* MAP.

Chinook A warm, dry SW wind that blows down the eastern slopes of the Rocky Mountains in the United States and Canada in winter and early spring.

Cirrus A type of high cloud with wispy and fibrous-looking bands and streamers. It is sometimes called 'mares'-tails' or 'banners'. *See* CLOUD.

City (i) A large and important town, where thousands of people live and work. (ii) An ancient town, which may be small, with a cathedral and bishop. (iii) In some countries, a town so designated by royal charter or letters patent.

Cliff A steep rock face.

Climate The general weather conditions over a fairly large area of the Earth's surface, observed during a period of years. Sunshine, rainfall, altitude and geographical location all affect climate. *See* WEATHER.

Climatology The scientific study of the Earth's various climates—their distribution and the way they affect human, animal and plant life. *See* CLIMATE; METEOROLOGY; WEATHER.

Cloud A mass of water vapour floating in the air; it is made up of condensed droplets of water or ice crystals. Clouds usually form high above the Earth's surface and adopt many different shapes. They are classified according to their height and shape. *See* CIRRUS; CONDENSATION; CUMULONIMBUS; CUMULUS; NIMBOSTRATUS; RAIN; STRATUS.

Coastline The shoreline of the continents and other land masses. Also, specifically, the highwater mark of medium tides or, sometimes, the mark reached by the biggest storm waves.

Col A high pass in the mountains; a saddle-like gap between two peaks or across a ridge.

Cold desert (i) A term for the polar regions and tundra, where plant life is limited or non-existent because of snow and ice. (ii) A high inland plateau such as the Gobi Desert, in Asia, whose altitude and distance from the sea make it very cold in winter and dry. *See* DESERT; TUNDRA.

Cold front The boundary between a mass of warm air and a mass of colder air that overtakes it and thrusts underneath it.

The Grand Canyon in Arizona, in the United States, was cut by the Colorado River. It extends for more than 200 miles (320 km) and is more than one mile (1.6 km) deep in places.

Compass An instrument for finding directions. It has a magnetized needle that turns freely on a pivot and always points to the magnetic North and South Poles. The needle is mounted on a card marked in degrees and showing the cardinal points. *See* CARDINAL POINTS; GYRO-COMPASS.

Condensation Water droplets formed when the air has more water vapour than it can hold. The amount it can hold depends on its temperature. *See* DEWPOINT; SATURATION.

Coniferous forest A forest whose trees are mostly evergreen and cone-bearing, with needle-like leaves. Such forests grow in a variety of climates from the subtropical to the subarctic, and yield valuable softwood timber.

Continent One of the seven major unbroken land masses of the Earth: Asia, Europe, Africa, North America, South America, Australia and Antarctica. *See* CONTINENTAL DRIFT.

Continental drift The theory that the continents, during millions of years, have shifted their positions by 'drifting'. It suggests that originally there was only one huge continent and that in time it broke into several pieces.

Continental shelf Shelf round the coasts of the continents. It slopes gently and is covered by shallow water. Its approximate edge is at the depth 100 fathoms (about 600 ft or 182 m).

Continental slope Slope leading from the outer edge of the continental shelf to the deep ocean floor. Its slope is steeper than that of the shelf. *See* CONTINENT; CONTINENTAL SHELF.

Contour line Line on a map joining all places at the same height above (or depth below) sea level.

Convection The upward flow of air that has been heated by contact with the Earth's surface. As it warms, it expands and rises. Cold air takes its place at the surface and is in turn heated and caused to rise. *See* ADVECTION.

Coral reef A ridge of rock in the sea that is composed chiefly of the skeletons of small animals called *reef-building coral polyps.*

Core The dense, innermost part of the Earth, with a diameter of about 4,320 miles (6,950 km). There is evidence of a molten *outer core* and a solid *inner core. See* CRUST; MANTLE.

Coriolis Force An effect, caused by the rotation of the Earth, on objects moving over the Earth's surface. Such objects appear to be pushed to the right in the Northern Hemisphere and to the left in the Southern Hemisphere.

Corrasion The wearing away of a river's bed by the abrasive action of solid materials carried in the water. *See* ABRASION; EROSION.

Corrosion The wearing away of rocks as the result of a chemical process. An example is the corrosion of limestone rocks by carbonic acid in solution. *See* EROSION.

Right: The white cliffs of England. Steep chalk cliffs rise above the waters of the stormy English Channel.

Below: The crater of Mount Ararat in Turkey. The volcano last erupted in 1840, causing great loss of life and property.

Crater (i) The funnel-shaped depression at the top of a volcano. (ii) A hollow in the Earth's surface where a meteorite has struck. *See* VOLCANO.

Crater lake A lake formed in the crater of a volcano. *See* CRATER; LAKE; VOLCANO.

Creek (i) A small stream. (ii) A small inlet or tidal estuary of a river.

Crust The outer shell of the Earth, about 10 to 30 miles (16 to 50 km) thick. It is also called the *lithosphere. See* CORE; MANTLE.

Cumulonimbus A type of dense thundercloud, often with a very dark base and sometimes with an anvil-shaped top.

Cumulus A white, detached, billowing type of cloud formed in upward, convection currents of air. *See* CLOUD; CONVECTION.

Current (i) The upward movement of an air mass. (ii) The distinct flow of water in the channel of a river or stream. (iii) The driving force of an ocean. Ocean currents are caused by the wind and the rotation of the Earth.

Cyclone A region of low atmospheric pressure in which winds spiral inwards towards the centre of lowest pressure. In the Northern Hemisphere, the winds blow anticlockwise; in the Southern Hemisphere, clockwise. In temperate regions cyclones are called *depressions*; in tropical regions they are more violent, and are called *typhoons* or *hurricanes. See* DEPRESSION; HURRICANE; TYPHOON.

Dam A barrier built across a river to hold back and control the flow of water.

Dead reckoning Navigation by calculating direction and distance travelled.

Debris Loose surface material from the break up of rocks; also called *detritus.*

Deciduous forest A cool temperate forest made up of broad-leaved trees such as oak, elm and beech that shed their leaves each winter and renew them the following spring. In monsoon forests, trees shed their leaves during the hot season.

Deep *See* TRENCH.

Degree (i) A measurement unit of temperature. It has different values, according to whether the scale used is Celsius or Fahrenheit. (ii) A unit of angular measurement equal to 1/360th part of a circle, used to measure latitude and longitude. *See* CELSIUS SCALE; FAHRENHEIT SCALE; LATITUDE; LONGITUDE; TEMPERATURE; THERMOMETER.

Delta A fan-shaped area of land where a river enters a body of standing water.

Denudation The wearing away of a land surface by erosion and weathering. *See* ABRASION; ATTRITION; CORRASION; EROSION; WEATHERING.

Deposition The laying down of rock particles carried by wind, water or ice. *See* EROSION.

Depression (i) A hollow or low-lying area in the Earth's surface. (ii) A centre of low pressure, also called a *cyclone*, producing unsettled weather. *See* CYCLONE.

Desert A large area of land that lacks water. A desert region is sometimes defined as one having less than 10 inches (250 mm) of rain a year. Deserts include *hot deserts*, such as the Sahara, and *cold deserts*, such as the Gobi. *See* COLD DESERT.

Dew point The temperature below which water vapour in the air begins to condense into droplets of water. *See* CONDENSATION; SATURATION.

Divide *See* WATERSHED.

Doldrums A low-pressure belt round the equator with ligh surface winds or calms. The doldrums are hot and often wet. Sailing ships avoided them because of the danger of being becalmed.

Down An upland tract of gently rolling pasture land, especially the chalk hills of southern England.

Drainage (i) The process of water loss from an area by means of streams that flow into a river. (ii) The drying-out of wet land in order to make it suitable for agriculture or building. Surplus water is led away in ditches or drainpipes. *See* BASIN; CATCHMENT AREA.

Drought An unusually long period of dry weather in a normally fertile area. As a result of a prolonged drought, crops may wither away and animals die.

Drowned valley A valley that has been filled with water usually as the result of a change in sea or land level.

Dune A hill or mound of sand formed by the wind blowing over beaches or deserts. Crescent or half-moon-shaped dunes are called *barkhans*; long dunes in parallel lines are called *seif-dunes*.

Dust Minute solid particles in the air, so small and light that they settle only when the air is still.

Dust bowl An area of drought or semi-drought and dust storms where the wind has blown or is blowing the topsoil away.

Dust storm A storm in which a strong wind sweeps up the topsoil from the land, producing choking clouds of dust. *See* DUST BOWL.

Earth A general term for the planet on which we live, the solid material of which it is made, or the loose material (soil) that lies on the surface over solid rock. *See* CORE; CRUST; MANTLE; SOLAR SYSTEM.

Earthquake A sudden rolling, shaking or trembling of the Earth's crust, caused by the breaking and movement of rocks or by volcanic action. Most earthquakes take place along belts where new mountains and volcanoes have been formed.

Easting Distance that a point on a map is eastwards from a given meridian, usually the meridian that is the origin of the map grid. *See* GRID; MERIDIAN; NORTHING.

Eclipse The obscuring of the whole or part of a heavenly body by another. A *lunar eclipse* occurs when the Earth lies between the sun and the moon; a *solar eclipse* occurs when the moon comes between the Earth and the sun.

Ecology The study of the relationship between living things (plants and animals) and their environment. It may be subdivided into human, animal, plant and bio-ecology (the effect of animal and plant life on each other).

The Kariba Dam on the River Zambezi in Africa provides hydro-electric power for Zambia and Zimbabwe. Lake Kariba, formed by the dam, is 175 miles (280 km) long.

Environment The sum total of the conditions affecting the lives of people, animals and plants. These conditions may be internal, such as body temperature and bacteria; or external, such as material surroundings and climate.

Epicentre A point or line on the Earth's surface directly above the point of origin of an earthquake. *See* EARTHQUAKE.

Epoch A subdivision of a Period in geological time. Only the Tertiary and Quaternary Periods are subdivided in this way. *See* GEOLOGICAL TIME.

Equator The (imaginary) great circle round the Earth at right angles to the Earth's axis, at latitude 0°, and midway between the North and South Poles. *See* AXIS OF THE EARTH; GREAT CIRCLE.

Equatorial rain-forest Dense, evergreen forest that grows in a region within about 7° N and S of the equator. This is an area of heavy rainfall, high temperatures and almost no difference between the seasons.

Equinox The time of the year when the sun is directly over the equator at noon. The word means 'equal night' and is a time when days and nights are of equal length in all parts of the world. The sun 'crosses the equator' twice a year so there are two equinoxes in a year: the *spring* or *vernal* equinox occurs about March 21 and the *autumn* or *autumnal* equinox about September 20.

Era A main division of geological time, subdivided into a number of Periods. The Eras are the *pre-Cambrian, Palaeozoic, Mesozoic* and *Cenozoic*.

Erg The sandy part of the Sahara.

Erosion The wearing down of earth and rock and the carrying off of the debris by wind, water or ice. *See* ABRASION; ATTRITION; CORRASION; CORROSION; DENUDATION; WEATHERING.

Escarpment A steep slope or inland cliff, produced usually by erosion of dipping sedimentary rocks.

Dune. Shifting wastes of sand are fringed by steppe lands and hills in the Gobi Desert of Mongolia and China.

Estuary The broad V-shaped mouth of a river, formed generally by the sinking or washing away of the coastal land. When the tide flows in, river and sea water mix. *See* BORE; DELTA; DROWNED VALLEY.

Evaporation The change of a liquid into a vapour (gas). The water vapour in the atmosphere results from the evaporation of surface water on the Earth. *See* CONDENSATION; DEW POINT.

Fahrenheit scale A scale used on thermometers, the freezing point of water being 32° and the boiling point 212°. To convert Fahrenheit into Celsius, first subtract 32 from the Fahrenheit reading, then multiply the result by 5/9. *See* CELSIUS SCALE; TEMPERATURE; THERMOMETER.

Fathom A unit of length used mainly to measure water depth; equal to 6 ft (1.8 m).

Fault An enormous crack or break in the rock strata of the Earth's crust, caused by horizontal or vertical movement.

Fauna The animal life of a specific geological period or of a region of the Earth. The corresponding term for plant life is *flora.*

Fertilizer Plant food, either natural or artificial, used as an additive to increase the fertility of the soil. Fertilizers contain nitrogen, phosphorus, potassium or other elements necessary for plant growth.

Fiord, Fjord A long, narrow inlet of the sea, usually enclosed by steep cliffs. The fiords of southern Norway are famous for their beauty. Probably, they are the result of erosion by glaciers moving seawards. *See* GLACIER.

Floe A horizontal sheet of floating ice that has become detached from the polar ice mass. *See* ICEBERG.

Flood The inundation by water of an area that is normally dry.

Flood plain Low-lying land by the side of a river that is regularly flooded when the river overflows its banks. The river deposits alluvium and gradually builds up the flood plain. *See* ALLUVIUM; FLOOD.

Flora The plant life of a specific geological period or of a region of the earth. The corresponding term for animal life is *fauna.*

Fluvial Of or belonging to a river.

Fog Low cloud that forms near the surface of the land or on a body of water. *Advection fogs* form when warm air touches cold areas of land or water. *Radiation fogs* occur when ground warmth is released quickly into the upper atmosphere and the air close to the ground is chilled as a result. In the International Meteorological Code, a fog exists when objects 1 kilometre away can no longer be seen. *See* CLOUD; CONDENSATION.

Föhn A warm and dry wind that blows down the lee slope of a mountain. It is common in the valleys of the northern Alps. In North America it is called a *chinook.*

Fold, Folding The bending of rock strata, usually molten strata, by pressures beneath the Earth's crust. Folding usually takes place along lines of weakness.

Foreland A headland jutting out into the sea. *See* CAPE; POINT.

Forest A large area of unbroken woodland. *See* CONIFEROUS FOREST; DECIDUOUS FOREST; EQUATORIAL RAIN-FOREST.

Fossil The impression or remains of an animal or plant that lived during some earlier age, preserved in sedimentary rocks, in ice, or in amber.

Freezing-point The temperature at which a liquid becomes cold enough to turn into a solid. Each liquid has its own freezing-point. Water freezes at 0°C (32°F). *See* CELSIUS SCALE; FAHRENHEIT SCALE; FROST; ICE; TEMPERATURE.

Frigid zone The climatic zone of cold winters and cool summers in polar regions. *See* PERMAFROST; TEMPERATE ZONE; TORRID ZONE.

Front The boundary between cold and warm air masses at the Earth's surface. *See* COLD FRONT; OCCLUDED FRONT; WARM FRONT.

Frontier A narrow track of country on either side of the boundary line between two nations. Sometimes it is confused with *boundary,* which is the actual line of demarcation.

Frost Thousands of tiny crystals of frozen moisture that form on solid objects when the temperature falls to 0°C (32°F) or below. *See* FREEZING-POINT; TEMPERATURE.

Fumarole A hole or vent in the ground in volcanic regions through which fumes or gases escape.

Gale A strong wind that blows at speeds of between 39 and 63 mph (62–101 kph)—numbers 8, 9 and 10 on the Beaufort Wind Scale. *See* BEAUFORT WIND SCALE; WIND.

Gap A break in a ridge through which a river sometimes flows. A dry gap is known as a wind-gap.

Geography The study of the Earth's surface as the home of mankind. It describes its physical features, resources, climate, soils, plants, animals and peoples and their distribution. It draws largely from the Earth sciences and social sciences, such as geology, meteorology, botany, zoology, economics and history.

Geoid A term—meaning 'Earth-shaped body'—that

refers to the general shape of the Earth. Although the Earth is spherical, it is not a true sphere (globe or ball) because it bulges slightly towards the equator.

Geological time The Earth's history described chronologically from the Pre-Cambrian Era, through the Palaeozoic and Mesozoic Eras, to the Cenozoic Era. The chief divisions of geological time are Eras, Periods and Epochs. Broadly speaking, each Era ended in a time of orogeny—active mountain building. *See* EPOCH; ERA; OROGENY; PERIOD.

Geology Earth science; the study of the history, build-up and materials of the Earth and of the forces that act upon it. More specifically, it deals with the history and structure of the Earth's crust.

Geosyncline An enormous downward curve or inverted arch in the Earth's crust, generally many miles wide and hundreds of miles long—a *syncline* on a very large scale. Marine sediments that settled at the bottom of the curve have resulted in the formation of thick deposits of sedimentary rock. *See* SYNCLINE.

Geyser A hot spring that periodically throws steam and hot water up into the air with explosive force. Geysers are common in the north island of New Zealand, in Iceland and in Yellowstone National Park in the United States. *See* HOT SPRING.

Glaciation The covering of part of the Earth's surface by an ice sheet or a glacier. Major glaciation occurred in northern Europe and northern North America during the Pleistocene Epoch. *See* ICE AGE.

Glacier A very slow-moving river of ice that creeps down a valley from a snowfield above the snow line. It moves under the influence of gravity.

Globe Another name for the Earth; a sphere with a map on it, representing the Earth.

Gradient (ii) A slope whose steepness is expressed either in degrees as an angle to the horizontal, or as a proportion between the vertical interval and the horizontal. (ii) The slope of the bed of a stream, expressed usually in ft per mile or cm per km.

Grassland A plain where grass is the natural and main vegetation. Grasslands do not receive enough rainfall for extensive tree growth, but are wetter than deserts.

Gravity, Gravitation The force that attracts masses of matter to each other. Gravitation holds the universe together. The moon's and sun's gravity causes the tides on Earth..

Great Circle An imaginary line drawn right round the Earth, with its plane passing through the Earth's centre. The equator is a great circle. Each two corresponding meridians together form a great circle.

Green belt A strip of open country surrounding a town or lying between towns, with the aim of preventing over-building. Green belts include parks, playing fields and unspoilt countryside and are kept free of houses, factories and other development.

Greenwich Mean Time (G.M.T.) A reckoning of time that is known as *standard time* in the British Isles and Western Europe. The sun's crossing of the meridian 0° is taken as 12 noon. *See* GREENWICH MERIDIAN; STANDARD TIME.

Greenwich meridian The meridian 0° that passes through Greenwich in England, the former site of the Royal Observatory. Longitude is expressed in degres E or W of this meridian. *See* GREENWICH MEAN TIME; STANDARD TIME.

Grid Horizontal and vertical lines drawn on a map to form a pattern of squares. Each line is numbered and any point on the map can be referred to by giving its *co-ordinates*—its position in relation to nearby lines. The vertical lines are called *eastings* and the horizontal lines *northings*.

Groyne A barrier or fence built out into the sea from the shore in order to stop the drift of sand or shingle.

Gulf A large, deep sea inlet extending far inland. *See* BAY.

Gyro-compass A compass based on the use of a spinning gyroscope. The axle of the gyroscope stands parallel to the Earth's axis, thus indicating true north. Unlike an ordinary compass, the gyro-compass makes no use of the Earth's magnetism. *See* COMPASS.

Left: Ayers Rock, a giant eroded outcrop of sand-stone in Central Australia, rises 1,110 ft (335 m) above the plain. Depending on the light, its colour varies from purple to red or orange.

Below: A rocky foreland, part of the bleak Cliffs of Moher on the coast of Clare, in Ireland.

Above: The Lauterhorn Glacier in Switzerland.

Centre: The Great Geysir, north of Reykjavik in Iceland, has given its name to all other geysers. Today, its activity is reduced, but it remains one of the largest of the dozens of similar hot springs nearby.

Below: The Monte Rosa glacier, on the borders of Italy.

Habitat The natural environment of any particular animal or plant.

Hail, Hailstones Hard globules of ice that fall from cumulonimbus clouds, often during thunderstorms.

Hamada Bare rocky areas of the Sahara. *See* DESERT.

Harmattan A strong, hot, dusty wind that blows over parts of western Africa from the Sahara. *See* WIND.

Haze A cloud of dust, smoke, salt or other particles that reduces visibility close to the Earth's surface. A haze is said to exist when visibility is less than 1.25 miles (2 km) but more than 0.6 miles (1 km). *See* FOG.

Headland *See* CAPE.

Headwater The source of a river or a river system.

Heath, Heathland A stretch of open, uncultivated country covered with low shrubs such as ling.

Heat wave A spell of weather that is unusually hot and dry for the area in which it occurs.

Hemisphere One half of a sphere or globe. On the Earth's surface, the half north of the equator is called the *Northern Hemisphere*; the half south of the equator, the *Southern Hemisphere.*

Highland Another term for a high upland. Sometimes the word is used in the plural as a proper name, as in the *Highlands of Scotland. See* UPLAND.

High seas The open sea or ocean outside any country's territorial waters.

High water The highest point reached by the tide; the time at which the tide is highest. *See* TIDE.

Hinterland Region lying inland from a coast or port.

Horizon The line where Earth and sky seem to meet and beyond which the observer cannot see because of the curvature of the Earth.

Horse Latitudes Subtropical zones of high pressure on either side of the equator close to latitudes 30°N and 30°S. They lie between the trade winds and the westerlies and are marked by light winds or calms and light rainfall. *See* TRADE WINDS; WESTERLIES.

Hot spring A continuous stream of hot water that bubbles to the surface from underground in volcanic regions. It differs from a geyser in that the water does not come out with explosive force. Such a spring is also called a *thermal spring. See* GEYSER.

Humidity The amount of water vapour in the air. The air can hold only a certain amount of water vapour at any given temperature. The warmer it is, the more it can hold. When it can hold no more it is said to be *saturated.* The amount of water vapour in the air compared with the amount the air would hold when saturated is the *relative humidity. See* CONDENSATION; HYGROMETER; SATURATION.

Hurricane (i) A severe tropical cyclone that blows occasionally over the West Indies and the Gulf of Mexico. Hurricanes originate in mid-Atlantic and generally move first westwards and then north-eastwards. (ii) A wind *Force 12* or more on the Beaufort Wind Scale. *See* BEAUFORT WIND SCALE; CYCLONE;

DEPRESSION; TYPHOON; WIND.

Hydro-electric power Electricity produced by water power. Waterfalls, natural or manmade, turn turbines and dynamos, thus generating electricity. *See* DAM.

Hydrological cycle *See* WATER CYCLE.

Hydrosphere The waters of the Earth, which cover two-thirds of its surface. The hydrosphere includes seas, lakes, rivers, ice and snow-fields, and the water vapour in the atmosphere.

Hygrometer An instrument that measures relative humidity.

Ice The solid state of water, formed when the air temperature falls to freezing-point or below. Ice floats on water because it is less dense. *See* FREEZING-POINT; FROST.

Ice Age A time when much of the Earth's surface was covered with ice. The term commonly refers to the latest Ice Age, the Pleistocene Epoch, which began about 1 million years ago. During this Epoch much of North America and northern Europe was covered with ice.

Iceberg A huge mass of ice floating in the sea after having broken off from an ice barrier or the end of a glacier. Only about one-ninth of an iceberg can be seen above the water. *See* FLOE.

Ice-cap A mass of ice covering land in polar regions. Huge ice-caps, such as those that cover Antarctica and much of Greenland, are sometimes called *ice-sheets*.

Ice-field A large continuous sheet of ice; a very large floe. *See* FLOE.

Igneous rock Rock that has solidified from molten rock or *magma*. It is one of the three main kinds of rocks that make up the Earth's crust. *See* MAGMA; ROCK.

Impermeable rock Non-porous rock that therefore does not allow fluids to seep through it.

Inland sea A sheet of water, larger than a lake, that is not connected with the open sea—for example, the Dead Sea. *See* LAKE.

Inlet A narrow bay. *See* BAY; GULF.

Insolation Radiant energy from the sun received by the Earth and its atmosphere in a spectrum of radiation. *See* RADIATION.

International Date-Line An imaginary line that roughly follows the 180° meridian in the Pacific Ocean. E and W of the Date-Line the date differs by one day. The line zigzags in places to avoid land areas. Each 15° longitude means one hour's difference in time (the earth turns through 360° in 24 hours). To make up for the total difference of 24 hours between one side of the line and the other, ships sailing E (or planes flying E) alter the date to gain a day, and those sailing W lose a day.

Inversion of temperature An increase of air temperature with increase in altitude, unlike the usual decrease in temperature with increasing height.

Irrigation The artificial channelling of water to dry land in order to make it suitable for agriculture.

Granite. A towering granite 'peninsula' on the coast of Victoria in Australia.

Island A piece of land, smaller than a continent, entirely surrounded by water.

Isthmus A narrow bridge of land joining two large land areas—for example, the Isthmus of Panama, which joins South America to Central America.

Jet stream A westerly current of air moving in the upper atmosphere at speeds of between 60 and 270 mph (100 and 430 kph). Such air streams occur at heights of about 40,000 ft (12,000 m).

Karst A rugged type of landscape made up of barren limestone, such as occurs in north-western Yugoslavia, the northern Pennines in England, and the Yucatan region of Mexico. Its main features are dry valleys, sinkholes, caverns and underground streams.

Kilometre A measure of length. It equals 1,000 metres or 0.621 statute miles.

Lagoon (i) A shallow stretch of water separated completely or partly from the sea by a narrow strip of land. (ii) A stretch of water enclosed by an atoll. (iii) A stretch of water lying between a coral reef and the mainland. *See* ATOLL; CORAL REEF.

Lake A body of water surrounded by land. Very large lakes are called *inland seas*. The hollows that fill with water to form lakes may be caused by erosion, deposition, volcanic action, or by faults in the Earth's crust. *See* CRATER LAKE; DEPOSITION; EROSION; INLAND SEA.

Land breeze A breeze flowing from land to sea at night as a result of the fact that land heats and cools more rapidly than the sea. Faster cooling of the land at night causes higher pressures over the land, so that air flows out to the sea.

Landslide, Landslip The slipping of large masses of rock and Earth down the side of a mountain or hill. It may result from the action of water seeping into the Earth, earthquakes, waves undermining a sea coast, or cracks in the rocks. *See* AVALANCHE.

Latitude Angular distance in degrees N or S of the equator of any point on the Earth's surface. Latitude is measured along an imaginary north-south line called a *meridian*. The latitude of the equator is 0°. *See* MERIDIAN; PARALLEL OF LATITUDE.

Lava White-hot molten rock or magma thrown out by a volcano. *See* MAGMA; VOLCANO.

Right: An island in the River Rhine, at Kaub in Germany. The toll-collecting castle of Pfalzgrafenstein was built in 1327.

Below: A 'dream island' in Polynesia, in the Pacific.

Lee, Leeward The side away from or protected from the wind.

Levee A ridge or bank along the edges of a river's channel. When a river floods, it deposits material on its banks.

Lightning Static electricity in the sky. It is an enormous electric spark caused by a discharge between positively-charged and negatively-charged areas within clouds, between clouds, or between a cloud and the earth. *See* THUNDERSTORM.

Light-year The distance light travels in one year—a measure often used in astronomy. It is equal to 5,880,000 million miles (9,463,000 million km).

Lithosphere *See* CRUST.

Littoral The seashore, or the stretch of land between high and low water marks.

Loch A Scottish lake, or a long, narrow arm of the sea such as is common along the coast of north-western Scotland.

Lode A thick vein of mineral ore in rock, or a number of closely parallel veins.

Loess Fine, clinging, yellowish dust that has been deposited as a soil by the wind. It comes from deserts and from areas that were once covered by ice sheets.

Longitude The angular distance measured in degrees east or west of the prime meridian—0°, the Greenwich meridian. *See* GREENWICH MERIDIAN; LATITUDE; MERIDIAN; PARALLEL OF LATITUDE.

Low, Atmospheric *See* DEPRESSION.

Lucerne A perennial fodder plant of the pea family. It is also called *alfalfa*.

Lunar day The period of time taken by the Earth to rotate once in relation to the moon. This is about 24 hours 50 minutes.

Magma Molten rock at a very high temperature and pressure lying under the surface of the Earth. Igneous rocks are formed from magma. Magma loses much of its water and gas when thrown out by volcanoes as *lava*. *See* IGNEOUS ROCK; LAVA; VOLCANO.

Magnetic Pole One of two poles of the Earth's magnetic field to which the free-swinging magnetic needle of a compass points. The North magnetic Pole is near Prince of Wales Island, in northern Canada. The South magnetic Pole is in Wilkes Land, in Antarctica. The magnetic poles are not true N and S. The North magnetic Pole is about 1,000 miles (1,600 km) from the North Pole; the South magnetic Pole about 1,500 miles (2,400 km) from the South Pole. *See* COMPASS; POLE.

Mantle A dense layer of rock lying between the Earth's crust and core. It is about 1,800 miles (2,900 km) thick. *See* CORE; CRUST.

Map A picture on a flat surface of part or all of the Earth's surface, drawn to a definite scale. Certain features (political, physical, etc.) will be shown and others left out, according to the purpose of the map. *See* MAP READING; PROJECTION.

Maritime climate Climate common on islands and on the coasts, particularly the western coasts, of the continents. The moderating influence of the sea gives a relatively mild winter and a warm summer. Rainfall is fairly heavy.

Marsh A low-lying region of soft, wet land, treeless and partly or completely under water. *See* BOG; SWAMP.

Massif A central mountainous mass, such as the Massif Central of France.

Meadow An area of grassland mown for hay. A *water-meadow* lies on the flood plain of a river and is regularly flooded. *See* PASTURE.

Mean sea level The average level of the sea. It is calculated by taking readings at various places over a long period.

Mediterranean climate A climate marked by hot, dry summers and moist, warm winters. It occurs on the western edges of continents, latitudes 30° to 40°.

Meridian A line of longitude; a half circle of the Earth's surface from the North Pole to the South Pole. *See* LONGITUDE.

Metamorphic rock One of the three main kinds of rocks that make up the Earth's crust. Originally igneous or sedimentary rock, it has been metamorphosed (transformed) by pressure, heat and chemical action. *See* IGNEOUS ROCK; ROCK; SEDIMENT.

Meteor A metallic or stony body that enters the Earth's atmosphere from outer space. Meteors become visible when friction with the air makes them glow. They are also called *shooting stars. See* METEORITE.

Meteorite A meteor that hits the Earth instead of burning up in the atmosphere. *See* METEOR.

Meteorology The scientific study of the Earth's atmosphere and weather. Among other things it enables the experts to forecast the weather. *See* WEATHER.

Mid-latitudes *See* TEMPERATE ZONE.

Midnight Sun The sun that is visible at midnight during the summer months in latitudes greater than 63.5°. It never sinks below the horizon and shines 24 hours a day.

Migration The mass movement from one region to another of certain birds and other animals, or people. Some animals (chiefly birds) migrate seasonally.

Mile A measure of length. (i) *Statute mile* equals 1,760 yards or 1.609 kilometres. (ii) *Geographical mile* equals one-sixtieth of a degree or one minute of latitude. This is taken to mean about 1,850 metres at the equator.

Millibar A unit of pressure used by meteorologists when preparing their weather charts; 1,000 millibars = 1 bar, or 29.53 inches (750 mm) as indicated on a barometer. *See* BAR.

Mineral (i) An inorganic substance that occurs in nature and that possesses distinct chemical and physical properties. Most minerals are crystalline in structure. Rocks consist of various mineral particles. (ii) Any substance—such as coal, or oil—that is mined.

Minute A unit of measurement. It may be one-sixtieth

Left: The 'dark lake', Scutari, between Albania and Montenegro in Yugoslavia.

Centre: The beautiful but fearsome sight of a river of lava, destroying everything that will burn.

Below: An artificial lake in Austria, formed by the building of dams.

of an hour, or one-sixtieth of a degree of angle.

Mistral A very cold wind that sweeps down the Rhône Valley from a high pressure area over the central plateau of France to a low pressure area over the Mediterranean.

Mole A breakwater or stone pier designed to shelter a harbour. *See* BREAKWATER.

Monsoon A wind system that reverses its direction from season to season. More specifically, the term applies to the seasonal winds over the Indian Ocean, blowing from the SW from April to October and from the NE from October to April. The SW winds bring torrential rains to India and neighbouring areas.

Moon The Earth's only natural satellite, which travels round the Earth once every lunar month. It is a dead world without water or atmosphere and its distance from the Earth varies between 221,000 and 253,000 miles (356,000 and 407,000 km). Its mass is less than one-eightieth that of the Earth and its diameter about one quarter (2,160 miles; 3,476 km).

Moorland A wild, open, upland area of land, often covered with grass and bracken.

Moraine Large quantities of rock debris carried along and deposited by a glacier. *See* GLACIER.

Mountain A high hill, or any defined mass of land that rises above the surrounding region and that has steep sides.

Right: Peace and beauty in a national park in Yugoslavia.

Below: A lion ignores intruders as it enjoys a meal in a national park in Kenya.

Nation A large group of people united by attributes they have in common—language, origin, culture or history.

National park An area of natural beauty or other special interest that has been set aside for protection. Some national parks have the object of preserving wild life. Usually, national parks are open to the public.

Navigation The art of finding one's way from place to place, keeping on a chosen course and knowing where one is all the time. Expert navigation is especially important for ships and aircraft. *See* CHART; COMPASS; DEAD RECKONING; SEXTANT.

Neap tide The tide with the smallest range between low water and high water. It occurs when the gravitational pull of the sun is at right-angles to that of the moon. *See* SPRING TIDE.

Nimbostratus A type of low cloud that forms dark grey, ragged sheets. It often brings heavy rain or snow. *See* CLOUD.

Nimbus A term that means a *rain cloud*. It forms part of the composite names of some clouds to indicate that they are rainbearing—for example, *cumulonimbus*. *See* CLOUD.

Northing The horizontal lines of a map grid, indicating the northerly progression from the point of origin of the grid. *See* EASTING; GRID.

Oasis A spot in a desert where there is water, and as a result vegetation grows and permanent settlement is possible.

Occluded front A front formed in a depression when a cold front overtakes a warm front so that the warm sector is lifted out of contact with the surface.

Ocean One of the great sheets of salt water that surround the continents and make up more than 70% of the Earth's surface. The five oceans are: the Pacific, Atlantic, Indian, Arctic and Antarctic or Southern oceans. *See* SEA.

Orbit of the Earth The path taken by the Earth round the sun. It is elliptical. On 4 July (called *aphelion*) the Earth is about 94 million miles (151 million km) from the sun; on 3 January (*perihelion*), about 91 million miles (147 million km). The Earth takes 1 year to complete its orbit at a speed of about 66,000 mph (106,000 kph).

Ore A mineral deposit that contains enough metal to make it economically valuable. *See* MINERAL.

Orientation Determining direction according to the points of the compass.

Orogeny A major phase of mountain building. It involves the formation of mountain ranges as a result of rocks being thrown up into folds or blocks. *See* FOLD; GEOLOGICAL TIME.

Orographic rainfall Rain caused by the rising of moisture-laden air over a mountain range. The air is cooled and releases some or all of its moisture as rain.

Outcrop An exposed piece of bedrock.

Pack ice Large masses of ice floating on the sea, the result of the breaking up of an ice-field by wind and wave. *See* ICE-FIELD.

Padi, Paddy Rice plant or rice grain in the husk. Rice is grown in a *paddy-field.*

Pampas The vast grasslands that surround the estuary of the Rio de la Plata in South America. *See* GRASSLAND; PRAIRIE; STEPPE.

Parallel of latitude A line on a map that is parallel to the equator and that joins all points the same distance N or S of the equator. *See* LATITUDE.

Pass A narrow way through or over a mountain barrier. It is usually the result of erosion. *See* COL.

Pasture Grassland on which domestic animals graze. *See* MEADOW.

Peak A mountain summit, usually formed by the erosion of surrounding rocks. *See* MOUNTAIN.

Pediment A gentle, smooth, sloping rockface at the base of a steeply eroded upland.

Peneplain A region made almost level by denudation.

Peninsula A piece of land projecting into the sea or into a lake.

Period A division of geological time that is shorter than an Era but longer than an Epoch. *See* EPOCH; ERA; GEOLOGICAL TIME.

Permafrost Subsoil and bedrock that is permanently frozen in polar regions. Even though the topsoil may thaw temporarily, the peramafrost sometimes goes down as far as 2,000 ft (610 m). *See* FRIGID ZONE; TUNDRA.

Permeable rock Rock that will let fluids seep through it, either because it is porous (coarse-grained, with open texture) or pervious (cracked and fissured).

Petroleum A mixture of hydrocarbons lying in the Earth's crust, valuable as a source of energy.

Piedmont A region of foothills.

Pipe (i) The vent of a volcano. (ii) A cylindrically-shaped mass of mineral ore. (iii) Vertical joints in chalk, filled with gravel and sand.

Plain A broad, flat area of land or one that is almost featureless. Generally, such areas are at low elevation. *See* GRASSLAND; PAMPAS; PRAIRIE; STEPPE.

Planet A heavenly body that revolves round the sun in a path called an *orbit*. The nine planets that, with the sun, make up the Solar System are Mercury, Venus, Earth, Mars, Jupiter, Saturn, Uranus, Neptune and Pluto. *See* SOLAR SYSTEM.

Planetary winds The general pattern of wind systems near the Earth's surface—such as the *doldrums, trade winds* and *westerlies. See* DOLDRUMS; TRADE WINDS; WESTERLIES.

Plankton A mass of tiny animal and plant life that floats in seas and oceans and provides valuable food for many marine animals.

Plantation A large estate, usually in tropical or sub-tropical regions, where a crop such as sugar or coffee is grown for profit.

Pampas. A gaucho—a cowboy of the pampas—prepares his midday meal.

Plateau A large, level area of land that, generally, rises above the surrounding land. It may have been formed by the upward thrusting of a plain, or by layers of hard rock resisting denudation. Some plateaux, however, lie between higher fold mountain ranges. Plateaux of this type are described as *intermontane.*

Point A narrow piece of land that juts out into the sea; a cape. *See* CAPE.

Polder A term used in the Netherlands to describe a stretch of land, at or below sea level, that has been reclaimed from the sea by the building of dykes.

Pole The N or S end of the Earth's axis. *See* AXIS OF THE EARTH; MAGNETIC POLE.

Port A place on the coast or on a river where ships can tie up and load and unload passengers and cargo. The term is also used in relation to aircraft.

Pot-hole A circular hole cut in the bed of a river by the action of pebbles carried by the swirling waters.

Prairie Huge treeless plains of tall grasses found in the interior of North America. They correspond to the *pampas* of South America and the *steppes* of Europe and Asia. *See* PAMPAS; STEPPE.

Precipice The vertical or very steep face of a mountain, cliff or rock.

Precipitation Moisture such as rain, snow or hail that falls on the Earth's surface as a result of condensation in the atmosphere. *See* HAIL; RAIN; SNOW.

Prevailing wind The wind direction most common in any particular place.

Projection A representation on a flat map of the surface of a geographical globe with its network of meridians and parallels. All projections are inaccurate in one respect or another and each kind aims at accuracy in some particular aspect, such as directions, shapes or areas.

Quarry A pit or open site where commercially valuable stone is cut or excavated.

Race (i) One of the main divisions of mankind, consisting of people that have a recognizable combination of physical traits handed on from one generation to another. (ii) A fast tidal current flowing through a narrow channel.

Radiation The process whereby radiant energy is emitted by a body in the form of short-waves or long-waves. The sun's emission of radiant energy is called *solar radiation*. The loss of heat from the Earth's surface at night by long-wave radiation is called *terrestrial radiation*. *See* INSOLATION.

Radioactive decay The continuous and increasing breakdown of certain radioactive elements in nature. *See* RADIOCARBON DATING.

Right: The port of Luanda in Angola. Ports are vital to the economies of countries that live by trade.

Below: Polder in the region of Ijsselmeer, formerly the Zuider Zee, in the Netherlands. Many of the windmills that are a feature of the Dutch landscape are water pumps, used for draining the land.

Radiocarbon dating A method of finding out the ages of long-dead organisms or substances derived from them. All living things contain (as well as carbon) carbon-14, a radioactive isotope of carbon. As soon as an animal or plant dies, it begins to lose its carbon-14 at the rate of 50 per cent every 5,730 years.

Rain Water droplets that fall to the Earth as a result of the condensation of water vapour in saturated clouds. *See* CLOUD; CONDENSATION; RAIN GAUGE; RAINFALL; SATURATION.

Rain gauge An instrument used to measure the amount of rainfall in a particular area in any given period of time. The simplest kind of gauge is a funnel fitted into a glass jar, which holds the water. Rainfall is usually measured in millimetres or inches. *See* RAIN; RAINFALL.

Rainfall The total amount of water that falls on a given region during a certain period of time, as recorded by a rain gauge. Melted snow and hail are also included in this total. *See* PRECIPITATION; RAIN; RAIN GAUGE.

Rain-forest *See* EQUATORIAL RAIN-FOREST.

Range (i) A chain of mountains. (ii) In some countries, grazing land for cattle, goats and sheep.

Rapids Part of a river where the current flows with great swiftness and the water is broken.

Ravine A long, deep, narrow valley.

Raw materials Natural substances that are manufactured into consumer goods. They may be raw substances such as iron ore or partly-processed substances such as wholemeal flour.

Reef (i) A ridge of rock at or near the surface of the sea, specifically one made of coral. (ii) In mining, a vein of gold or some other metal or metal ore. *See* ATOLL; BARRIER REEF; CORAL REEF.

Relief map A map that shows the *relief*—differences in height—of any part of the Earth's surface.

Representative fraction The ratio of the distance between any two points on a map and the actual distance between them on the ground. It indicates the *scale* of the map.

Reservoir A place where a large volume of water is stored for hydro-electric or irrigation purposes, or to form the water supply of a populated area.

Ridge of high pressure A long, narrow area of high atmospheric pressure wedged in between two areas of low pressure. It is often marked by a temporary period of fine weather. *See* ATMOSPHERIC PRESSURE; TROUGH.

Rift valley A long, narrow depression formed when the land sinks between two parallel faults. The most famous is the Great Rift Valley that stretches some 3,000 miles (4,830 km) from Syria to eastern Africa and is only 20 to 30 miles (30 to 50 km) wide. *See* FAULT.

River A freshwater stream that flows in a definite channel to the sea, a lake or another river. The point where it originates is called its *source*. Lesser rivers that contribute water to it are called *tributaries*. *See* WATERSHED.

Roaring Forties A southerly region of the oceans, located between latitudes 40° S and 60° S, where there are no land masses and where, as a result, the prevailing NW to W winds blow fiercely and regularly. *See* WESTERLIES.

Rock The hard mass of mineral matter that forms most of the Earth's crust. *See* IGNEOUS ROCK; METAMORPHIC ROCK; SEDIMENT.

Rotation of the Earth The spinning of the Earth on its axis from W to E. One rotation takes approximately 24 hours. The speed of rotation at the equator is about 1,050 mph (1,690 kph). *See* AXIS OF THE EARTH.

Saddle A type of wide, level col located in a ridge between two peaks. *See* COL.

Salinity The relative saltiness of sea-water. The amount of salt in the Earth's seas as a whole averages 34.5 parts per thousand.

Salt flat The bed of a former salt lake, now dried up and forming a flat salt-crust, often many miles long.

Sandbank A band of sand in the sea or in a river; formed by tides and currents and sometimes visible at low water. A low sandbank is called a *shoal.*

Saturation (i) The condition of the atmosphere when it can hold no more water vapour. (ii) The state of rocks whose pores are filled with water. *See* CONDENSATION; WATER TABLE.

Savanna Tropical grassland with some scattered trees. The word is also spelt *savannah* or *savana. See* GRASSLAND.

Scale The ratio between distance on the ground and the length that represents it on a map. It may be stated in words, shown as a graduated line, or given as a representative fraction. *See* REPRESENTATIVE FRACTION.

Scrub The vegetation that grows in semi-arid conditions, on poor, stony soils. It is made up largely of coarse grasses, bushes, and low, stunted trees.

Sea A general term for the mass of salt water that covers most of the Earth's surface. Specifically, it refers to lesser divisions of the oceans such as the North Sea and the China Sea. *See* OCEAN.

Sea level *See* MEAN SEA LEVEL.

Season Within the tropics the sun is always relatively high at noon and seasons do not depend on temperature, but on rainfall which gives a wet season and a dry season in many areas. On the equator there are virtually no seasons. In high latitudes the tilt of the axis and the annual orbit round the sun gives a warm season (summer) and a cool season (winter) with spring and autumn between. *See* AUTUMN; SPRING; SUMMER; WINTER.

Seaway A ship-canal or inland waterway that is navigable by seagoing ships. A famous example is the St Lawrence Seaway in North America.

Second In the measurement of angles, one-sixtieth of a minute.

Sediment Material that has been carried by wind, water or ice and eventually deposited. Hence *sedimentary rock* is a type of rock made up of sediments that have been laid down in layers and compacted or cemented.

Left: Some comparatively minor rivers can carry a great volume of water. The vast Lake of the Ozarks in Missouri, in the United States, was formed by the damming of the Osage, a tributary of the Missouri.

Below: Budapest on the Danube. Most of the world's great cities are built on rivers.

Seismology The scientific study and measurement of earthquakes.

Selva Specifically, a Brazilian word for the equatorial forest of the Amazon basin; generally, any region of similar vegetation. *See* EQUATORIAL RAIN-FOREST.

Sextant An instrument used for measuring angles, especially the angle of the sun or some other star. In this way a navigator is able to fix his position.

Shelf A platform or ledge under the sea. *See* CONTINENTAL SHELF.

Shield A large area of very old, crystalline, Pre-Cambrian rock, exposed at the Earth's surface. Well-known examples are the Canadian Shield and the Deccan in India.

Sial *See* CONTINENTAL DRIFT.

Sierra A Spanish term for a mountain range with jagged, saw-like peaks.

Sima *See* CONTINENTAL DRIFT.

Sirocco, Scirocco A warm, dusty, southerly wind that blows from the Sahara across northern Africa and on to Sicily and southern Italy. It normally lasts for one or two days, and, though dry over Africa, may be very humid by the time it reaches Sicily.

Right: Salt Flat. A thick deposit of salt covers the desert in the Danakil Depression, in northern Ethiopia. The Depression, some 400 ft (120 m) below the level of the nearby Red Sea, once held a salt lake that has now evaporated.

Below: By contour ploughing, following the natural contours of the land, a farmer can prevent soil being washed away and improve irrigation.

Smog A word coined from 'fog' and 'smoke' to indicate a condition of fog polluted with smoke. It is a particular nuisance in industrial areas.

Snow Precipitation in the form of delicate, feathery ice crystals.

Soil The mixture of rock particles, decomposed organic matter, aqueous solution of chemicals and gases that makes up the loose surface-layer of the Earth. Agriculturally, the most important part of the soil is the thin layer of fertile topsoil.

Soil erosion The removal, mainly by wind and water, of the topsoil. Man often speeds this process by over-grazing, burning, or deforestation. *See* DUSTBOWL; EROSION.

Solar System The system to which the Earth belongs. It is made up of the sun together with the nine planets, the asteroids, the meteoroids, the comets and the various particles that revolve round it in almost the same plane. It is part of the galaxy called the *Milky Way. See* PLANET; SUN.

Solar year A period of slightly more than 365 days, the time taken by the Earth to make one complete orbit of the sun. *See* ORBIT OF THE EARTH; YEAR.

Solstice The time of year when the difference between lengths of days and lengths of nights is at its greatest. At this time the sun reaches its maximum declination from the equator. The summer solstice in the Northern Hemisphere is about June 21 and the winter solstice about December 22. In the Southern Hemisphere these dates are reversed.

Sound (i) A stretch of water, usually broader than a strait, that links two larger areas of water. (ii) An inlet of the sea. *See* INLET; STRAIT.

Spit A narrow, low-lying tongue of sand, shingle or gravel sticking out into the sea from the coast.

Spring (i) One of the four seasons of the year, between winter and summer. In the Northern Hemisphere it begins about 21 March; in the Southern Hemisphere, about 23 September. (ii) A natural flow of water gushing from the ground, usually because the water table cuts across ground level. *See* HOT SPRING; WATER TABLE.

Spring tide A very high and very low tide that occurs twice a month. It occurs when the Earth, sun and moon are in the same, almost straight line.

Spur A small ridge jutting out from a mountain or a hill.

Stalactite An icicle-shaped deposit hanging from the roof of a cave. It is formed by drops of water containing calcium carbonate dripping from the roof and evaporating. *See* STALAGMITE.

Stalagmite A stumpy, icicle-shaped mineral deposit rising from the floor of a cave. Like the stalactite, it forms from a solution of calcium carbonate dripping from the roof.

Standard time The time adopted in a particular longitudinal zone with reference to the mean time of a standard meridian. Most standard times differ from Greenwich Mean Time by an exact number of hours or half hours. Standard time is also called *zone time. See* GREENWICH MEAN TIME; INTERNATIONAL DATE LINE.

Steppe Temperate grassland, as a rule level and tree-less. Specifically, steppes are found in the mid-latitudes of Europe and Asia. They correspond to the *prairies* of North America and the *pampas* of South America. *See* PAMPAS; PRAIRIE.

Storm (i) Force 11 on the Beaufort Wind Scale. (ii) Any violent disturbance of the atmosphere accompanied by wind, rain, snow, thunder or other manifestation. *See* BEAUFORT WIND SCALE.

Strait, Straits A narrow stretch of water that connects two larger bodies of water. *See* SOUND.

Stratum A single, generally distinct, horizontal layer or bed of sedimentary rock (plural *strata*). *See* SEDIMENT.

Stratus A type of low cloud that stretches in a grey, unbroken layer. *See* CLOUD.

Subsidence Sinking of the Earth's crust. It may be a small local settling of the Earth produced perhaps by the collapse of old mine workings, or a huge depression across part of a continent, such as the Great Rift Valley. *See* RIFT VALLEY.

Summer One of the four seasons of the year, between spring and autumn. In the Northern Hemisphere it begins about 21 June; in the Southern Hemisphere, about 21 December. *See* SEASON.

Sun The star round which the Earth orbits and from which the Earth receives its heat and light. It is about 93 million miles (150 million km) from the Earth. Its mass is about 330,000 times greater than the Earth's mass. Scientists believe that the sun has a liquid centre, which is surrounded by flaming gas. All life in the Solar System depends for its existence on the sun's radiation. *See* RADIATION; SOLAR SYSTEM.

Survey The precise measuring and estimating of the location, size, shape and features of any part of the Earth's surface in order to map it.

Swamp An area of permanently waterlogged soil and vegetation. *See* BOG; MARSH.

Syncline A trough or downfold in a layer of rock. *See* ANTICLINE; GEOSYNCLINE.

Tableland A flat-topped upland or plateau with steep sides.

Taiga Coniferous forests in the high latitudes of the Northern Hemisphere. Tundra lies to the north and steppes to the south.

Temperate Zone A climatic zone lying between the Torrid Zone and the Frigid Zone.

Temperature As an element of climate, the degree of heat in the atmosphere, in the land, or in the sea.

Terrain The physical characteristics and features of any stretch of country.

Theodolite An instrument used in surveying for measuring angles and directions. It is mounted on a tripod and is made up of a telescope and a spirit-level with graduated horizontal and vertical scales. *See* SURVEY.

Thermal A rising current of warm air, caused when an area of the Earth's surface is heated by the sun's rays.

Thermometer An instrument for measuring temperature. It consists of a glass tube marked with a graduated scale. The base of the tube is a bulb containing mercury, alcohol or some other liquid that expands or contracts with changes in temperature. The resultant changes in the level of liquid in the tube are read off on the scale. *See* CELSIUS SCALE; FAHRENHEIT SCALE; TEMPERATURE.

Thunderstorm A violent storm in the atmosphere accompanied by lightning, thunder, rain and sometimes hail.

Tidal wave The rocking motion of the water body round the edges of oceans and seas, produced by tides.

Tide The rise and fall in the level of the oceans and seas caused chiefly by the gravitational pull of the moon and the sun. *See* NEAP TIDE; SPRING TIDE.

Timber line, Tree line The upper limit of tree growth on a mountainside, or a limit of latitude beyond which trees do not grow. The timber line varies from place to place, depending on climate, soil, slope, species of trees and other factors.

Topography A detailed description of the physical features of an area, or their representation on a map, called a *topographical map*.

Cypress Gardens, Florida, in the United States, a swamp that has been turned into a pleasure ground. Oaks and cypresses rise above quiet waterways and rare and beautiful plants.

Tornado A small, extremely violent whirlwind moving destructively across country at 20 to 40 mph (30 to 60 kph).

Torrid Zone The climatic zone lying between the Tropic of Cancer and the Tropic of Capricorn, where the weather is almost always hot and the sun shines directly overhead twice each year. *See* FRIGID ZONE; TEMPERATE ZONE.

Town *See* CITY.

Trade route A land, sea or air route that has become an established route for commerce.

Trade winds Fresh, dry, steady winds that blow from high pressure areas in subtropical regions towards the low pressure region of the equator. In the Northern

The seemingly limitless coniferous taiga forests of Siberia, in the U.S.S.R.

Hemisphere they blow from the NE and in the Southern Hemisphere from the SE. The trade wind belts vary a little in latitude during the course of the year.

Tree line *See* TIMBER LINE.

Trench One of the deepest parts of the ocean, usually a steep-sided depression in the ocean floor. The Marianas Trench in the western Pacific has the greatest known ocean depth—36,198 ft (11,033 m). Such ocean valleys are also called *deeps* or *troughs*.

Triangulation A method of surveying large areas based on the mathematical principle that if at least one side and two angles of a triangle are known, the other two sides and the third angle can be calculated. The angles are measured by a theodolite. *See* SURVEY; THEODOLITE.

Tributary A stream or river that flows into a larger one. *See* RIVER.

Tropic *See* CANCER, TROPIC OF; CAPRICORN, TROPIC OF.

Trough (i) A trench or deep in the sea bed. (ii) A U-shaped valley, sometimes made by a glacier. (iii) A long narrow region of low pressure in the atmosphere sandwiched between regions of higher pressure. (iv) At sea, a hollow between two successive waves. *See* TRENCH.

Right: Typical valley scenery in the high mountain ranges of southern Africa, with their thick vegetation, stark cliffs and deeply-cut canyons.

Below: Mount Etna, on the eastern coast of Sicily, is one of the few active volcanoes in Europe. It is known to have erupted more than 80 times. Its snow-clad upper slopes overlook shady woods, vineyards and orange groves.

Tsunami A great sea wave caused by a submarine earthquake.

Tundra A treeless Arctic plain that has no corresponding region in the Southern Hemisphere. It lies between the northern polar ice and the northern coniferous forests.

Typhoon A violent tropical storm that occurs in the western Pacific Ocean and the China Sea. It is accompanied by revolving winds that exceed 100 mph (160 kph), thunderstorms and very heavy rain. It is also called a *tropical cyclone*, and is the same kind of storm as a hurricane. *See* CYCLONE; HURRICANE.

Upland Land that is higher than the surrounding region. *See* HIGHLAND.

Valley A long narrow hollow in the Earth's surface formed usually by a glacier or river but sometimes by folding and faulting. Often a river flows through it. *See* FAULT; FOLD.

Veering A clockwise change in the direction of the wind—for example from N to NE. The opposite of *backing*. *See* BACKING.

Vegetation The plant life or flora of the Earth. *See* FLORA.

Vein A long narrow crack in a rock mass containing deposits of crystalline matter. Veins are often the source of metallic ores. *See* LODE.

Vent The pipe or the opening in the Earth's crust through which magma rises in a volcano. *See* MAGMA; VOLCANO.

Visibility The distance it is possible to see clearly. *See* HORIZON.

Volcano An opening in the Earth's crust through which magma is forced up to the surface. The magma that flows out is called *lava*.

Wadi An Arabic term for a steep-sided rocky stream bed in a desert.

Warm front The boundary line between a mass of cold air and an overtaking mass of warm air. The approach of a warm front is marked by rising temperatures and an increase in cloud formations. *See* COLD FRONT; FRONT; OCCLUDED FRONT.

Water cycle The continuous circulation of water from the Earth's surface into the atmosphere, and back to the surface.

Waterfall A sudden steep fall of the water in a river because of an abrupt drop in the river's bed.

Watershed The crest of the surrounding uplands that form the boundary of a river basin. *See* BASIN; CATCHMENT AREA; DRAINAGE; RIVER.

Waterspout A tornado over the sea. A funnel of cloud spirals down to the surface of the sea and scoops up a twisting column of water. *See* TORNADO.

Water Table The junction between saturated and non-saturated layers of rock. Water trickles through many rocks until the lower layers become saturated. After heavy rain the water table rises; after drought it sinks.

Wave, Ocean A surging motion of the ocean surface caused by the friction of wind blowing over it. The water moves in a kind of circle from trough up to crest and down and back into trough again. It does not usually travel forward in deep water.

Weather The day-to-day or hour-to-hour condition of the atmosphere at any particular place. The factors to be taken into account include temperature, precipitation, clouds, sunshine, wind and visibility. *See* CLIMATE.

Weathering The breaking down of parts of the Earth's crust by exposure to the atmosphere. It is one of the principal processes of denudation. *Mechanical* or *physical weathering* is caused by changes in temperature and by frost. *Chemical weathering* results from natural chemical processes such as oxidation (combining with oxygen) and solution (dissolving in a liquid). *See* ABRASION; ATTRITION; CORRASION; CORROSION; DENUDATION; EROSION.

Westerlies Winds that blow frequently and fairly regularly in belts that lie between the Horse Latitudes and the polar regions in both hemispheres. In the Northern Hemisphere they blow from the SW; in the Southern Hemisphere they blow from the NW and are known as the *Roaring Forties*. *See* HORSE LATITUDES; ROARING FORTIES.

Whirlpool A strong eddy or circular current in a body of water. It is usually caused by the meeting of two currents, but sometimes a whirlpool results when a fast current enters an obstructed channel.

Whirlwind A small, violent current of air rotating rapidly round an area of low atmospheric pressure. *See* CYCLONE; TORNADO; WATERSPOUT.

Wind Air currents moving across the surface of the Earth. Wind speeds are usually classified according to the Beaufort Wind Scale. Wind direction is the direction from which the wind blows; for example, an east wind is one blowing from the E.

Winter One of the four seasons of the year. It is the coldest season. Winter is the period between December 22 and March 21 in the Northern Hemisphere; and June 21 and September 22 in the Southern Hemisphere. *See* AUTUMN; SEASON; SPRING; SUMMER.

Year The time taken by the Earth to make a complete revolution round the sun. There are various ways of measuring this time, but for practical purposes a *calendar year* equals 365 days—beginning on 1 January and ending on 31 December. To make the calendar more accurate, a *leap year* of 366 days is introduced every fourth year. *See* SOLAR YEAR.

Zone (i) Any belt or region with its own distinct characteristics. (ii) One of the five climatic regions of the Earth; the northern and southern *frigid* zones; the northern and southern *temperate* zones; and the *torrid* zone. *See* FRIGID ZONE; TEMPERATE ZONE; TORRID ZONE.

Waterfall. Several of the world's highest waterfalls spill in filmy veils down the sides of Merced Canyon, in Yosemite National Park, California. One, the Yosemite Falls, is 2,425 ft (740 m) high. The Park lies in the Sierra Nevada mountains.

The effects of two thousand years of weathering can be seen on the ruined columns of Hatra, in Iraq.

The Seven Continents

The continents are the world's great land masses. Together, they cover about one quarter of the Earth's surface. Europe is usually described as a separate continent but—geographically—it is really a peninsula of Eurasia, a land mass made up of Europe and Asia.

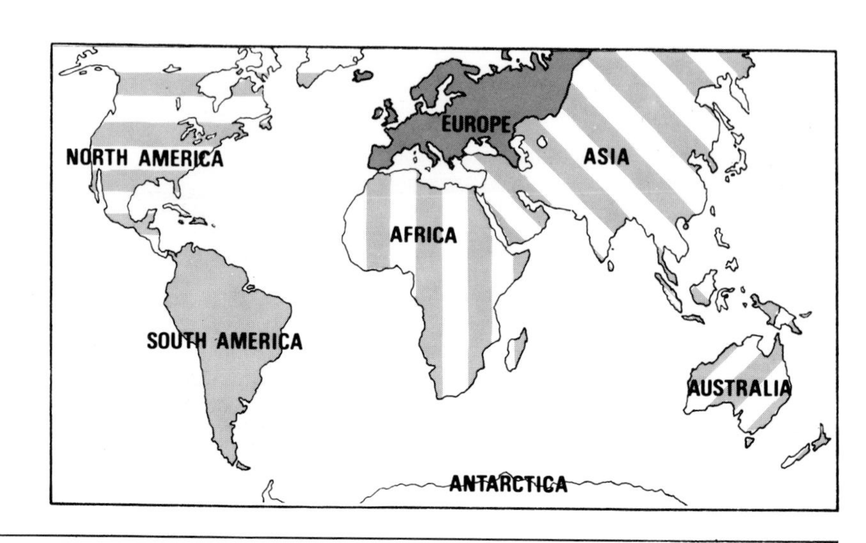

EUROPE

The River Rhine at Kaub, in Germany. An ancient toll-collecting castle stands on an island in the river.

Area: 4,036,000 sq miles; 10,453,000 sq km.
Greatest Distances: N to S, 4,000 miles (6,400 km); W to E, 3,000 miles (4,800 km).
Population: 702 million.
Europe is the smallest of the continents, but has influenced world history in a way that is out of all proportion to its size. It is bounded on the N by the Arctic Ocean, on the S by the Black Sea and the Mediterranean Sea, and on the W by the Atlantic Ocean. On the E, the Ural Mountains separate it from the continent of Asia.

ASIA

A floating market in Bangkok, the capital of Thailand. The city has a network of busy canals.

Area: 17,336,000 sq miles; 44,900 sq km.
Greatest Distances: N to S, 5,400 miles (8,700 km); W to E, 6,000 miles (9,700 km).
Population: 2,666 million.
Asia is the largest of the continents. It also has the largest population, and is the richest in natural resources. It is bounded on the N by the Arctic Ocean, on the E by the Pacific Ocean, and on the S by the Indian Ocean. On the W, the Ural Mountains form its boundary with Europe.

AFRICA

A village in the highlands north of Nairobi, in Kenya. It belongs to the large Kikuyu tribe.

Area: 11,700,000 sq miles; 30,300,000 sq km.
Greatest Distances: N to S, 5,000 miles (8,000 km); W to E, 4,700 miles (7,600 km).
Population: 480 million.
Much of Africa was not fully explored until the 1900s because deserts, jungles and unnavigable rivers made travel extremely difficult. The continent is bounded on by the N by the Mediterranean Sea, on the E by the Red Sea and the Indian Ocean, and on the W by the Atlantic Ocean.

NORTH AMERICA with Central America

Contour farming in the United States. By following the land's natural contours, soil erosion is prevented.

Area: 9,600,000 sq miles; 24,900,000 sq km.
Greatest Distances: N to S, 4,500 miles (7,200 km); W to E, 4,000 miles (6,400 km).
Population: 371 million.
Although much of it was almost uninhabited until Europeans 'discovered' it 500 years ago, North America is today the richest and most powerful of the continents. It is bounded on the N by the Arctic Ocean, on the E by the Atlantic Ocean and on the W by the Pacific Ocean. A narrow 'bridge' joins it to South America.

SOUTH AMERICA

Banana and coffee plantations in Bolivia. The coffee berries are harvested laboriously by hand.

Area: 6,772,000 sq miles; 17,540,000 sq km.
Greatest Distances: N to S, 4,750 miles (7,640 km); W to E, 3,200 miles (5,150 km).
Population: 243 million.
There may be places in the vast forests of South America where no human being has ever walked. The continent still challenges explorers. It is bounded on the N by the Caribbean Sea and the Atlantic Ocean, on the E by the Atlantic, and on the W by the Pacific Ocean. Its southern tip approaches Antarctica.

AUSTRALIA

Herding sheep on a ranch in Australia. More than a quarter of the world's wool comes from Australia.

Area: 2,968,000 sq miles; 7,687,000 sq km.
Greatest Distances: N to S, 2,000 miles (3,200 km); W to E, 2,400 miles (3,900 km).
Population: 15 million.
Australia is the only continent that is also a single country. It is bounded on the N by the Timor Sea and the Arafura Sea, on the E by the Pacific Ocean, and on the W by the Indian Ocean. The term *Australasia* refers to a region comprising Australia, New Guinea, New Zealand, and other smaller islands nearby.

ANTARCTICA

Penguins in Antarctica. Their feathers and a layer of fat under the skin protect them from the cold.

Area: 5,000,000 sq miles; 13,000,000 sq km.
Greatest Distance: Wilhelm II Coast to Eights Coast, 2,700 miles (4,350 km).
Most of Antarctica, the freezing, desolate continent surrounding the South Pole, lies beneath a mass of ice and snow that is over a mile thick in places. The lowest known temperatures on Earth have been recorded in Antarctica. No animals live in the inland areas of the continent. Animals can find the means of existence only near the coasts or in surrounding waters. A few plants exist around the edges of Antarctica.

Europe

COUNTRIES OF EUROPE
POPULATIONS AND MONETARY UNITS

Country	*Population*	*Money*
Albania	2,795,000	Lek
Andorra	34,000	Franc, Peseta
Austria	7,502,000	Schilling
Belgium	9,863,000	Franc
Bulgaria	8,890,000	Lev
Czechoslovakia	15,355,000	Koruna
Denmark	5,119,000	Krone
Finland	4,811,000	Markka
France	53,963,000	Franc
German Democratic Rep.	16,736,000	Mark
Germany, Federal Rep. of	61,713,000	Deutsche Mark
Gibraltar	30,000	Pound
Greece	9,707,000	Drachma
Hungary	10,708,000	Forint
Iceland	231,000	Króna
Ireland, Rep. of	3,440,000	Pound
Italy	56,244,000	Lira
Liechtenstein	26,000	Franc
Luxembourg	364,000	Franc
Malta	360,000	Pound
Monaco	26,000	Franc
Netherlands	14,293,000	Florin (Guilder)
Norway	4,105,000	Krone
Poland	36,062,000	Zloty
Portugal	9,931,000	Escudo
Romania	22,457,000	Leu
San Marino	21,000	Lira
Spain	37,654,000	Peseta
Sweden	8,326,000	Krona
Switzerland	6,366,000	Franc
Turkey (Europe)	4,500,000	Lira
U.S.S.R. (Europe)	212,640,000	Rouble
United Kingdom	55,619,000	Pound
Vatican City	1,000	Lira
Yugoslavia	22,646,000	Dinar

Europe is part of the same land mass as Asia, but is always considered a separate continent. It is the second smallest continent and its many countries have often been devastated by war between themselves.

Europe is bounded on the N by the Arctic Ocean, and on the W by the Atlantic Ocean. The Mediterranean and Black Seas wash its southern shores. On the E it is linked to Asia, with the Ural Mountains and the Caspian Sea marking the boundary between the two continents. In the SE, from the Black Sea to the Caspian, stretch the Caucasus Mountains. On the other side of the Mediterranean Sea lies Africa. At the Strait of Gibraltar, where the Mediterranean is joined to the Atlantic, Europe and Africa are only 9 miles (14 km) apart.

The area of Europe is 4,036,000 sq miles (10,453,000 sq km). The greatest distance N

Top: The Alps stretch in a great arc for nearly 700 miles (1,100 km) across southern Europe. Farming is confined to the valley floors and to the high summer pastures—called alps—above the tree line.

Centre: The temple of Olympian Zeus, near the Acropolis in Athens. Parts of it date from 515 B.C., but it was completed by the Roman Emperor Hadrian in A.D. 129. Only 15 of its 104 Corinthian columns still stand.

Below: Saint-Tropez, on the Côte d'Azur in southern France. Yachts and fishing-boats share the crowded harbour. Once a quiet Provençal fishing village, Saint-Tropez became known as a haunt of painters and was soon a fashionable tourist resort.

Above: 'The Mouth of Hell', a cave formation on the storm-swept Atlantic coast near Cascais in Portugal.

to S is about 4,000 miles (6,400 km), and W to E 3,000 miles (4,800 km).

Physical Features and Climate

Europe consists of three main regions: the northern mountains, the central plains, and the southern mountain chains.

The Northern Mountains lie in Scotland and northern England, and in Norway and Sweden. The Ural Mountains in Russia are also part of the region. Finland is not mountainous, but it is formed from the same rock system. During the Ice Age much of this region was affected by the action of glaciers. Among the results of this glacial action are the fiords of W Norway—deep, steep-sided sea inlets. Finland's thousands of lakes were also formed by glacial action.

Below: Farmland in England, carefully tended for hundreds of years.

The Central Plains cover by far the largest part of Europe. They stretch from Ireland in the W as far as the Ural Mountains. The greater part of the central plains is less than 500 ft (150 m) above sea-level. The eastern section of the plains is in Russia, and geologists often call it the *Russian Platform.*

The Southern Mountain Chains are a complex series of mountain ranges. The chief mountains are the Alps, which curve in a high, rugged arc that cuts off the Italian peninsula from the rest of the continent. The highest point of the Alps is Mont Blanc, whose peak is 15,781 ft (4,810 m) above sea-level. This and the other high peaks of the Alps are always covered in snow above the 8,000 ft (2,500 m) level. The Alps extend eastwards and are continued in the Balkan Peninsula as the Balkan Mountains. The Caucasus Mountains, between the Black and Caspian Seas, are also a part of the same mountain system. To the W, the Pyrenees, lying across the northern neck of the Iberian Peninsula (Spain and Portugal), are another extension of the Alpine chain.

These mountain chains form barriers between the coastlands bordering the Mediterranean Sea and the plains of N and central Europe. But there are several breaks in the chains, such as the valley of the River Rhône in France, and these serve as major routes from N to S.

Other mountain systems are found in Spain (the Meseta), central France, southern Germany and Czechoslovakia. They are not so high as the Alpine mountains. The upper slopes of many of them are thickly forested.

Europe has several inland seas. The North Sea, between Great Britain and the continental mainland, is a wide arm of the Atlantic. In northern Europe lies the Baltic Sea, with its two arms, the Gulf of Bothnia and the Gulf of Finland. It separates Norway, Sweden, and Finland from the central part of the continent. The Mediterranean Sea, to the S, is almost entirely landlocked. But it has a narrow link with the Atlantic in the W through the Strait of Gibraltar. The Tyrrhenian, Adriatic, Ionian, and Aegean Seas are arms or sections of the Mediterranean. Two narrow straits, the

Dardanelles and the Bosporus, with the little Sea of Marmara between them, link the Mediterranean with the Black Sea. The Sea of Azov is an almost entirely landlocked arm of the Black Sea. The Caspian Sea, although it is salt, is really a huge lake.

Europe's main rivers are the Danube, flowing eastwards through central Europe to the Black Sea, the Rhine, flowing northwards to the North Sea, and the Volga, flowing southwards through Russia to the Caspian Sea.

Most of Europe has a relatively mild climate. Northern, central, and eastern parts have warm summers and cold winters. Western parts, including the British Isles, have winters that are less cold. The countries around the Mediterranean have hot, dry summers and mild, wet winters. The extreme N lies within the Arctic Circle, but even in these extreme latitudes the warming action of the Gulf Stream can be felt.

Resources and Industry

Many parts of Europe are rich in mineral resources. The continent produces about 40 per cent. of the world's coal, and a third of its iron ore. Petroleum deposits lie in SE Europe, and also in the W near the Urals. Further large deposits of both petroleum and natural gas lie under the North Sea.

Other minerals produced in large amounts include bauxite, copper, lead, mercury, platinum, potash, salt, sulphur, and zinc. There are smaller deposits of a large number of minerals.

The world's greatest manufacturing region has developed around the coalfields of Belgium, Britain, northern France, and Germany. Its industries produce iron, steel and all kinds of metal goods, as well as textiles—silk, cotton, wool, linen, and artificial fabrics—chemicals, pottery and glass. In Russia the main manufacturing districts are in the neighbourhood of Moscow, in the basin of the River Don, and in the foothills of the southern Urals.

The motor-car factories of Britain, France, West Germany and Italy make almost a third of the world's vehicles. Machine-tools are produced in Britain and West Germany, and watches and similar precision machinery are a speciality of Switzerland. Heavy engineering industries are highly developed in West Germany and the U.S.S.R.

Although the first industries grew up around coalfields, and close to iron deposits, many modern industrial centres are based on the availability of hydro-electric power, particularly in northern Italy, Switzerland, and Norway and Sweden.

Much of Europe's prosperity depends on

Left: A village in the beautiful valley of the River Möll in Carinthia, southern Austria. In the background is part of the Austrian Alps.

Below: The low hills and rolling uplands of Sardinia. Many of the people of this Mediterranean island pursue an age-old way of life, jealously preserving traditional customs.

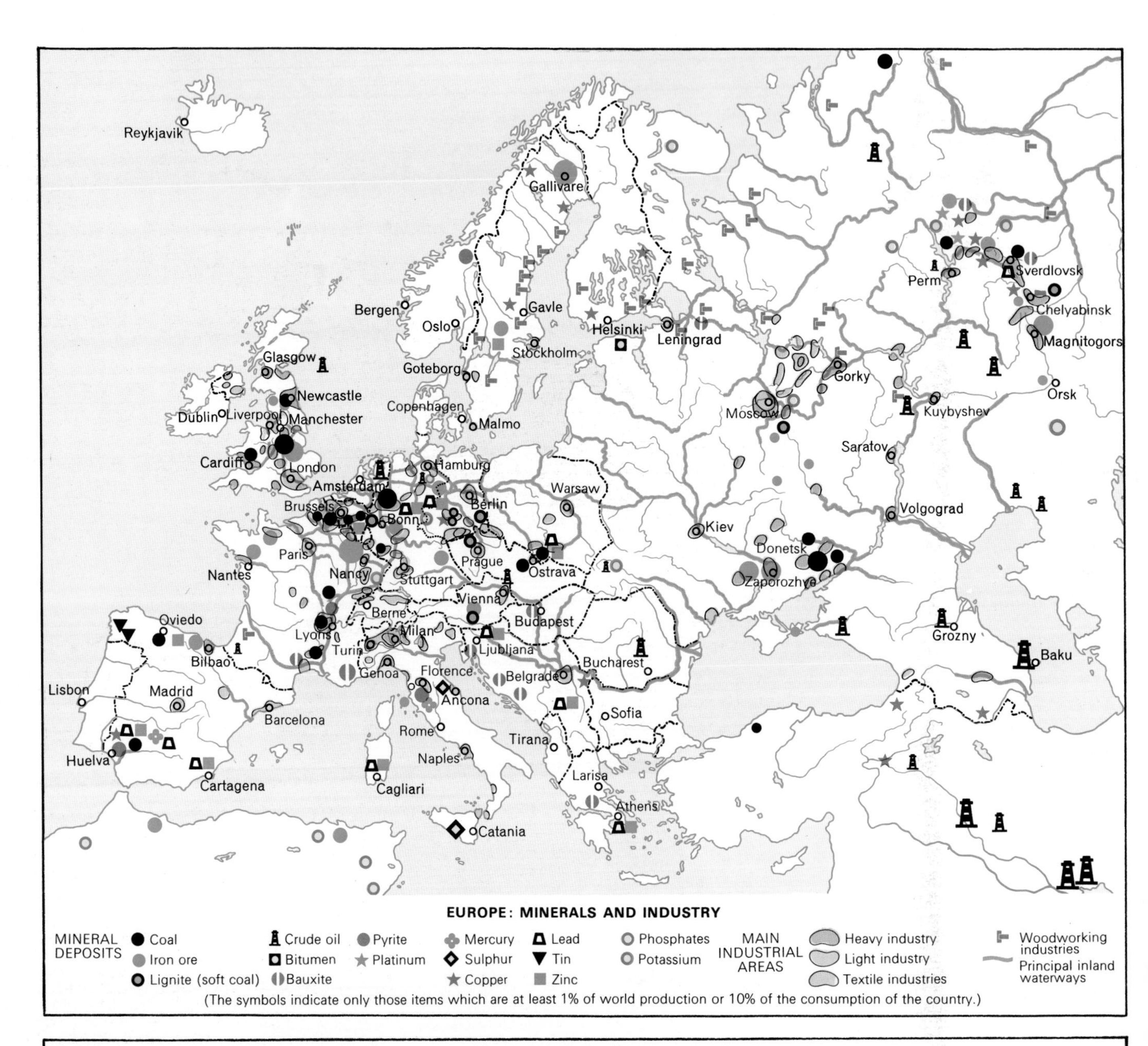

EUROPE: MINERALS AND INDUSTRY

THE EUROPEAN COMMUNITY

The European Community consists of ten associated countries of Western Europe: Belgium, Denmark, France, the Federal Republic of Germany (West Germany), the Republic of Ireland, Italy, Luxembourg, the Netherlands, the United Kingdom and Greece. It has about 80 per cent. of Western Europe's population, and only the United States has greater economic strength.

The aim of the member countries is to create a unified economy: to combine their economic resources—finance, industry and agriculture—for the benefit of all.

The Working of the Community

The European Community has its own civil service and courts, as well as a parliament and ministers.

The Council of Ministers is the lawmaking body of the Community, but it acts only on proposals put to it by the Commission. It consists of one representative of each of the ten member countries, and meets in Brussels or Luxembourg.

The Commission proposes new laws or policies to the Council.

The Assembly is the parliament of the Community. It controls the budget, and has power to dismiss the Commissioners. It meets in Strasbourg or Luxembourg.

The Court of Justice consists of eleven judges who have the task of settling legal disputes on Community matters. It meets in Luxembourg.

History

The idea of European unity is an old one, but it was given new life by the disastrous results of disunity in the 20th century: the two World Wars.

At first, in 1959, the Community had six members. In 1973, they were joined by Denmark, Ireland and the United Kingdom. Norway decided, after a referendum, not to apply for membership. Greece joined in 1981.

trade. Because of its long coastline, no part of western Europe is far from a seaport. Ships carry products from the factories of Europe to all parts of the world. A large proportion of the world's financial business, including international banking transactions and insurance, is carried on in the continent's major cities, particularly in London and Zurich. But Europe is not only a large exporter. It has to import huge amounts of raw materials, particularly petroleum and other minerals of which it has insufficient quantities for its industrial needs, and food.

Europe's major natural resources include the fertile soils of the central plains, where much of the continent's food is grown. Over the centuries, forests have largely been cut down, but wide tracts of Germany and other countries of central Europe are still heavily wooded. The pine forests of Scandinavia provide timber for building and furniture, and wood pulp for making paper.

The wildlife of Europe has been greatly reduced by man's activities—by hunting and by the destruction of wild woodland—but a number of large animals are still found in the more remote areas. They include bears, wild boar and wolves. Herds of reindeer roam the Arctic regions of northern Europe—an area known as *Lapland* that includes parts of Norway, Finland and Russia. The European bison, also called the wisent, survives in parks and nature reserves. Some animal species, such as the European beaver, are increasing in number again after having been almost wiped out. The continent is rich in birds, many species of which migrate S to Africa for the winter months.

Communications

Communications in Europe are more highly developed than anywhere else in the world. The continent is criss-crossed with a dense network of railways and roads and a busy system of waterways.

Railways run over many thousands of miles, connecting up all parts of the continent except N Scandinavia, Russia and areas of SE Europe. Some routes serve only local needs, but there are several main trunk routes radiating from Paris.

The first trunk route links Paris to Calais, and then—by way of ferries to Dover—to London and other British cities. The second trunk route runs from Paris around the western end of the Pyrenees to Madrid and Lisbon.

The third route runs S, via Dijon to Marseilles, on the Mediterranean coast. The fourth route branches off at Dijon and goes down to Italy through long tunnels under the Alps. The fifth route is eastwards to Strasbourg, Munich, Vienna, Budapest, Belgrade, and Sofia. The sixth line is via Berlin

Symbol of the Atomic Age. The Atomium, a steel and aluminium model of a metal molecule, in the park of Laeken in Brussels. A relic of the World Fair of 1958, it is 335 ft (102 m) high and houses a scientific exhibition.

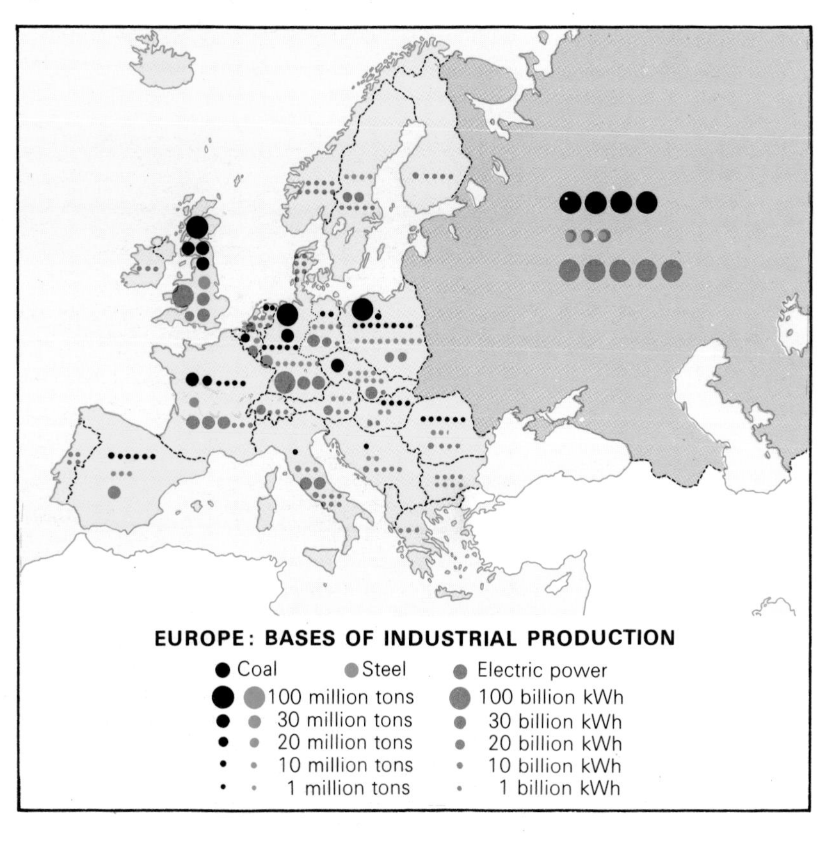

Top: The ancient town of Bellinzona in Switzerland. Few towns and villages are without their petrol and service stations, successors to the blacksmiths and stables of the past.

Centre: Europe's' Largest Port. Rotterdam, in the Netherlands, is 15 miles (24 km) from the North Sea, but a canal system enables it to be used by even the biggest ships. Because of its position on an arm of the River Rhine, it is also a great inland port.

Below: Tuna fishing, formerly a major item in the economies of Sicily, Sardinia, and other Mediterranean islands. Deep sea fishing is an important industry in several countries on the Atlantic seaboard.

and Warsaw to Moscow. Though Russian railways run on broad gauge rails, some Russian carriages can be adapted to run on the standard gauge rails of other European countries. From Moscow, the Trans-Siberian Railway crosses Asia to Vladivostok.

Roads. The tremendous increase in the number of road vehicles in recent years has been paralleled by improvements in roads and highways. Today, fast motorways connect all the main centres of population. The first of these major highways were the German *Autobahnen* constructed in the 1930s. Much heavy freight that once would have been carried by rail is now transported on the road in huge lorries popularly called 'juggernauts'. Increasingly, too, passengers—particularly tourists—travel by long-distance bus instead of by train.

Air Services link the major cities of Europe with one another and with the rest of the world. But there are still relatively few local airports.

Rivers and Canals provide cheap and efficient transport in the central and western low-lying regions. The main waterways are the Rhine and the Danube with their tributaries and linking canals in western Europe; and the Volga, Don and Dnieper in Russia.

Railways, roads, and waterways all serve the continent's ports, most of which remain ice-free all the year round. The only ports that are closed in winter are Archangel on the White Sea, those along the Baltic coasts, and some Black Sea ports—which, however, can be kept open by icebreakers.

Other forms of transport and communication are also plentiful. Networks of pipelines carry water, petroleum and gas. Telephones link nearly all populated areas. Europe's newspapers have some of the world's largest circulations.

Agriculture, Fisheries and Land Use

Because of Europe's temperate climate and good rainfall, most of its farms are very productive. Even so, it has to import about one fifth of its food from other continents. Some heavily-populated countries, such as the United Kingdom, produce much less food than they need. Others, such as the Netherlands, produce more than they need and export the surplus.

The continent has three important farming

Inland Waterways. A heavily-laden barge on the Rhine, at Bacharach in West Germany. The Rhine, one of the great trade arteries of Europe, provides an outlet to the sea at the port of Rotterdam in the Netherlands. It passes through several countries, and is linked by canal to many other rivers. The huge Rhine barges carry mineral ores, oil and other heavy cargoes. Often a barge is 'home' to its crew, which may be a family. The family car is sometimes carried on deck for use in port.

regions: the lands around the Mediterranean Sea; the western part of the plains; and the eastern lowlands of Russia.

The Mediterranean lands have dry, sunny summers. There, farmers can grow large crops of such fruits as almonds, apricots, lemons, olives, oranges, peaches and pomegranates. Wheat, barley, maize and rice are grown too. There are few good pasture lands for cattle, but goats and sheep graze on the scanty vegetation of the mountainsides.

The western plains have fertile soil, and the most efficient and intensive farming takes place in this region. Traditionally farmers have had mixed farms on which they have raised animals and grown crops. Crops have usually been grown by rotation, that is, a different crop being grown each year, often on a four-year cycle. In recent years there has been a tendency for farms to become bigger and more specialised.

Small fields have become larger as hedgerows have been torn up so that machinery is more easily used. Chemical fertilisers give heavy crop yields with less need for crop rotation. However, there is a danger that the quality of the soil may be reduced in the long run. Wheat, barley, oats and rye are the chief cereal crops. Sugar beet and potatoes are important root crops.

In this region the commonest farm animals are daily cattle. Some European cattle breeds, such as Jersey, Hereford, Friesian and Brown Swiss, have become famous as stock all over the world. Belgium, Denmark and the Netherlands are the chief exporters of dairy produce—milk, butter and cheese. The British Isles has much hill land and therefore has more sheep farming than other countries.

The eastern lowlands of Russia also have good soil, but the farming methods used are generally less efficient than in the W. As a result, the yields of crops are often lower. The main crops are rye and flax in the central areas, and wheat on the rich black earth of the steppe-land farther S.

In Alpine regions, transhumance is frequent—that is, farmers drive their cattle, sheep and goats up to the high pastures during the warm summer months. Then they bring them down again to the lowlands in the cold weather when the highlands are covered with snow.

Fisheries are another important part of Europe's food industry. The leading fishing country is Norway, which has about 60,000 fishermen. Other leading fishing countries

Top: Terrace Farming. A village in central Italy, built on a mountain slope. Over many laborious years, the villagers have constructed terraces on the rocky slopes to preserve the thin soil and to hold rainfall.

Below: The windmills of Kinderdijk in the Netherlands in a region of 'polder' land, mostly below sea-level, that has been reclaimed from the sea by the building of dikes. Many of the famous Dutch windmills were pumps for draining the land; others were used for grinding corn. Today, the windmills of Kinderdijk are no longer in use, but are preserved as relics of a day that has gone.

COUNTRIES OF EUROPE
AREAS

Country	*Sq Miles*	*Sq Kilometres*
Albania	10,630	27,532
Andorra	191	495
Austria	32,373	83,846
Belgium	11,774	30,495
Bulgaria	42,858	111,002
Czechoslovakia	49,377	127,886
Denmark	16,620	43,045
Finland	130,165	337,126
France	211,200	547,000
German Democratic Rep.	41,660	107,900
Germany, Federal Rep. of	95,980	248,590
Gibraltar	2.25	5.8
Greece	50,944	131,944
Hungary	35,919	93,030
Iceland	39,709	102,846
Ireland, Rep. of	27,135	70,279
Italy	116,303	301,223
Liechtenstein	62	161
Luxembourg	999	2,587
Malta	122	316
Monaco	1	1.8
Netherlands	13,500	34,965
Norway	125,181	324,217
Poland	120,360	311,730
Portugal	34,240	88,680
Romania	91,700	237,500
San Marino	24	62
Spain	194,885	504,750
Sweden	173,400	449,110
Switzerland	15,940	41,284
Turkey (Europe)	9,121	23,623
U.S.S.R. (Europe)	2,151,000	5,571,000
United Kingdom	94,216	244,018
Vatican City	108 acres	44 hectares
Yugoslavia	98,750	255,760

are Britain and Iceland. Some of the most important fishing grounds are in the North Sea, but many fishing vessels go out into the North Atlantic in search of cod. Russian fishermen catch sturgeon in the Black and Caspian Seas.

Four countries of Europe are much more densely populated than any others. They are Belgium, England, West Germany and the Netherlands, which exceed an average of 500 inhabitants to each sq mile (more than 190 per sq km). Most of the rest of Europe has a population of between 33 and 256 per sq mile (12 and 98 per sq km). Most of the areas of high population lie around the regions of the coalfields, where heavy industries grew up and many of the biggest cities are.

Other major areas of high population lie in N Italy; around Oporto and Lisbon in Portugal; around Madrid and Barcelona in Spain; and around Moscow. The valleys of

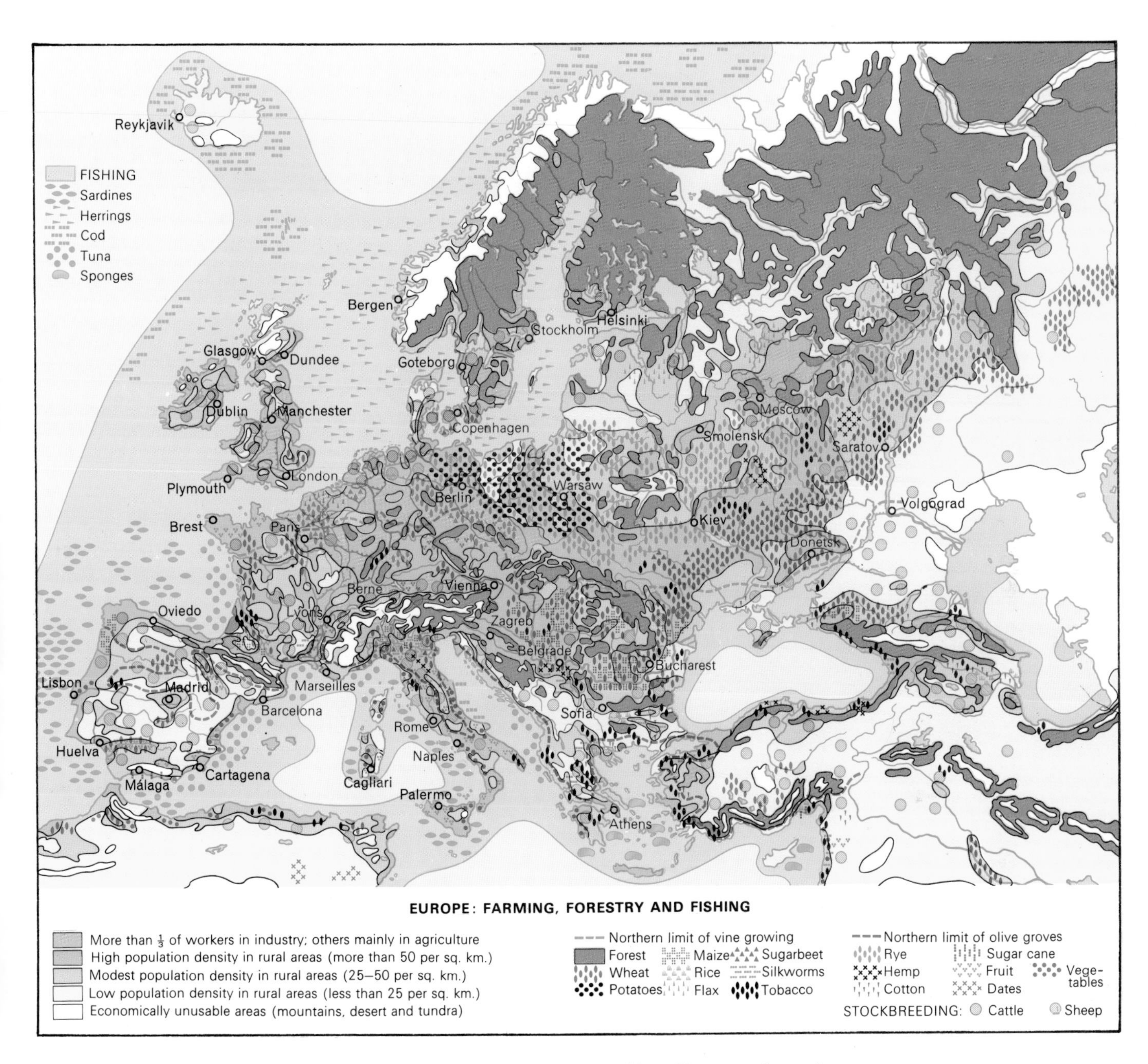

EUROPE: FARMING, FORESTRY AND FISHING

the Rhine and Rhône rivers are also heavily populated, partly because of the high level of agriculture in these regions.

People and Ways of Life

Nobody knows quite how long people have lived in Europe, but fossilized skeletons suggest that the continent has been inhabited for at least 300,000 years. Cave paintings that still exist in France and Spain were made at least 30,000 years ago. The Altamira caves in Spain and the Lascaux caves in France are among the most famous of these.

The present population of Europe consists mainly of members of the Caucasian race—commonly called 'white people'. Their ancestors moved into Europe a long time ago from the plains of central Asia. There are three main groups of European peoples, and though they have intermarried over the years each one of these types is still more common in certain areas than the others are.

Nordic people have long skulls, fair hair, fair complexions, and blue eyes. They are generally tall. They are still found in north-western Europe—in Scandinavia, northern Germany, and in smaller numbers the Netherlands.

Mediterranean people have long skulls too, but they have dark hair and complexions and brown eyes. They are shorter than Nordic people. They are found chiefly in countries

Above: Lincoln Cathedral in England, a splendid example of Early English Gothic, stands as testimony to beliefs and aspirations that have played a large part in shaping European life.

Right: The old houses in a market place in Bergen, Norway, provide a background to shoppers and sightseers.

Far right: The village of Portofino, on the Gulf of Genoa in northern Italy. Tourism has become a major local industry in many parts of Europe. Holiday-makers find in such spots as Portofino their dreamed-of holiday combination: the sun, the sea and the feeling of being 'away from it all'.

bordering the Mediterranean Sea.

Alpine people have broader skulls than the other two types, have hair and complexions that are medium to dark, and a sturdy build.

Two other major groups are the *Dinaric* people, who have round heads and are found in the Balkans, and the *East Baltics*, who have broad, round heads, and are found in eastern Europe, including parts of Russia.

Today the peoples of Europe are grouped in different 'nationalities'. The continent is divided politically into 35 countries, and the people of those countries speak about 60 different languages.

The *Germanic* languages include Danish, Dutch, English, German, Norwegian and Swedish. The *Romance* or *Latin* languages include French, Italian, Portuguese, Romanian and Spanish. The *Slavic* languages include Bulgarian, Czech, Serbo-Croatian, Polish, Russian and Ukrainian. Other language groups include the Celtic tongues—Gaelic, Welsh and Breton—and Greek and Albanian groups.

The people of Finland and Hungary speak languages akin to those of central Asia. The Basques, a people who live on the borders of France and Spain, have a language whose connexions are not clear.

All these variations of language have helped to keep the nations of Europe divided. For hundreds of years Latin—the language of ancient Rome—was used as a universal language by scholars, and was in daily use as an international language by the Catholic Church until the 1970s. Today, English, French, German and Russian are the most useful languages for a European. The most widely spoken language in Europe is German.

Since the end of World War II many peoples have emigrated to Europe from other continents. The United Kingdom, for example, now has a large number of Negro, Indian and Pakistani citizens, many people from Asian Turkey are found in West Germany and many North Africans in France.

Today, the nations of Europe often tend to group themselves into the non-communist 'Western' countries and the communist 'Eastern' countries. A few, such as Austria and Switzerland, remain 'neutral'. Ten countries of western Europe form the *European Community*, an economic alliance, still in its formative stages, which many hope will become a political union.

IMPORTANT DATES IN THE HISTORY OF EUROPE

3000–4000 B.C. The Minoan civilization developed on Crete.
1194–1184 B.C. According to tradition, the Helladic peoples (the Achaens) fought against Troy, the war celebrated in Homer's *Iliad*.
490 B.C. The Greeks defeated the Persians at Marathon.
461–431 B.C. The Golden Age of Athens, Greek civilization reaching its peak.
146 B.C. Greece became a Roman province.
55–54 B.C. Julius Caesar invaded Britain.
27 B.C. Augustus became the first Roman emperor.
100s A.D. Rome attained its greatest power.
293 The emperor Diocletian divided the Empire.
330 Constantine, the first Christian emperor, moved the capital to Byzantium.
476 The 'barbarians' under Odoacer deposed the last Western Roman emperor, Romulus Augustulus.
800 Charlemagne, king of the Franks, crowned emperor of the Romans by the Pope.
962 Otto of Germany founded the Holy Roman Empire.
1096–1099 The First Crusade sought to take the Holy Land from the *Saracens*, the Muslims; it established a Christian kingdom of Jerusalem.
1147–1291 Several later Crusades ended in failure.
1337–1453 The Hundred Years' War between England and France. At its end only Calais was left in English hands.
1300s–1500s The Renaissance followed the medieval period, and with it came a new 'humanist' attitude to art and knowledge and a new interest in Europe's past.
1453 The Turks captured Constantinople.
1492 The last Muslim kingdom in Spain destroyed by the Spaniards.
1492 Christopher Columbus 'discovered' the New World.
1517 The Protestant Reformation began with Martin Luther's protest at the sale of indulgences.
1545–1563 The Council of Trent reformed the discipline of the Catholic Church, and began the Counter-Reformation.
1555 The Peace of Augsburg left the German lands half Catholic and half Protestant.
1571 The 'Turkish threat' ended by the defeat of the Turkish fleet by Don Juan of Austria at Lepanto.
1683 On land, the Turkish advance was ended by the victory of John Sobieski, king of Poland, at Vienna.
1618–1648 The Thirty Years' War limited the power of the German emperor and increased the influence of France.
1770s–1800s The Industrial Revolution began in England.
1783 End of the American War of Independence. England lost the American colonies.
1789–1799 The French Revolution began. The rise of Napoleon.
1815 Final defeat of Napoleon at Waterloo.
1848 The Year of Revolutions.
1870–1871 The Franco-Prussian War ended in the defeat of France and the creation of a new German Empire.
1914–1918 The First World War. Defeat of the Central Powers by the Allies, destruction of Austria-Hungary and abdication of the German emperor.
1917 The October Revolution in Russia led eventually to the victory of the Bolsheviks.
1933 Adolf Hitler and the National Socialists in power in Germany.
1939–1945 The Second World War. Defeat of Germany, Japan and supporters by the Allies. Europe left in ruins.
1949 North Atlantic Treaty Organization created.
1959 Treaties of Rome establish the European Economic Community with six members.
1973 Denmark, Ireland and UK joined EEC.
1981 Greece joined EEC.

THE COUNTRIES OF EUROPE

Albania

Shqiperia, 'Land of the Eagle'. Republic in the Balkan Peninsula, on the E shore of the Adriatic Sea. The Albanian Alps cover most of the country, but there are lowlands in the W. The climate is Mediterranean in the W, severe in the mountains. There is little industry, but some deposits of petroleum, coal, oil and copper. Chief towns: Tiranë (the cap.), Shkoder (Scutari), Korcë, Durrës (Durazzo). Area 10,630 sq miles (27,532 sq km).

Austria. The castle of Werfen in the valley of the Salzach, south of Salzburg. Many of the Alpine valleys have similar medieval fortresses built in commanding positions.

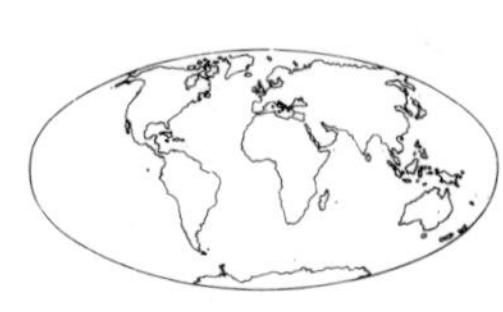

Andorra

Tiny republic in the E part of the Pyrenees, between France and Spain. The President of France and the bishop of Urgel in Spain are joint suzerains. The country consists of mountains and high valleys. Its river, the Valira, has a number of branches. Cattle and sheep graze the hills and crops are grown in the valleys. Tourism is important. The people speak Catalan, French and Spanish. Andorra la Vella, a small town, is the cap. Area 191 sq miles (495 sq km).

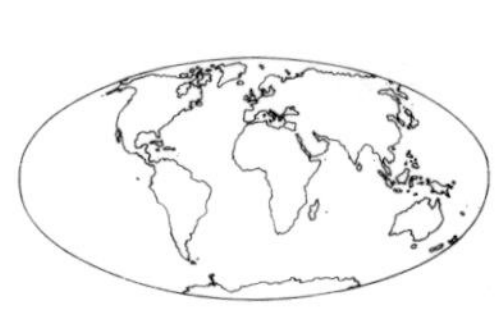

Austria

Republic of central Europe, until 1918 the most important part of the Austro-Hungarian Empire. Austria has no sea-coast and is one of Europe's most mountainous countries. The Alps, in the W and S, extend across two-thirds of the country. The highest peaks are Grossglockner (12,461 ft; 3,798 m) and Wildspitze (12,383 ft; 3,774 m). N of the Alps is the plain of the River Danube, with the wide Vienna plain at its E end. Beyond the Danube, the land rises northwards towards the mountains. The Danube, one of Europe's largest rivers, enters Austria from West Germany and flows E to Czechoslovakia. The Danubian plain has cold winters and warm summers. In Alpine areas the weather is often very cold and wet. In the valleys, most people live by agriculture. They raise cattle, sheep and goats and, where the land is suitable, grow fruits and cereals, including rye and wheat.

Manufacturing industries are sited chiefly in the low-lying areas of the N and E. The factories produce iron and steel goods, textiles, leather goods, chemicals, paper and processed foods. The main source of power is hydro-electricity. A national referendum decided against the introduction of nuclear power stations. The country has valuable mineral resources, including iron, copper, coal, lignite, petroleum, natural gas and graphite. Tourism is also important. In winter, some Alpine resorts are thronged with people who enjoy the winter sports. Music lovers from all parts of the world visit the music festivals in Salzburg. Vienna is the cap. and largest city. Other major cities are Graz, Linz, Salzburg and Innsbruck. Even mountain areas have good road and rail communications. The Danube is an important international trade route and is navigated by heavy barges. The heaviest concentration of population is in the low-lying N and E. About a quarter of the people live in Vienna. Almost the whole population speaks German, except in some border districts. Area 32,373 sq miles (83,846 sq km).

Belgium. The Grand-Place in Brussels, one of the finest squares in Europe. Its superb flamboyant buildings are intricately carved and gilded.

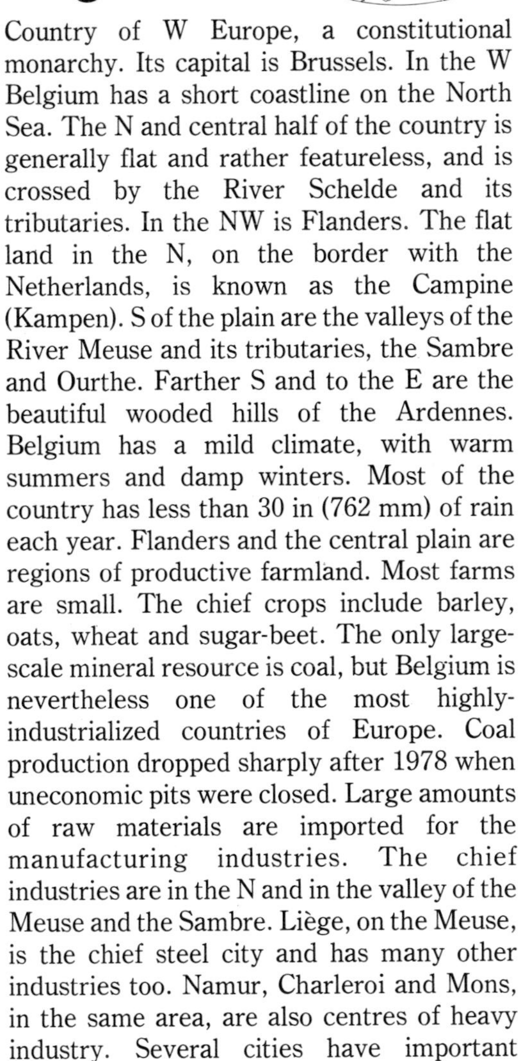

Belgium

Country of W Europe, a constitutional monarchy. Its capital is Brussels. In the W Belgium has a short coastline on the North Sea. The N and central half of the country is generally flat and rather featureless, and is crossed by the River Schelde and its tributaries. In the NW is Flanders. The flat land in the N, on the border with the Netherlands, is known as the Campine (Kampen). S of the plain are the valleys of the River Meuse and its tributaries, the Sambre and Ourthe. Farther S and to the E are the beautiful wooded hills of the Ardennes. Belgium has a mild climate, with warm summers and damp winters. Most of the country has less than 30 in (762 mm) of rain each year. Flanders and the central plain are regions of productive farmland. Most farms are small. The chief crops include barley, oats, wheat and sugar-beet. The only large-scale mineral resource is coal, but Belgium is nevertheless one of the most highly-industrialized countries of Europe. Coal production dropped sharply after 1978 when uneconomic pits were closed. Large amounts of raw materials are imported for the manufacturing industries. The chief industries are in the N and in the valley of the Meuse and the Sambre. Liège, on the Meuse, is the chief steel city and has many other industries too. Namur, Charleroi and Mons, in the same area, are also centres of heavy industry. Several cities have important textile industries. Other major industries include brewing, chemicals and glass. Antwerp is the largest city and chief port. Other large cities are Brussels, Ghent, Liège, Mechelen (Malines) and Schaarbeek. The road and railway systems carry much international traffic. Barges on the canals carry raw materials from Antwerp to the manufacturing centres and take back goods for export. The country is thickly populated, the heaviest concentrations being in the N and centre. There are two distinct groups of people: the Flemish-speaking northerners and the French-speaking Walloons of the south. Area 11,774 sq miles (30,495 sq km).

Belgium. Brugge (Bruges) in Flanders, known for its romantic buildings and bridges and its quiet canals, was once a great commercial city.

Bulgaria

The People's Republic of Bulgaria lies in the E half of the Balkan Peninsula and has a coastline on the Black Sea. The centre, W and S of the country are mountainous. The Balkan Mountains run E-W across the centre, and the wild and rugged Rhodope Mountains across the SW part. Between them lies the broad valley of the River Maritsa. To the N, the great River Danube forms most of the boundary between Bulgaria and Romania. The plains in the N are bitterly cold in winter and hot in summer. In the mountains, the climate is also severe, but it is generally mild in the Maritsa valley. Bulgaria is primarily an agricultural country. On the Danubian plain, farmers grow barley, maize, rye and wheat, and raise sheep and other livestock. In the Maritsa valley, with its milder climate, tobacco and fruits are also grown. Many manufacturing industries are related to agriculture, but there are also heavy industries producing chemicals, metals and machinery. The country has sizeable deposits of coal, iron, lead, zinc, copper, manganese and oil. Tourism is important. Visitors are attracted by the spectacular mountain scenery and the beautiful beaches of the Black Sea coast. The people of Bulgaria come from two main stocks: Bulgar and Slav. Their language, Bulgarian, is Slavic. The chief cities are Sofia (the cap.), Plovdiv, Varna, Burgas, Dimitrovo (Pernik) and Ruse. Area 42,858 sq miles (111,002 sq km).

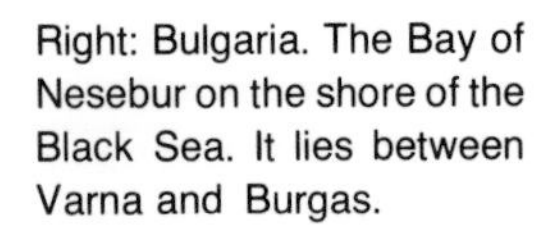

Right: Bulgaria. The Bay of Nesebur on the shore of the Black Sea. It lies between Varna and Burgas.

Below: Czechoslovakia. The Tatra Mountains in the eastern half of the country are part of the Carpathian Range. There are numerous lakes in the valleys. Ski-ing and other winter sports attract many visitors.

Czechoslovakia

Country in central Europe formed after World War I from territories that were formerly part of Austria-Hungary. It is bordered to the W by West Germany, to the N by East Germany and Poland, to the E by the U.S.S.R. and to the S by Hungary and Austria. The country has three natural regions, which also correspond to national groupings. The western, Czech, part of the country is Bohemia. It is a basin-shaped plateau, fringed by mountains. In the low-lying centre is the city of Prague, on the Vltava River. To the NW is the mountain range of the Erzgebirge (Krŭsné Hory in Czech) and in the SW are the wooded hills of the Bohemian Forest. Moravia, the central section of the country, is a fertile region of low hills. It is drained by the Morava River. The eastern part of Czechoslovakia is Slovakia. It is largely occupied by ranges of the Carpathian system. The High Tatra, famous for its beautiful scenery, includes Czechoslovakia's highest mountain. This is the Gerlachovka, which rises to 8,737 ft (2,663 m). The southern part of Slovakia is extremely fertile. It lies in the Danube basin. The Danube, called the *Duna*, forms part of the country's southern border. On it is Bratislava, the chief city of Slovakia and an important river port. Czechoslovakia has, in general, hot summers and cold winters.

There is heavy rainfall in most of the mountainous regions. Deposits of coal, iron and uranium are the principal mineral resources. Nuclear power stations are already in use. The country also has china clay, gold, graphite, lead, manganese, mercury, silver, zinc, and a silica sand suitable for making glass. But it has to import much of its raw materials. The main industries are metal refining, heavy engineering, glass-making and textile manufacture. Farms in the lowland areas produce rich crops of cereals, potatoes and sugar-beet. The chief languages are Czech, spoken in the W, and Slovak, spoken in the E. A few people speak German or Magyar. Chief towns: Prague (the cap.), Brno, Bratislava, Ostrava, Kosiče and Plzen. Area 49,377 sq miles (127,886 sq km).

Left: Denmark. The city hall in Copenhagen. Its clock tower is a well-known landmark.

Below: Czechoslovakia. The 14th-century Charles Bridge in Prague, leading from Hradcany Castle to the Old Town on the east bank of the Vltava (Moldau) River. It is named in honour of the Holy Roman Emperor Charles VI (1316–78).

Denmark

The small country of Denmark in NW Europe is a constitutional monarchy. It is one of the oldest independent countries in the world. Denmark consists of a large number of islands and an irregular peninsula that extends northwards from the N coast of West Germany. The peninsula, Jutland, is separated from the Scandinavian Peninsula by the Skagerrak on the NW and the Kattegat on the NE. These channels form part of the waters linking the North Sea with the Baltic Sea. The largest of Denmark's islands is Zealand (Sjaelland), which is separated from Sweden by a narrow passage called the *Sound* (Öresund). Copenhagen, the country's capital, is on Zealand. Between Zealand and Jutland is another large island, Fünen (Fyn). It is separated from Zealand by a passage called the *Great Belt*, and from Jutland by the *Little Belt*. The next largest island is Lolland, to the S of Zealand and Fünen. Altogether, Denmark has about 500 islands, of which 100 or so are inhabited. Jutland is mostly low-lying. There are low hills in the centre, but they rise only some 500 ft (150 m) above sea-level. The W coast, on the North Sea, has sandy beaches and lagoons. The country inland from them is largely unproductive. The climate is temperate and the rainfall is low, but because of its position Denmark is subject to gales sweeping in from the Atlantic. The country has almost no mineral resources. But it is one of Europe's leading dairy-farming countries, renowned for its butter and cheese. It is also a leading producer of high-quality bacon and exports eggs and poultry to many countries. Much of its arable land is given over to the production of feeding-stuffs for livestock. In places, beet and potatoes are cultivated. The fishing industry is an important source of revenue. There are some manufacturing industries, chiefly based on engineering. They rely on imported raw materials, especially petroleum and its products. The people speak Danish, a language related to Swedish and Norwegian. The chief towns are Copenhagen (the cap.), Aarhus, Odense and Aalborg. Area 16,620 sq miles (43,045 sq km).

Denmark's territory includes the Faeroe Islands, lying between Iceland and Scotland, and the great barren land of Greenland, the largest island in the world. Most of Greenland—which is only a few miles from the coast of Canada—lies within the Arctic Circle. Despite its size, 840,000 sq miles (2,176,000 sq km), only about 60,000 people live there.

Finland

Finland, situated between Scandinavia and the U.S.S.R, is a land of lakes and forests. In the W it has a long coastline on the Gulf of Bothnia. Nearly all of the country is low-lying, but there are some mountains in the N. The N regions form part of Lapland. About one-tenth of Finland consists of lakes, most being in the S. Many form lake-systems—winding, irregular sheets of water, containing numerous islands. The largest lakes are Saimaa (680 sq miles; 1,761 sq km) and Päijänne (608 sq miles; 1,575 sq km). There are many short rivers, some with cataracts. The waters off the S coast are dotted with islands, including a major island group, the Åland Archipelago. About a third of the country is within the Arctic Circle and has a severe climate. Elsewhere, summers are warm though winters are long and cold. Some two-thirds of Finland is forested, and timber is the basis of the chief industries, which include paper manufacture, chemicals and furniture. Forest products form 40% of exports. There are also textile, metal and electrical industries. Though only a tenth of the country is farmland, cattle are raised and some cereals are grown. Swedish is recognized as a second official language. The chief cities are the cap. Helsinki (Helsingfors), Tampere (Tammerfors), Turku (Åbo) and Lahti. Area 130,165 sq miles (337,126 sq km).

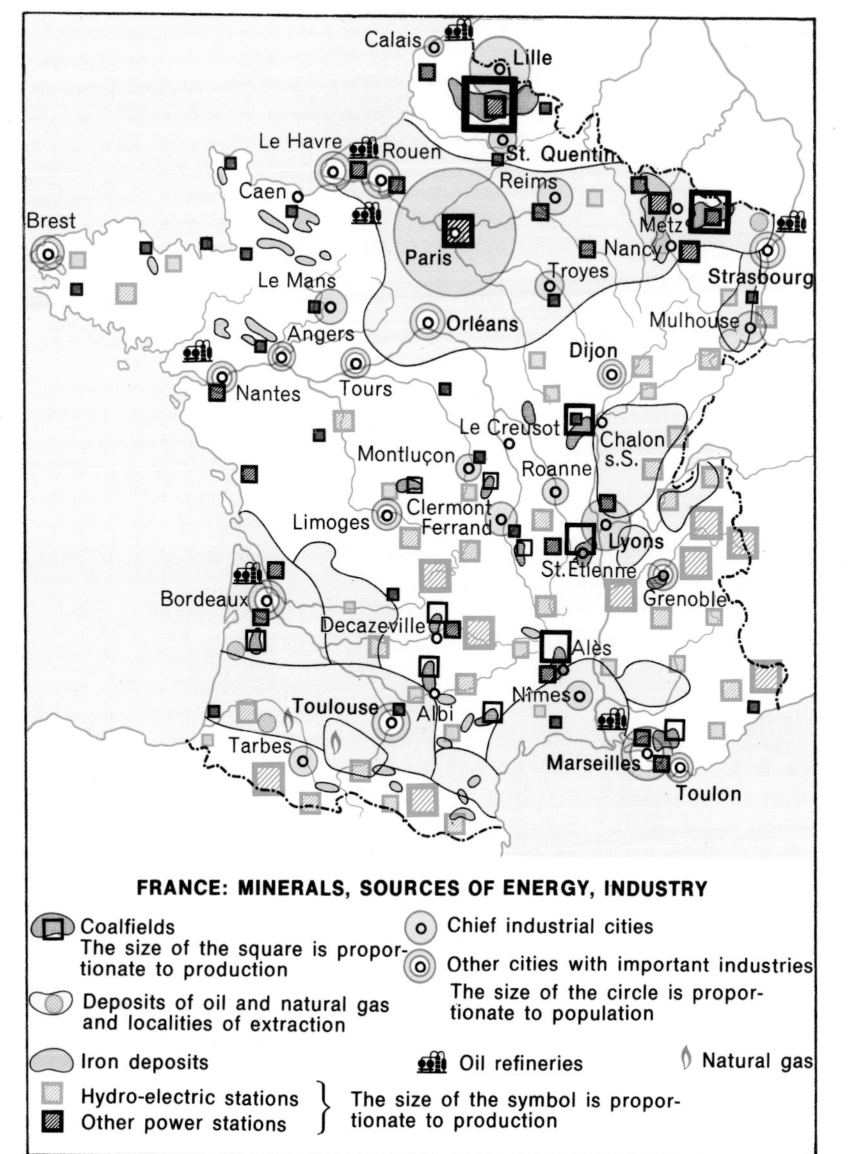

FRANCE: MINERALS, SOURCES OF ENERGY, INDUSTRY

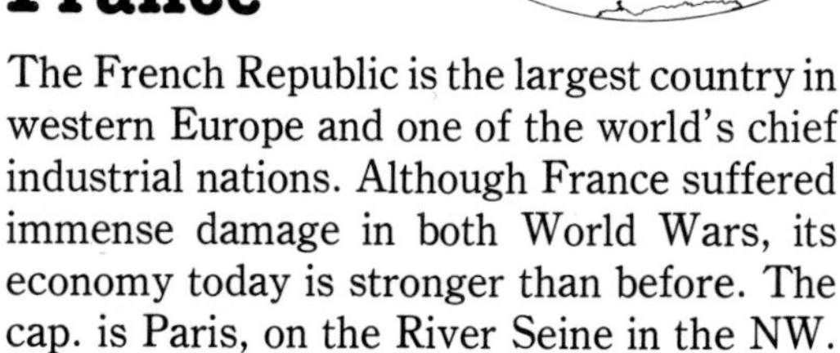

France

The French Republic is the largest country in western Europe and one of the world's chief industrial nations. Although France suffered immense damage in both World Wars, its economy today is stronger than before. The cap. is Paris, on the River Seine in the NW. Area 211,200 sq miles (547,000 sq km).

Land and Climate

On the NE it has borders with Belgium and Luxembourg, and on the E with West Germany, Switzerland and Italy. In the SE it has a coastline on the Mediterranean Sea. France's southern boundary consists of the high, wild Pyrenees Mountains, which separate it from Spain. On the W and NW France has a long, irregular coastline on the Atlantic. The large island of Corsica in the Mediterranean is part of France.

The various parts of the lowlands have their distinctive features. Gascony, in the SW, has marsh and grasslands. To the N of it is the rich region watered by several great rivers. Farther N are the broad plains of NW France, divided into two principal regions: the basin of the River Loire, and the *Paris basin*, whose heart is the Île de France.

The highlands are even more varied. In the NW, the hills of the Armorican Massif occupy Brittany and part of Normandy. The Massif Central in the centre and SE is a vast plateau that occupies about one-sixth of the whole country. To the NE are the wooded hills of the

Jura, the Vosges and the Ardennes. The Alps rise along the borders of Switzerland and Italy in the SE. The highest Alpine peak is Mont Blanc (15,781 ft; 4,810 m). In the extreme S, the Pyrenees form a high, relatively narrow barrier to the Iberian Peninsula.

France's longest river is the Loire, which rises in the Massif Central and flows W for 650 miles (1,050 km) to the Bay of Biscay. The Garonne rises in Spain and flows NW to join the River Dordogne near Bordeaux. Another great river, the Rhône, enters France from Switzerland and flows W then S to the Mediterranean. In the N the chief rivers are the Seine and its tributaries.

Western and northern France has cool summers and mild winters. In the Massif Central, winters are often very cold. Southern regions have a Mediterranean climate, with hot, dry summers.

France. The Château of Langoais on the Loire, built in about 1465 by one of the ministers of Louis XI, still has the menacing air of a feudal fortress. It contains fine tapestries and paintings.

Agriculture and Industry

France is Europe's most productive agricultural country. Cereals are grown in most regions, but principally on the relatively large farms of the Paris basin and the areas running up to the Belgian border. The chief cereal crop is wheat, but barley, maize, oats and rye are also grown. Rice is cultivated in the marshy Rhône delta, called the *Camargue*.

One of the most important crops is sugar-beet. There are also high yields of potatoes and fodder crops. Market-gardening is common, particularly in the south. Normandy, Brittany and Maine grow apples and pears, part of the crop being made into cider.

The great vineyards of Bordeaux, the Loire, Champagne, Burgundy, Beaujolais, the Rhône and Alsace give France its pre-eminent position among the world's wine-producing countries.

The chief industries are the production of iron and steel; and the manufacture of motor vehicles, aircraft, precision instruments, electrical goods, plastics and drugs. There are major rubber and oil industries. The chief mineral resource is coal. Others include iron, aluminium, zinc and salt. A great industrial centre has grown at Fos, east of Marseilles.

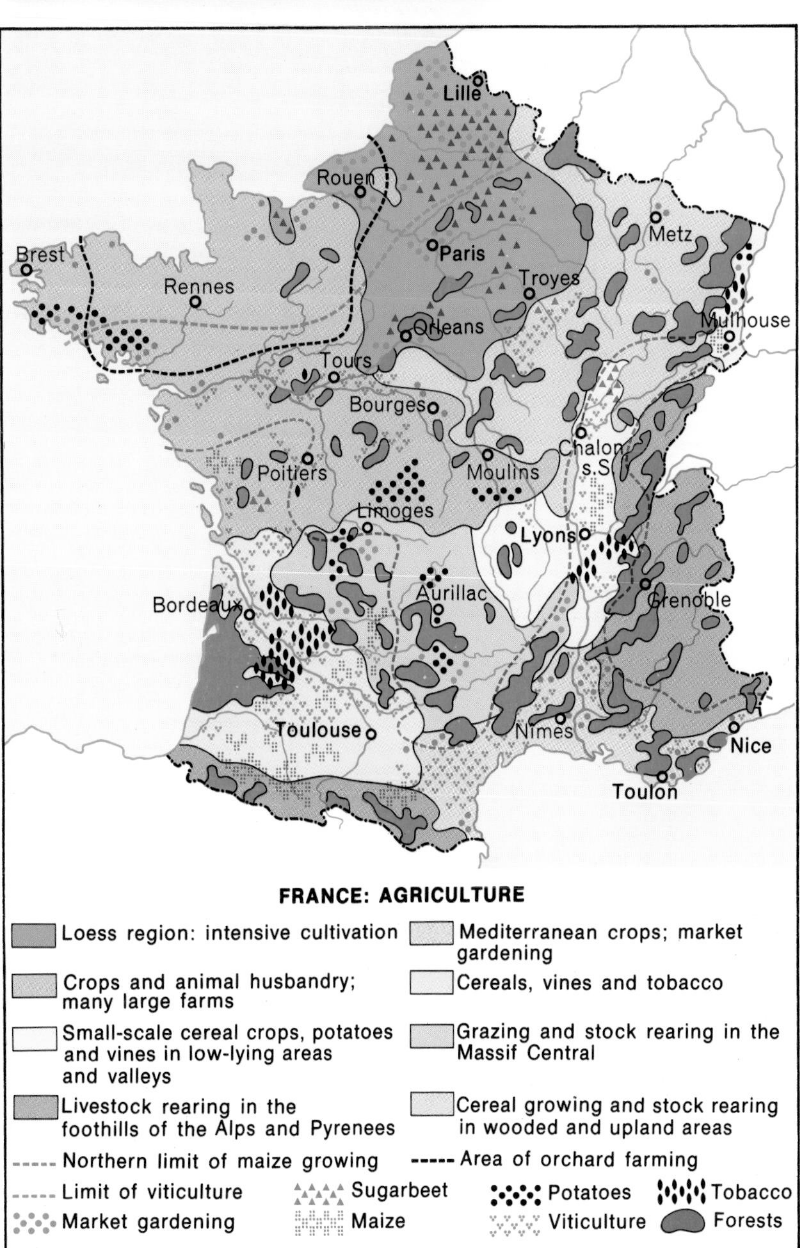

Cities and Communications

France's largest and most famous city is Paris, the capital. Other large cities include Marseilles, Lyons, Toulouse, Nice,

Above: The French Renaissance château of Chenonceaux in the Loire Valley, spanning the River Cher. Its great gallery is nearly 200 ft (60 m) long. The château once belonged to Catherine de Medici.

Bordeaux, Nantes, Strasbourg, St-Étienne, Le Havre and Lille. All major cities are linked by road and rail and there are some 5,000 miles (8,000 km) of waterways. There are several international airports. A high-speed train now links Paris with Lyons.

The People

The French vary in ancestry. Generally, the people are mostly of Celtic origin in Brittany; of Norse descent in Normandy; Germanic in the E; and those in the S are often of Italian descent. The Basques, who live in the Pyrenees, have their own language whose origin is unknown.

Below: The town of Braunfels on the River Lahn, near Wetzlar in Germany. The castle overlooking the town, formerly the seat of the princes of Solms-Braunfels, dates partly from the 1300s.

German Democratic Republic

The German Democratic Republic (East Germany) is the part of Germany that formed the Russian occupation zone after World War II. East Germany is a NE segment of pre-war Germany and includes Brandenburg, Saxony, Mecklenburg and Thuringia. Although the city of Berlin is within its territory, only part of the city, *East Berlin*, belongs to East Germany. Area 41,660 sq miles (107,900 sq km).

Land and Climate

On the N, East Germany has a coastline on the Baltic Sea. Its E boundary follows the line of the rivers Oder and Neisse; beyond them is Poland. On the SE, the border with Czechoslovakia lies along the mountains of the Erzgebirge. To the SW and W is West Germany. The greater part of East Germany is low-lying. Its northern two-thirds is part of the great N European plain, which slopes generally downwards towards the N and NW.

The southern part of the country is mountainous. It includes, in the SW, the thickly-wooded hills of the Thuringian Forest and in the SE the 'Ore Mountains', the Erzgebirge. On the W, on the border with West Germany, are the lower Harz Mountains.

The chief river is the Elbe, which flows SE–NW across the country. Most of East Germany has a mild climate. But winters in the E are severe.

Agriculture and Industry

Almost all agricultural land has been allocated to state farms or to collectives. Consequently, the farms are large by the standards of most European countries. In the central areas, where the soils are relatively poor, there is mixed farming. The chief crops are rye and potatoes, and great numbers of pigs are reared. Cattle, sheep and poultry are also kept. The richest land in the country is in a belt running SE from the Harz Mountains. There, farmers produce large harvests of sugar-beet and wheat. Wheat is grown, too, in the N. In these Baltic lands, dairy farming is important and consequently fodder crops are cultivated extensively.

East Germany is a major industrial country. Its principal industries are concerned with engineering, such as the manufacture of machinery and machine tools, and shipbuilding. Some of the cities of the S have long-established reputations for precision engineering and the manufacture of optical equipment and electrical goods. There are also highly-developed chemical and textile industries. Mineral deposits are few.

Cities and Communications

East Berlin has a population of more than one million. Other important cities are Leipzig, Dresden, Karl-Marx-Stadt (Chemnitz), Magdeburg, Halle, Erfurt and Rostock. There are extensive networks of road and rail communications.

The People

The population of East Germany was swelled after World War II by the influx of refugees from the former German lands E of the Oder-Neisse line. But it later dropped as millions of people left to seek a better life in West Germany. The closing of the frontier halted this decline. But East Germany still suffers from a labour shortage. In recent years, there has been a growing sense of pride among the people because of their achievements in rebuilding their country with little outside aid.

Germany, Federal Republic of

The Federal Republic of Germany (West Germany) is one of the most powerful industrial countries in the world. It is composed of the areas of Germany that formed the British, American and French zones of occupation after World War II, and is more than twice as large as East Germany. The federation consists of 11 *Länder* (states). One *Land* is West Berlin, which is within the territory of East Germany. But since West Berlin is still under military occupation, federal laws do not automatically apply there. The capital of West Germany is Bonn. Area 95,980 sq miles (248,590 sq km).

Land and Climate

On the N, West Germany has coastlines on the North Sea and the Baltic Sea, as well as a short border with Denmark in the Jutland Peninsula. In the E, it borders on East Germany and Czechoslovakia, and in the S on Austria and Switzerland. On the W, it has frontiers with France, Luxembourg, Belgium and the Netherlands.

The northern lowlands of West Germany are part of the vast N European plain, which stretches right across the continent. Its most fertile areas are those to the S. The coastal lands bordering the North Sea are marshy in many places, but some bogland has been reclaimed by the building of dikes. The Baltic coast has a quite different character, with many long inlets.

The central uplands form a much broken region of hills, valleys and plateaux. This region has some of the most varied scenery in Europe. In the W are the small hills and round, crater lakes of the Eifel. Farther S and E are the most beautiful sections of the Rhine and Mosel (Moselle) river valleys, with their steep, vine-covered and picturesque, castle-crowned sides.

The several mountain ranges of the southern highlands include the Schwarzwald (Black Forest) in the SW. To the extreme S are the Alps. Although the highest peak of the Bavarian Alps (Zugspitze) rises only to 9,720 ft (2,960 m), the scenery is among the most spectacular in Europe.

West Germany is a land of large rivers. The chief one is the Rhine, Europe's principal waterway. Other rivers that flow to the North Sea include the Ems, Weser and Elbe. In the NE of the country, the Oder empties into the Baltic. In the S, the Danube flows SE into Austria, eventually reaching the Black Sea.

Above: An old street in Bad Wimpfen, on the River Neckar in Baden-Württemberg. The Blue Tower in the background is part of the ruined medieval palace of the Hohenstaufens.

Below: The Baroque church of the former Benedictine abbey of Zwiefalten, in the Swabian Jura.

The climate of West Germany is generally mild. But the Alpine areas often have severe winters.

Agriculture and Industry

Most German farms are small, generally run as family holdings. This is particularly true of the central uplands where the soil is poor in many sections and mixed farming is common. The lands just N of the uplands are extremely fertile, however, and produce large crops of wheat and barley, as well as sugar-beet and vegetables. Rye is another important crop.

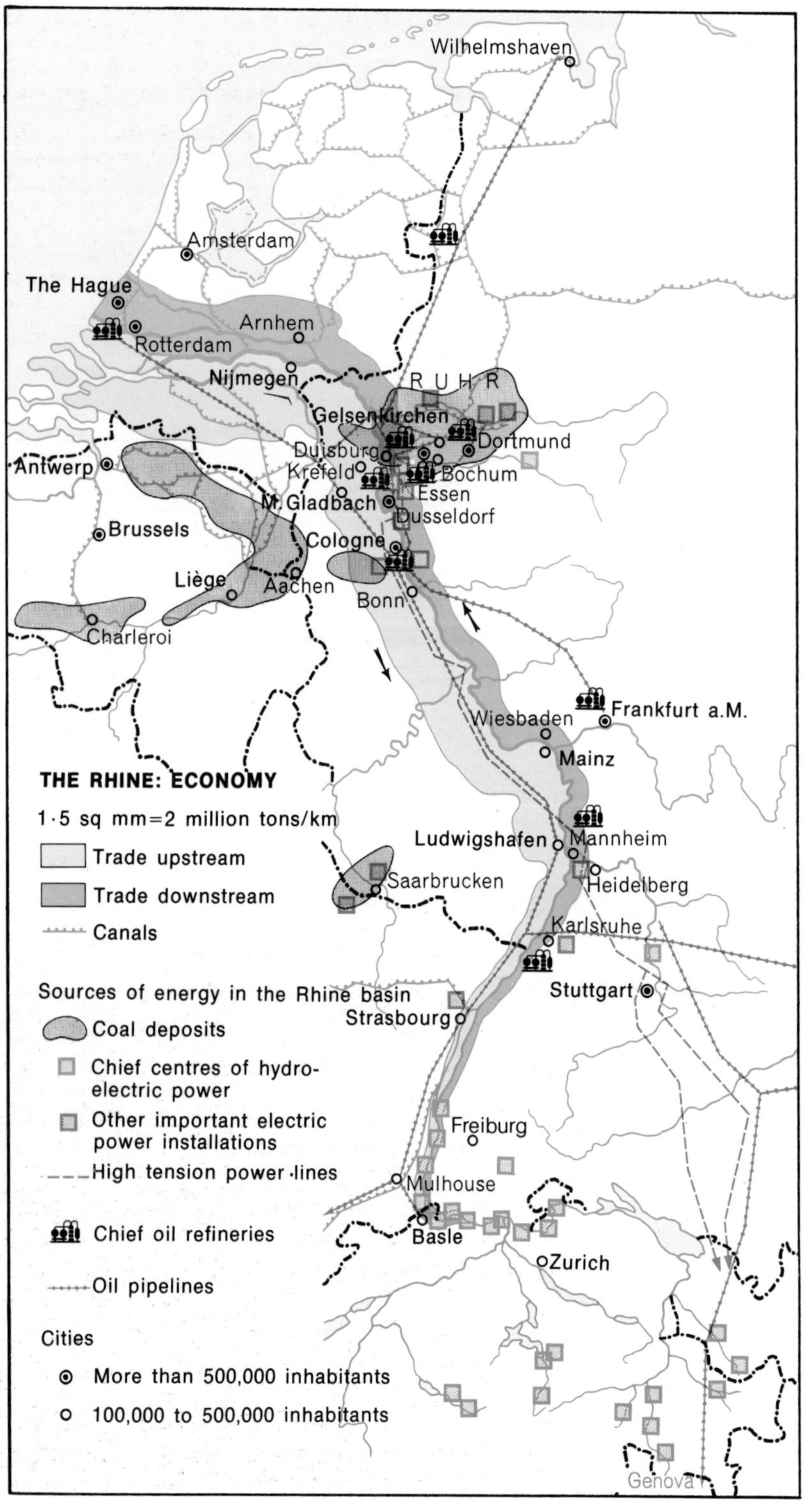

In the NW, large areas of grassland are given over to stock raising. The land in this part of the country is not very suitable for cultivation.

Livestock raising is also of major importance in the southern highlands, where the dairy industry has a high reputation. The valleys of the Rhine and its tributaries are known for their vineyards.

German industry is famous for its high productivity and its inventiveness. The chief industrial area is the *Ruhrgebiet*, a complex of industrial cities in the rich mining area of the Ruhr valley; the Ruhr is a NW tributary of the Rhine. West Germany's industries include almost all kinds of processing and manufacturing: steel-making; the manufacture of motor vehicles, engines, machine tools, electrical goods, precision instruments, drugs, fertilizers, dyes, textiles, plastics, optical instruments, furniture, toys and ceramics; brewing and food processing; and printing.

Some of Germany's industry is based on its mineral resources. These include coal, potash, iron, oil, silver, copper and lead.

Cities and Communications

West Germany's largest cities are West Berlin, Hamburg, Munich, Köln (Cologne), Essen, Frankfurt am Main, Dortmund, Düsseldorf and Stuttgart. The country's transport and communication systems are among the most highly-developed in the world.

Autobahnen (fast motor roads) link the major cities and are complemented by efficient railways. A complex waterway system, with the Rhine as one of its major components, plays a large part in industrial transport. There are several important ports in the N and a number of international airports.

The People

The largest concentrations of population are on the northern seaboard and in the Rhine and Ruhr valleys. The people are about 50 per cent. Protestant and 45 per cent. Roman Catholic.

Left: Bernkastel-Kues, a famous wine village in the Mosel (Moselle) valley in Germany. Among its buildings is the Cusanusstift, a refuge and hospital founded by the humanist Cardinal Nicholas Cusanus in the 1400s; its work still continues today.

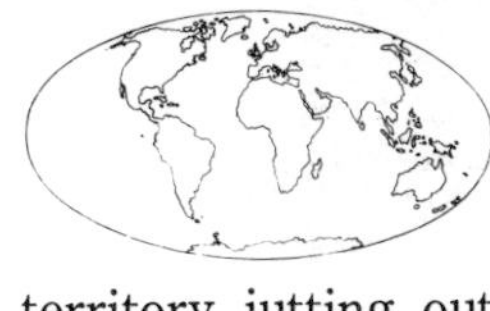

Gibraltar

Gibraltar is a British territory jutting out from the Spanish mainland. Africa is about 14 miles (22 km) distant across the Strait of Gibraltar. The chief feature is 'the Rock'—the limestone promontory that rises 1,408 ft (429 m) above the Mediterranean Sea. Gibraltar has been a British fortress and naval base since the early 1700s, but the Spanish government claims it as part of the territory of Spain.

Greece

Greece is a mountainous republic in the eastern Mediterranean. It became a member of the EEC in 1981. The largest part of its territory is in the Balkan Peninsula, but it also includes a great number of islands. Altogether Greece has about 400 islands, most of them rocky and many of them vine-covered on their S-facing slopes. The low-lying coastland of NE Greece is the most densely-populated part. The N of the peninsula is mountainous. Mount Olympus, N of Thessaly, rises to 9,558 ft (2,914 m) above sea-level. Towns and farmlands lie in valleys between the hills. Rivers flow fast in winter, but tend to dry up in summer. Greece has hot summers. In winter the coastlands have mild weather but in the mountains it is very cold. Greece has some mineral resources. There are large deposits of soft coal and small amounts of various metals. Industry is increasing, but the country is predominantly agricultural. Farmers keep goats and sheep, and grow olives, grapes, currants, cotton and tobacco. The language of nearly all the people is Greek. The chief towns are Athens (the cap.), Salonika, Patras, Piraeus, Vólos, Larissa and Iráklion (Crete). Area 50,944 sq miles (131,944 sq km).

Below: A village in the Peloponnese, the southern peninsula of the Greek mainland. The villages of this part of Greece have a simple beauty, but life is hard for those who have to wrest a living from the sparse areas of arable land.

Hungary. Budapest, on the Danube, has two distinct sections: Buda on the hilly west bank of the river and Pest on the flat east bank. The Margaret Bridge, in the background of the picture, connects the Margitsziget (Margaret Island) to the rest of the city. Budapest is the centre of Magyar culture.

Hungary

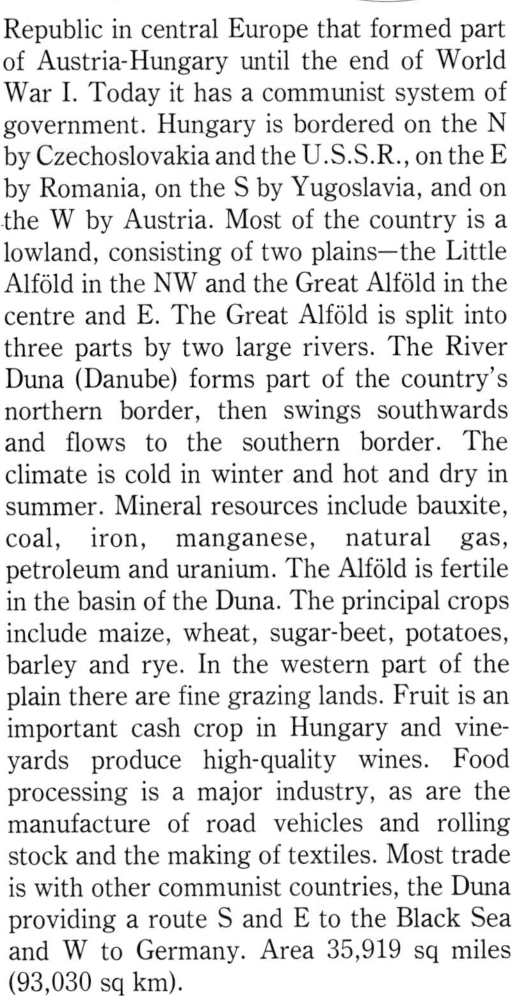

Republic in central Europe that formed part of Austria-Hungary until the end of World War I. Today it has a communist system of government. Hungary is bordered on the N by Czechoslovakia and the U.S.S.R., on the E by Romania, on the S by Yugoslavia, and on the W by Austria. Most of the country is a lowland, consisting of two plains—the Little Alföld in the NW and the Great Alföld in the centre and E. The Great Alföld is split into three parts by two large rivers. The River Duna (Danube) forms part of the country's northern border, then swings southwards and flows to the southern border. The climate is cold in winter and hot and dry in summer. Mineral resources include bauxite, coal, iron, manganese, natural gas, petroleum and uranium. The Alföld is fertile in the basin of the Duna. The principal crops include maize, wheat, sugar-beet, potatoes, barley and rye. In the western part of the plain there are fine grazing lands. Fruit is an important cash crop in Hungary and vineyards produce high-quality wines. Food processing is a major industry, as are the manufacture of road vehicles and rolling stock and the making of textiles. Most trade is with other communist countries, the Duna providing a route S and E to the Black Sea and W to Germany. Area 35,919 sq miles (93,030 sq km).

Iceland

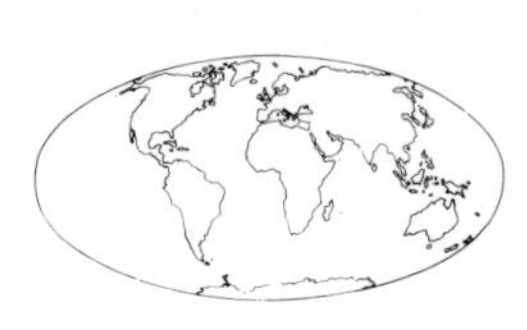

Island republic in the North Atlantic, the N tip of Iceland lies inside the Arctic Circle. Much of the island is a high, ridged plateau, with more than 200 volcanoes. The best known is the active volcano Mt Hekla (4,747 ft; 1,447 m) in the S. A new volcano on the island of Heimaey resulted in the evacuation of all 5000 inhabitants for several months. Clouds of steam drift from hot springs and geysers: the word *geyser* comes from the Great Geysir in the SW. Ice sheets cover one-eighth of the land. Elsewhere there is desert, scrubby moorland and bog. Rivers carry away the water from melting glaciers, passing over waterfalls on their way to the sea. Because of the moderating effect of the Gulf Stream, Iceland's climate is less severe than might be expected. The chief industries are fishing and the processing of fish. To protect these industries, Iceland claims a 200 mile extended fishing zone. In the limited areas which are suitable for farming, cattle, sheep and horses are raised, and hay and vegetables are grown.

The people of Iceland are descended from Viking settlers. Despite the rigours of life on the island, their standard of living is high. Half of the population lives in Reykjavik, the cap. and the only sizeable town, but there are a few other towns, too. Area 39,709 sq miles (102,846 sq km).

Ireland, Republic of

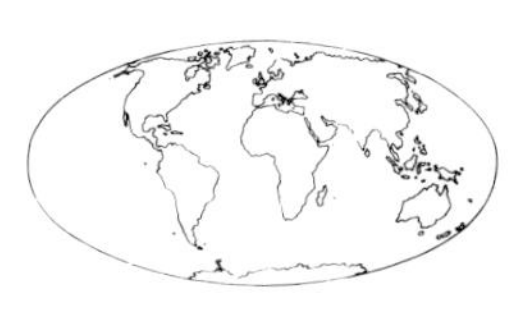

The Republic of Ireland comprises five-sixths of Ireland, the second largest island of the British Isles. In Gaelic, its name is *Eire*. The rest of Ireland, the six NE counties of Ulster, form *Northern Ireland*, a part of the United Kingdom. But some of Ulster, including the most northerly part, is in the Republic. On the E, the Irish Sea and St. George's Channel separate the country from Great Britain. On the W and N is the Atlantic Ocean. The Republic of Ireland has a wide central lowland plain, consisting of good pasture and arable land and peat bogs—called turf bogs. Many low mountain ranges fringe the plain. Those in the W and S are particularly beautiful. MacGillycuddy's Reeks in the SW include Carrantuohill (3,414 ft; 1,040 m), the highest mountain in Ireland. The W coast is irregular, with many bays and long, narrow inlets. the River Shannon flows for 215 miles (346 km) from its source in the NW to its estuary on the Atlantic in the SW. It is the longest river in the British Isles. It passes through several lakes, including Lough Ree and Lough Derg. Among other lakes are the famous Lakes of Killarney in the SW and the remote, dark loughs of Connemara in the W. In the E the River Liffey flows into the Irish Sea at Dublin. S of Dublin are the Wicklow Hills, popular with climbers and walkers. Ireland's climate consists of mild winters and warm summers. There are some mineral deposits: copper, lead, zinc and coal. Turf (peat) is in widespread use as fuel, either in its natural or in processed form. It is used as fuel in some power stations.

Ireland's most important natural resource is its rich farmland. Agriculture is the chief industry, the raising of cattle and the production of milk, butter and cheese taking the leading place. Pig-breeding is also a major farming activity, and the country is celebrated for its horses. The principal crops include cereals and vegetables. The official languages of the country are Gaelic and English. The chief cities are Dublin (the cap.), Cork, Limerick, Dun Laoghaire, Waterford and Galway. Area 27,135 sq miles (70,279 sq km).

Italy

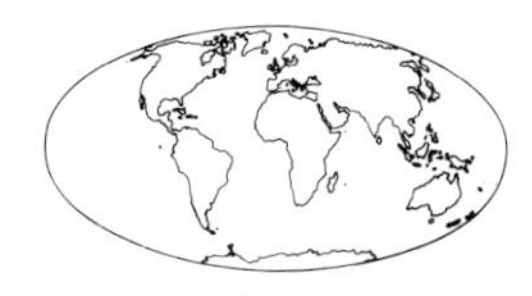

The Republic of Italy includes the islands of Sardinia and Sicily. Its capital is Rome, on the River Tiber in central Italy, near the W coast. Area 116,303 sq miles (301,223 sq km).

Land and Climate

Right across the N of Italy is the great barrier

Left: Italy. The Rialto Bridge in Venice, built in the 1500s, the oldest of the bridges across the Grand Canal. The bridge, a single marble span, has arcades lined with shops. It is named after the Rialto Islands on which part of the city is built.

Below: The deserted beauty of the western seaboard of Ireland, with its dark hills, lakes, moors and boglands. Everywhere, ruined castles and monasteries tell of a turbulent past.

Above: Modern buildings in Milan seen across the pinnacles of the vast Gothic cathedral.

Right: A legacy of classical times at Tivoli, near Rome. It was here that the Emperor Hadrian built his luxurious villa, said to have covered 7 sq miles (18 sq km).

of the Alps. To the E of the country is the Adriatic Sea, lying between the peninsula and Yugoslavia and Albania. To the S is the Ionian Sea, with its northern inlet, the Gulf of Taranto, forming the instep of the Italian 'boot'. To the W Italy has a long coastline on the Tyrrhenian Sea and in the NW on the Ligurian Sea.

Peninsular Italy—that is, the mainland of the country—has several distinct natural regions:

The Alpine Region includes a series of spectacular mountain ranges, with many peaks rising above 13,000 ft (3,900 m). The lower Alpine slopes are terraced for growing fruit, vines and vegetables; and on the higher slopes there are Alpine meadows in many areas.

The Valley of the Po is a broad triangular lowland between the Alps and the Apennines. It is the most densely populated region of Italy, has many of the largest cities and has much of the country's industry. It is also the principal agricultural region.

The North Adriatic Lowland extends NE from the Po valley. It includes part of the stony region known as the *Karst*, most of which is in Yugoslavia.

The Apennines, Italy's backbone, are more than 700 miles (1,100 km) long. They tower above the Ligurian coast and then, crossing the peninsula, follow the line of the Adriatic coast. On this coast, their foothills are much serrated. The average height is about 4,000 ft (1,200 m), but there are higher peaks, and Monte Corno in the Gran Sasso d'Italia rises to 9,650 ft (2,914 m).

The Western Coastal Plain lies along the Tyrrhenian Sea from La Spezia in the N to Naples in the S. It includes the hilly land of Tuscany. The Pontine Marshes and other unhealthy marshy areas near Rome have been reclaimed and turned into good agricultural land.

Rivers and Lakes. Most of Italy's rivers have their sources in the Apennines, but the two great rivers of the N, the Po and the Adige, flow from the Alps.

Climate. Italy's climate is varied. The N tends to have hot summers and cold winters, with much rain. In a large part of the peninsula and the islands, the winters are warm and there is only moderate rainfall. Some places suffer from drought.

Agriculture and Industry

Italy has many minerals, but few of the deposits are large. They include coal, manganese, bauxite (a recent discovery may be the largest in Europe) and sulphur. In the basin of the Po, wheat and maize are important crops and sugar-beet is grown, particularly in the W. Wheat is also produced on a lesser scale in the centre and the S. Other major crops include olives, citrus and other fruits, and vegetables. Vineyards are widespread: Italy produces some of the world's most celebrated wines. Fodder crops are grown to support the dairying industry.

Much of the country's industry is centred in the NW, in the famous Milan-Genoa-Turin triangle. But some new industry has been established in the S, sometimes supplanting older industries of the N. The most important industries are based on steel and engineering technology. Of primary importance is the manufacture of motor vehicles and components. Italy is also one of Europe's leading nations in the textile and chemical industries.

Cities and Communications

The chief cities are Rome (the cap.), Milan, Naples, Turin, Genoa, Palermo (Sicily), Bologna and Florence. Communications are highly developed. To a large extent, the railways are electrified. Italy's *autostrade* (motorways) are among the finest highways in the world. There are many airports.

Italy. The art and architecture of the Baroque. The Piazza Navona in Rome, with Bernini's superb 'Fountain of the Rivers'.

Liechtenstein

The tiny principality of Liechtenstein lies between Austria and Switzerland. It has a postal and customs union with Switzerland and uses Swiss money. The W half of the country is a plain bordering the Rhine. The small farms and orchards grow mixed crops. The people are mainly Austrian in origin and speak German. There is only one town, Vaduz, the cap. Area 62 sq miles (161 sq km).

Luxembourg

Independent grand duchy in W Europe. Most of the country lies on the plateau of Lorraine, but in the N there are some of the beautiful wooded hills of the Ardennes. In the E, the canalized River Moselle runs along the border with West Germany. The country has a mild climate, except in winter. Although one of the smallest countries in Europe, it is one of the most prosperous. It has valuable iron deposits in the S which are the basis of its iron and steel industry. Other industries include leather tanning and the manufacture of textiles and chemicals. Beer and wine are produced. Except in the N, the farms are productive, and cattle and other animals are raised. Area 999 sq miles (2,587 sq km).

Valletta, on the north-east coast of the island of Malta, is one of the best anchorages in the Mediterranean, and was for long an important naval base. The city is built on a rocky peninsula between two deep harbours. It has many reminders of the Knights of St. John of Jerusalem—the Knights of Malta—who ruled the island for more than 250 years.

The city of Luxembourg, capital of the Grand Duchy, built on steep rocky heights above the River Alzette. The Pétrus gorge, crossed by bridges, separates the old town from the new.

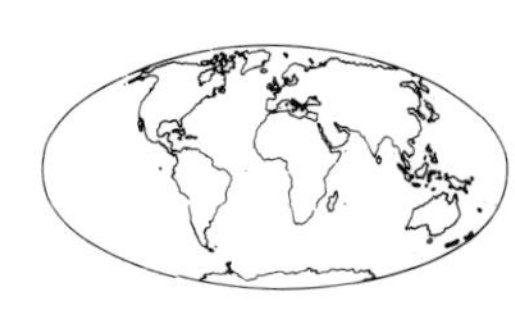

Malta

Island country in the Mediterranean Sea, 58 miles (93 km) S of Sicily. It consists of the islands of Malta and Gozo and the smaller islands of Comino, Cominotto and Filfla. From 1814 to the 1960s it was a British colony and an important naval base. During World War II the island was awarded the George Cross by King George VI of the United Kingdom for the people's bravery in the face of air attack.

The country is generally low and rocky, with thin soil retained by terracing. There are no rivers or lakes. The climate is hot in summer and mild in winter, with moderate rainfall. Apart from the dockyard, industry is small, though growing. Most people live by agriculture or tourism. Malta exports some fruit and vegetables. The official languages are Maltese and English. Area 122 sq miles (316 sq km).

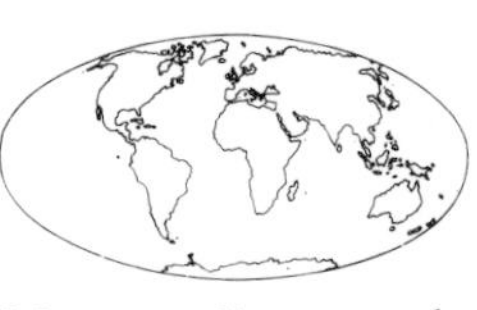

Monaco

The tiny country of Monaco lies on the Mediterranean Sea at the NE end of the French Riviera. It is an hereditary principality, and consists of a short strip of coast behind which rise foothills of the Maritime Alps. Almost the whole area is heavily built up. The chief source of income is tourism. The principality is famous for its gambling casino at Monte Carlo, for its motor sport events—the Monaco Grand Prix and the Monte Carlo Rally—and for its yachting harbour. The people are called *Monégasques*. Area 0.7 sq miles (1.8 sq km).

Netherlands

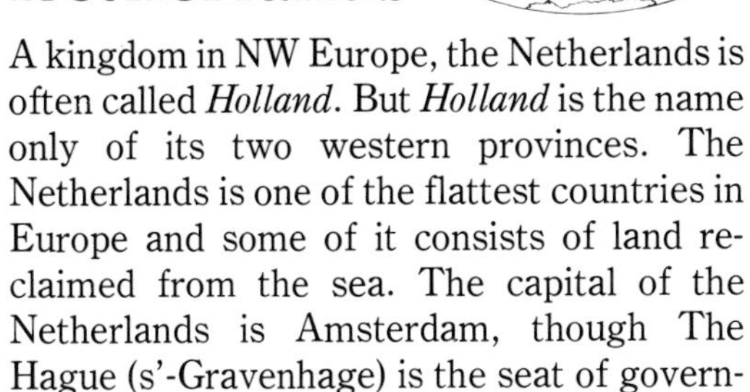

A kingdom in NW Europe, the Netherlands is often called *Holland*. But *Holland* is the name only of its two western provinces. The Netherlands is one of the flattest countries in Europe and some of it consists of land reclaimed from the sea. The capital of the Netherlands is Amsterdam, though The Hague (s'-Gravenhage) is the seat of government. Area: 13,500 sq miles (34,965 sq km).

Land and Climate

The Netherlands has the deltas of three of Europe's largest rivers, the Rhine, Maas (Meuse) and Scheldt. It is very low-lying, its highest point—in the extreme SE—being only 1,057 ft (322 m) above sea-level. Many parts of the country are as much as 20 ft (6 m) below sea-level, and are protected from the sea by dikes. The Netherlands has a long, low coastline, largely consisting of sand dunes that have been reinforced artificially. Just inland from the coastal dunes are the *polders*, the huge areas of land reclaimed from the sea, or from fresh-water lakes. The largest reclamation area is around the IJsselmeer, formerly called the *Zuider Zee*. The IJsselmeer was once a bay of the North Sea, but is now a lake. Reclamation of further areas continues. Almost half the country is polder and two-thirds of the polders are below sea-level.

Left: The Gothic tower of the cathedral of Utrecht in the Netherlands, generally considered the most beautiful tower in the country. It rises some 370 ft (112 m), and its topmost stage is octagonal. It was built between 1321 and 1382.

Below: The Voorburgwal in Amsterdam, lined with houses of many styles and periods. In the background is the Oude Kerk ('old church') of St. Nicolas, which dates from the 1300s. Amsterdam, sometimes called 'the Venice of the North', has about 50 miles (80 km) of canals, crossed by more than 400 bridges.

The crowded 17th- and 18th-century houses of the old part of Amsterdam, with their sharply-pointed gables.

BENELUX: ECONOMY

Chief urban areas
Industrial areas
Coalfields
Iron deposits
Coal-mining centres
Metallurgical centres
Textile centres
Centres of mixed industry
Transport and other centres
Internal waterways
Railways

Haarlem
Amsterdam
The Hague
Utrecht
Europoort
Rotterdam
Essen
Antwerp
Cologne
Bonn
Brussels
Luxembourg
Paris

The many rivers of the Netherlands are mostly delta arms of the Rhine and the Maas, linked by a network of canals that serve both for transport and for drainage. The terrible floods of 1950 led to the construction of a series of flood control barriers under the Delta Plan. The country's largest lake is the IJsselmeer. The climate of the Netherlands consists of mild to cold winters and cool summers, with moderate rainfall.

Agriculture and Industry

The Netherlands has very rich deposits of natural gas, and good deposits of coal and petroleum. Reclaimed polder lands provide fertile soil for growing crops and for providing good grazing. The main crops are wheat, sugarbeet and vegetables, plus fodder for livestock. In the SW, hothouses produce tomatoes, lettuce and cucumbers for export. Nearby, the sandy soil is planted with the bulbs for which the Netherlands is famous.

Dairy products are of primary importance. Condensed and powdered milk, butter and cheese are processed for export, and the country is Europe's largest exporter of eggs. Fisheries provide fish for the home market.

Metal refining and engineering are the principal industries, with chemical manufactures third. Shipbuilding is another major industry. The Dutch also specialize in cutting and polishing diamonds. A further major source of wealth is the large volume of trade for central Europe that passes through the Netherlands or is carried in Dutch ships. The Dutch also provide banking services for international trade.

Cities and Communications

The chief cities are Amsterdam, The Hague, Rotterdam, Utrecht, Eindhoven, Haarlem, Groningen, Tilburg and Nijmegen. Road and rail communications are highly developed. So is the waterway system: there are more than 4,000 miles (6,400 km) of canals and other waterways. A great port with industries has been developed at Europoort, downriver from Rotterdam.

The People

The Netherlands is one of the most heavily populated countries in Europe. The people are mainly of Germanic origin. Their language is Dutch.

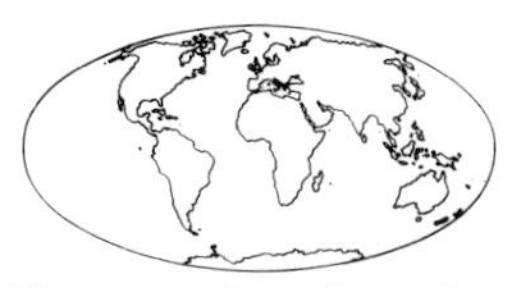

Norway

Kingdom of northern Europe, forming the western part of the Scandinavian Peninsula. In the Arctic Ocean, 400 miles (650 km) N of peninsular Norway, is the large archipelago of Svalbard (Spitzbergen). High mountains rise sharply from the western coast, which is indented with many fiords. Thousands of islands fringe the coast, but most of them are very small and are uninhabited. They help to protect coastal shipping from Atlantic storms. The chief island group, the Lofoten Islands, lies N of the Arctic Circle. The highest mountain in Norway, Galdhöpiggen (8,140 ft; 2,481 m), is in the Jotunheimen Mountains in the S. The climate is cold in the mountains and the N, but warmer along the W coast as a result of the Gulf Stream. Summers are short and mild. Rainfall is heavy. In winter the mountains are covered with snow. Norway's dense forests of evergreen and hardwood trees are exploited for their timber, but replanting is practised. The main mineral resource is iron, and there are also some other metals. Large deposits of petroleum lie off the W coast. Dairying is the main agricultural activity. Norway has a large fishing fleet. Area 125,181 sq miles (324,217 sq km).

Left: Geiranger Fiord, in south-west Norway, one of the innumerable long and narrow inlets along the coasts. Some fiords penetrate far inland and ships can sail up them for many miles.

Below: The Lofoten Islands, off the north-west coast, in the Norwegian Sea. Though they lie within the Arctic Circle, the islands have a relatively moderate climate because of the warming effect of the North Atlantic Drift.

The Palace of Culture in Warsaw, Poland, showing the post-war influence of the modern Russian grandiose style.

The 13th-century castle in the walled village of Obidos, Portugal.

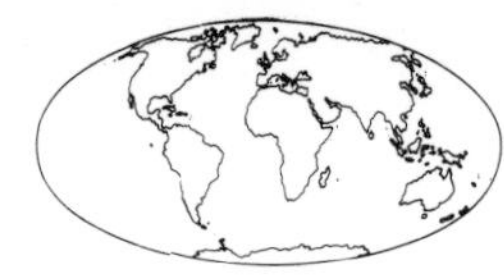

Poland

Poland lies on the N European plain and has a long coastline in the N on the Baltic Sea. On the E it is bordered by the U.S.S.R., on the S by Czechoslovakia, and on the W by East Germany. Like all of these countries it has a communist form of government. There have been serious disturbances in recent years resulting from steep increases in food prices. Most of Poland is flat, but the land rises slowly towards the Carpathian and Sudeten mountains in the S. The chief river is the Vistula, which rises in the Carpathians and flows northwards for 652 miles (1,050 km) to Gdańsk on the Baltic. Two other large rivers, the Oder and the Neisse, form the boundary with East Germany. The Polish climate tends to be warm and wet in summer, and cold in winter. Some rivers are frozen for several months a year. Poland is rich in minerals, including coal, lead, zinc, copper, sulphur, iron and potassium. Many of the country's farms are small family holdings. The chief crops are potatoes, rye, wheat, oats, barley and sugar-beet. Livestock includes pigs and cattle, as well as the horses for which Poland has long been famous. Among industries, the manufacture of steel and the making of various kinds of textiles and other clothing materials are important. Engineering works produce heavy machinery; Polish shipbuilding has a high reputation. The people of Poland are mainly Slavs. Area 120,360 sq miles (311,730 sq. km).

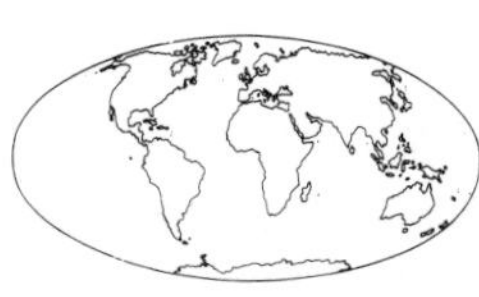

Portugal

Republic in the W of the Iberian Peninsula, with a long coastline on the Atlantic. The high central plateau of Spain extends into the northern half of Portugal. Along the Atlantic coast and in the S the land is low-lying and much of it is very fertile. Long stretches of Portugal's coast have fine sandy beaches popular with tourists. In the lowland regions the climate is warm and equable, but in mountain areas the winters are severe. Portugal has some mineral resources of importance, including wolframite, coal, lead, copper and antimony. Its forests are a major source of revenue; Portuguese cork oaks produce about half of the world's cork. The country is predominantly agricultural, the main farming regions being in the central highlands and the W. Crops include rice, wheat, maize, barley, potatoes and beans. Olive trees are found in most areas. In the S, citrus fruits, figs and almonds are grown. Portugal is a leading wine-producing country. Portuguese fishermen catch cod, tuna and sardines. There are factories producing textiles, steel and ceramics. Area 34,240 sq miles (88,680 sq km). The Madeira Islands and the Azores are part of Portugal.

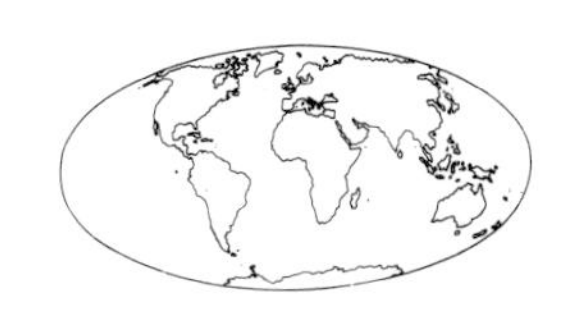

Romania

Romania is a republic of SE Europe, with a coastline on the Black Sea. Its name means 'land of the Romans'. It has a communist form of government. Romania is bordered on the N by the U.S.S.R., on the S by Bulgaria, and on the W by Yugoslavia and Hungary. The country is extremely mountainous: the Carpathian Mountains and the Transylvanian Alps form a great arc from N to W. They enclose the plateau of Transylvania in the NW. The wide plain in the S of the country lies in the basin of the River Danube, which runs along the country's southern boundary. Summers are warm, except in the high mountain regions, but winters tend to be severe.

Romania has rich mineral resources, particularly petroleum and natural gas. The famous oilfields of Ploeşti and Buzau are in the SE. Other valuable natural products are coal, iron and rock salt, and timber from the great forests. On the vast plains of the S, farmers grow wheat and maize. Romania also produces sugar-beet, fruit and potatoes. The country is also known for its wines, some of them of high quality. Among the products of its fishing industry is caviare—sturgeon roes. The government is trying to extend the manufacturing industries, particularly those based on metal and chemical technologies.

The early inhabitants of the land that is now Romania were the Dacians. Today's Romanians are descended from both the Dacians and all the many peoples of S and E Europe who arrived later either as invaders or settlers. The chief cities are Bucharest (the cap.), Braşov, Ploeşti and Iaşi. Area 91,700 sq miles (237,500 sq km).

Romanian villagers 'dressed for the occasion'. The people of the country's rich farmlands have a strong sense of tradition and faithfully preserve the customs of their ancestors.

San Marino

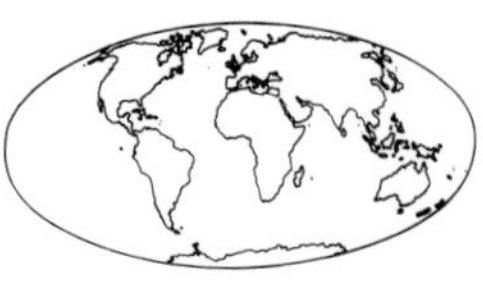

San Marino is the world's smallest republic. It is in northern Italy, on the E slopes of the Apennine Mountains, near the Italian town of Rimini. The republic's capital, a walled town also named San Marino, is built at the top of a steep, rocky, three-peaked hill, 2,424 ft (737 m) high. The chief sources of revenue are tourism and the sale of postage stamps, but there are several small industries, including the manufacture of cloth, dyes and ceramics.

Quarries produce stone for building. The people make oil from olives, and grow fruit and some cereals. There are many vineyards, in which dessert grapes are grown as well as grapes for wine-making. Area 24 sq miles (62 sq km).

Spain

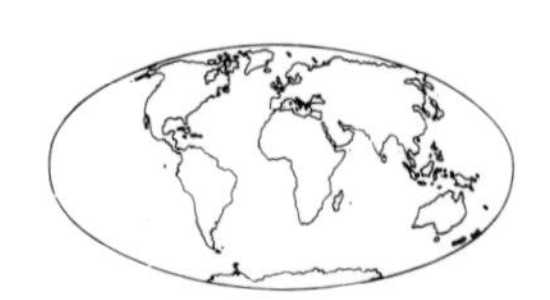

Spain occupies the greater part of the Iberian Peninsula in SW Europe. Its extreme southern tip on the Strait of Gibraltar is only 9 miles (14 km) from Morocco, in Africa. The Canary Islands in the Atlantic and the Balearic Islands in the Mediterranean are ruled as part of the country. The capital is Madrid. Area 194,885 sq miles (504, 750 sq km).

Land and Climate

On the E, Spain has a long coastline on the Mediterranean Sea, and on the W and N an even longer coastline on the Atlantic. The Pyrenees in the NE form the boundary with France. Three-quarters of the country is a vast plateau, the *Meseta*, parts of which are as

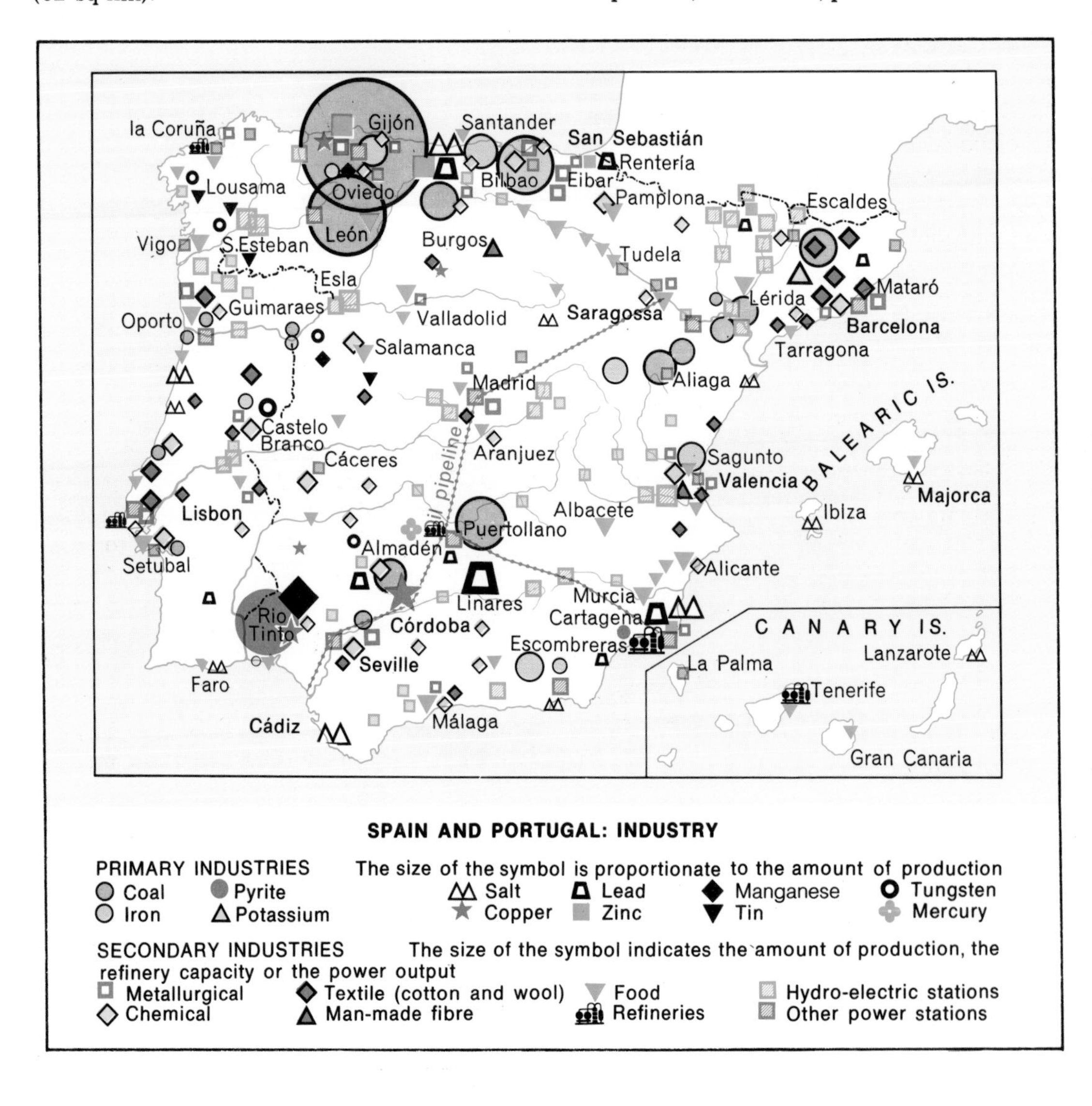

SPAIN AND PORTUGAL: INDUSTRY

high as 6,000 ft (1,800 m) above sea-level. The S and E coastal areas have mild winters and warm summers. The rest of the country has greater extremes of climate.

Agriculture and Industry

On the plateau, agriculture suffers from lack of water. In the plains, the most important crop is wheat, and some upland areas have oats and rye. Other cereal crops grown are rice, maize and barley. Olives contribute to the livelihood of many farmers. There are thousands of sq miles of vineyards. The E and S of the country are famous for their citrus fruits, almonds, peppers, figs, tomatoes and onions. Spain has valuable mineral deposits—mercury, coal, iron, copper and salt. The chief manufacturing industries are in the N. They include textile, engineering and paper industries. Tourism is important to the country's economy.

Cities and Communications

Spain has many beautiful and historic cities. The largest city is Madrid, the capital. Barcelona is an important industrial city and port. Other large towns are Valencia, Seville, Zaragoza, Malaga and Bilbao. In general, communications are not highly developed, because the country has few natural routes.

The People

The Spanish people are descended from the Iberians and from the Celts, Carthaginians, Romans, Moors and other peoples who later settled in the peninsula. Most of them speak Spanish. In the NE, around Barcelona, Catalan is spoken, and in the NW, Galician, a language similar to Portuguese, is used. In the western Pyrenees, the people speak Basque.

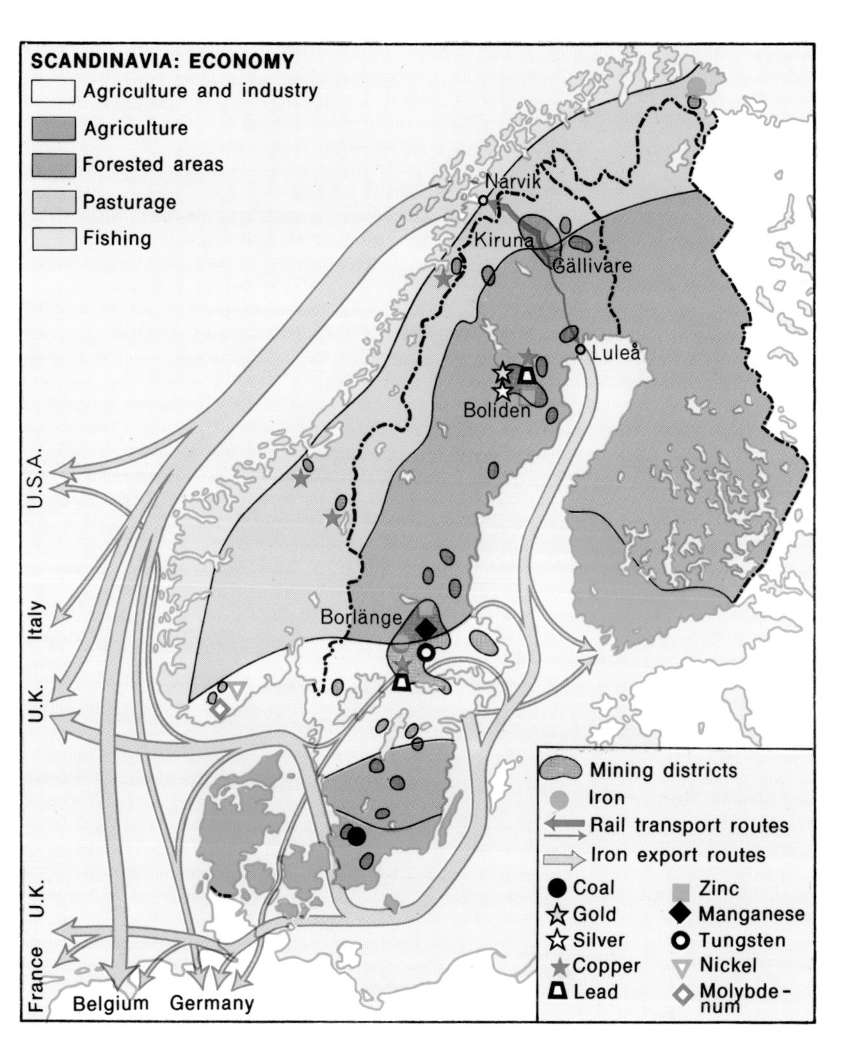

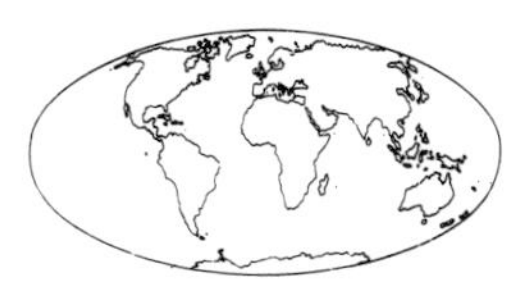

Sweden

The people of Sweden, the largest country in the Scandinavian Peninsula, enjoy one of the highest standards of living in the world. The country is a representative monarchy. On the N, it has boundaries with Norway and Finland. On the E and S it has a long and greatly broken coastline on the Gulf of Bothnia and the Baltic Sea. On the SW, the Kattegat and Skagerrak separate it from Denmark. And on the W it is flanked by Norway. The northern part of the country, called *Norrland*, lies partly within the Arctic Circle; the frozen northern lands belong to *the land of the midnight sun*. The mountains that rise along almost the entire western side of Sweden slope gradually down to the coastal plains of the E and S. The N of the country has many long, narrow lakes in the mountain valleys. The S has even more lakes, some of which, such as Lake Vanern (2,140 sq miles; 5,540 sq km), cover large areas. Some offshore islands in the Baltic are part of Sweden. Despite its northerly location, much of the country has a temperate climate. However, Sweden is colder than the coast of Norway in winter and trade is hampered by the freezing of the Baltic Sea. Most Swedish iron ore is exported via Narvik in Norway. Sweden makes good economic use of timber from the forests that cover some two-thirds of its area. Other valuable natural resources are iron, zinc, lead and copper. The greater part of the country is unsuitable for agriculture, but there are some rich farmlands in the S, where dairy farming is important. The fishing industry, also important, is concentrated chiefly in the SW. Partly because of abundant hydro-electric power, manufacturing industries are highly developed. Swedish products that have a high reputation for quality include aircraft, motor vehicles, heavy machinery, ships, precision instruments, paper, glass and foodstuffs. The people are related to the peoples of Denmark and Norway. The Swedish language is similarly related. The chief cities are Stockholm (the cap.), Göteborg and Malmö. Area 173,400 sq miles (449,110 sq km).

The monument of Gustavus II, Adolphus (1594–1632) in Hälsingborg, Sweden. This soldier-king led Sweden's armies through Europe and was killed at the battle of Lützen, although his troops were victorious.

Switzerland

Switzerland has been at peace for over a hundred years. It is a federation of 22 little states called *cantons*. Switzerland has no coastline. It is bordered by West Germany on the N, Austria and Italy on the E, Italy on the S, and France on the W. Three-quarters of the country consists of mountain ranges. The Jura Mountains extend in an arc in the NW. The Alps stretch across the whole of the southern part. Between is the plateau of the Mittelland. Two of Europe's largest rivers rise in the Swiss Alps. The Rhine originates as two headstreams that join and flow into Lake Constance. The Rhône has its source in a glacier and flows westwards through Lake Geneva to France. There are many waterfalls, some of them having falls of nearly 1,000 ft (300 m). The climate varies. The higher regions have much snow, and avalanches are a common hazard. Places on S-facing slopes have in general more equable climates than those facing N. In spring, the Mittelland has a prevailing warm SW wind called the *föhn*. Dairy farming is the chief form of agriculture; Swiss cheeses and other dairy products are known all over Europe. In the Alps, the cattle are driven up into the mountain meadows to graze in summer. Some cereals are grown, particularly in the Mittelland. On the warmer hill slopes there are vineyards and fruit trees. Swiss industry is highly developed, specializing in products that require skilled workmanship: watches, optical instruments, musical instruments and engines. The country is the centre of international banking. Another major source of revenue is tourism. The people speak German, French or Italian, according to the locality in which they live. Some also speak a language called *Romansh*. The chief cities are Zurich, Basel, Geneva, Bern (the capital), Lausanne and Winterthur. Area 15,940 sq miles (41,284 sq km).

The Marktgasse in Bern, the federal capital of Switzerland. The clock-tower was part of the medieval city walls. Its 16th-century clock has puppet figures that parade each hour.

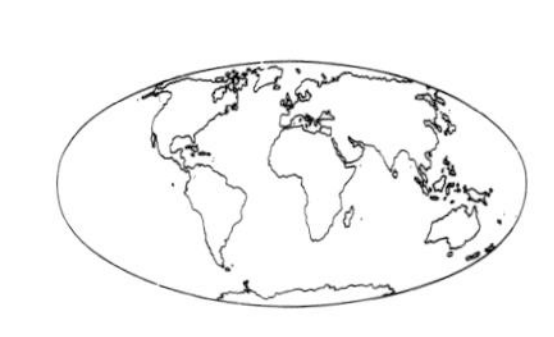

Turkey

Turkey, which lies between the Mediterranean Sea and the Black Sea, is mainly an Asian country. But in the NW its territory extends into the Balkan Peninsula, which is part of Europe. European Turkey, called *Thrace*, contains two of the country's chief cities, Istanbul and Edirne. The two parts of Turkey are separated by the Dardanelles, the Sea of Marmara and the Bosporus. For the main article on Turkey, see page 145.

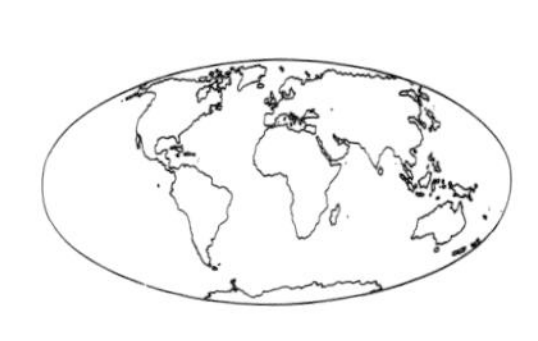

U.S.S.R.

The Union of Soviet Socialist Republics is the largest country in the world. Three-quarters of its territory is in Asia. But Moscow—the country's capital—and the majority of its other important cities are in Europe.

The U.S.S.R. is made up of 15 *union republics*. Only the United States rivals it in industrial and military power. The U.S.S.R. is the leading communist country, though its leadership is challenged by China.

The U.S.S.R. is often called the *Soviet Union* because its constitution provides for government by a system of councils known as *soviets*.

The distance across the U.S.S.R. from E to

Left: The church of St. Basil in Moscow, on the southern side of Red Square. It was built in the mid-1500s, during the reign of Ivan the Terrible. Each of its eight multi-coloured 'onion' domes has a different shape.

Below: The former Winter Palace of the czars in Leningrad. The huge Baroque building now houses part of the Hermitage Museum.

Left: Red Square, in Moscow, site of the great parades that take place on May 1 (May Day) and November 7 (the anniversary of the October Revolution).

W is 5,250 miles (8,450 km). The country's area is 8,649,000 sq miles (22,400,800 sq km). The area in Europe is 2,151,000 sq miles (5,571,100 sq km) and in Asia 6,498,000 sq miles (16,829,700 sq km).

Land and Climate

In the N the U.S.S.R. has a long coastline on the Arctic Ocean, and in the E a long coastline on the Pacific. At its most north-easterly point it is only 56 miles (90 km) from Alaska in North America, across the Bering Strait.

The East European Plain occupies most of the European part of the country. To the E it extends as far as the Ural Mountains. In the S are the Caucasus Mountains and in the SW the Carpathians. The SW part of the plain, bordering on the Carpathians and the Black Sea, is called the *Ukraine*.

The Turanian Plain lies around the Aral Sea in the S. Some of it consists of plateaux and grasslands, but it also includes two

Spring sown wheat, sugar beet and cattle farming
Winter sown wheat, maize and other cash crops
Flax, fodder and stock rearing
Cereals and stock rearing
Mainly forest, with some farming in river valleys
Fruit and vineyards
Arid and semi-arid grazing lands
Winter sown cereals, sugar beet and other cash crops
Intensive fruit and vegetable farming with cattle
Tea and citrus fruits
Upland pastures (for sheep) with mixed farming in valleys
Cattle rearing with some grain farming
Sheep in Central Asia with cotton, fruit and cereals
Reindeer herding in the Tundra region
Cattle and horse breeding
Industrial regions

Right: A new town in Siberia, the Asian part of the Soviet Union. Siberia has rich natural resources, now being exploited.

barren, sandy deserts, the Kara Kum ('Black Sands') and the Kyzyl Kum ('Red Sands'). There are also steppes. In the so-called Caspian Depression, on the northerly shores of the Caspian Sea, is the lowest point in the U.S.S.R, some 400 ft (120 m) below sea-level.

The Central Asian Mountains are S and E of the Turanian Plain, along the borders of Afghanistan and China. The country's highest peak, Mt. Communism (24,590 ft; 7,495 m), is in the Pamirs.

The West Siberian Plain lies between the Ural Mountains in the W and the Yenisey River in the E. This region is the greatest stretch of continuous plain in the world, and has an area of about one million sq miles (two and a half million sq km).

The Central Siberian Plateau extends eastwards from the Yenisey to the Lena River.

The East Siberian Highlands consist of several mountain ranges, deep valleys and plateaux between the Lena River and the Pacific Ocean. The Kamchatka Peninsula, its

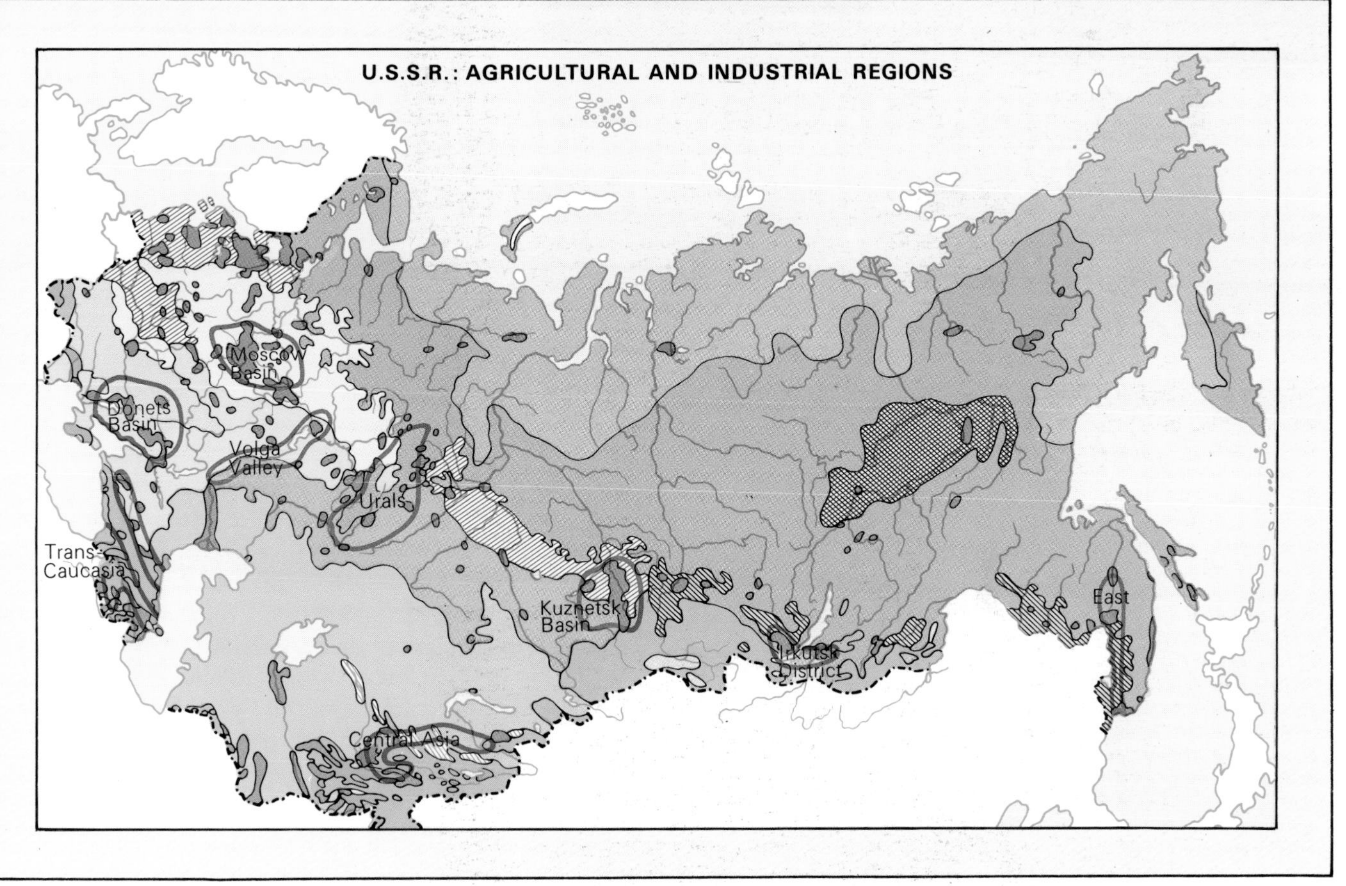

most easterly part, has more than 20 active volcanoes.

Rivers. The great rivers of the European part of the U.S.S.R. include the Volga, the Don, the Dnieper and the Ural. All these rivers drain the steppes, the temperate grasslands extending across the U.S.S.R.

The Siberian regions have several great rivers. The largest are the Ob, the Yenisey and the Lena. The basin of the Ob is practically the whole of W Siberia. Its banks are marshy, and flood regularly.

Inland Seas and Lakes. The Caspian Sea is the world's largest inland body of water, and has an area of about 170,000 sq miles (440,000 sq km). Its waters are salt. The nearby Aral Sea, also salt, has an area of about 26,000 sq miles (67,000 sq km). Lake Baykal, on the edge of the Central Siberian Plateau, has an area of about 12,000 sq miles (31,000 sq km). The largest freshwater lake in Asia, it is also the deepest body of fresh water in the world, with depths of some 5,700 ft (1,700 m).

The Climate of the U.S.S.R. varies greatly. In general, the country has a continental climate with hot summers and cold winters.

Golden fields of crops in Georgia, U.S.S.R. In the background is Mount Elbrus (18,481 ft; 5,633 m), the highest peak in the Caucasus Mountains and in Europe.

Agriculture and Industry

Only about a quarter of the land of the U.S.S.R. is farmed, and of this less than half is cultivated. Nevertheless, the total area of farmland is greater than in any other country. Nearly all the land is organized either in collectives or in state farms. The chief crops are cereals. The U.S.S.R. is the world's greatest producer of wheat and rye. Most of the wheat is grown on the steppes. Barley and maize are also important. Other major crops are potatoes, sugar-beet, fodder, flax and tobacco. Away from the great grain lands, mixed farming is common, and even in central and eastern Siberia there are productive pockets of agriculture. The raising of livestock plays a large part in the country's economy.

Fishing is a major industry. Fishing fleets operate in the domestic seas—the Baltic, Black and Caspian—and along the coasts. Russian deep-sea trawlers are a familiar sight in many of the world's fishing grounds.

Timber from the thousands of square miles of forests is one of the most valuable natural resources. But the U.S.S.R. also has extremely rich mineral deposits. The chief of these are coal, iron, manganese and petroleum. But there are also many other minerals of economic significance, including copper, aluminium, asbestos and tungsten. Agreement has been reached for the supply of natural gas by pipeline to Western Europe.

In manufacturing industry, the government has until recently concentrated on the production of heavy industrial goods—steel, rolling stock, motor vehicles, industrial plant, chemicals and so on—as well as armaments. But an increasing proportion of industry is now devoted to consumer goods to raise the living standard of the people. Nearly three-quarters of industry is concentrated in the European part of the country.

Cities and Communications

Moscow is the most heavily populated city as well as the capital. It and Leningrad are among the largest cities in Europe. Other cities with populations of more than one million are Gorki, Novosibirsk, Kuybyshev, Sverdlovsk, Kiev, Kharkov, Minsk and Baku. In the European regions, communications are reasonably well developed. Elsewhere they are poor. Railways are the most important means of transport. One famous line, the Trans-Siberian Railway, which links Moscow with the Pacific coast, is more than 5,000 miles (8,000 km) long. Several new railways are helping to open up Siberia.

The People

More than 70 per cent. of the people of the U.S.S.R. live W of the Urals. The Russians are by far the largest national group in the population. The Ukrainians are the next largest. Other large national groups include Belorussians, the Uzbeks, Kazakhs, Tartars, Azerbaijani, Armenians and Georgians. Each of these numbers several million people. There are also numerous smaller groups. The official language of the U.S.S.R. is Russian, which is taught in all schools. But about 60 different languages are spoken in various parts of the country.

United Kingdom

The United Kingdom of Great Britain and Northern Ireland comprises the whole of the British Isles except for the part of Ireland that forms the Republic of Ireland. Included in the U.K. are the Isle of Man, in the Irish Sea, and the Channel Islands, off the coast of Brittany in France. The U.K. is a constitutional monarchy. Its principal parts are the kingdoms of England and Scotland, the principality of Wales and the province of Northern Ireland. It is the centre of the Commonwealth of Nations, the world-wide association of countries that has succeeded the former British Commonwealth and Empire. The U.K. still has a few overseas possessions. In 1982 a task force recaptured the Falkland Islands, which had been invaded by Argentina.

The capital of the U.K. is London, in SE England. Three other cities have capital status, too: Edinburgh for Scotland; Cardiff for Wales; and Belfast for Northern Ireland. The area of the U.K. is 94,216 sq miles (244,018 sq km).

Land and Climate

The British Isles lie in the Atlantic Ocean on the continental shelf of the European mainland. Great Britain, the largest island in Europe, is separated from the mainland by the North Sea and the English Channel.

The Atlantic Ocean has a dominating effect on the climate. The prevailing SW winds bring much rain to the higher districts of the W. But the warm currents of the North Atlantic Drift help to moderate temperatures. The lowest temperatures are

Left: The ruins of Whitby Abbey in Yorkshire, a Benedictine foundation built in 1078 on the site of an abbey dating from 657.

Below: The chalk uplands of Salisbury Plain, in Wiltshire, cover 300 sq miles (780 sq km). The plain is renowned for its beef cattle and has long been used as a military training area.

Right: Today, most of the container traffic in the Thames plies in and out of Tilbury Docks in Essex.

experienced in the mountains and on the east coast. The SE generally has the warmest summers.

England, the largest of the four countries of the U.K., has an area of 50,334 sq miles (130,364 sq km). In general, it is high in the W and lower in the E.

Its chief mountain range is the Pennine Chain, extending from the Cheviot Hills on the Scottish border as far S as Derbyshire. The Pennines are often called 'the backbone of England'. Their highest point is Cross Fell (2,930 ft; 893 m) in the N. At their southern end is the Peak District, noted for its limestone caverns.

In the NW, between the Pennines and the coast, is the Lake District. There, in the Cumbrian Mountains, is Scafell Pike (3,210 ft; 978 m), the highest peak in England.

Farther to the S are the Midlands, a region of smooth hills and valleys. To their E are the Fens, inland from the wide, shallow bay called the *Wash*.

The low plateau of East Anglia occupies the 'bulge' S of the Wash. To its SW is the valley of the River Thames. The Thames Flood Barrier, designed to prevent flooding in London at times of exceptionally high tides, became operational in 1982. Between the Thames and the English Channel lies the chalk country of the Downs. This continues W to Salisbury Plain. Beyond lies the beautiful rocky peninsula of Devon and Cornwall.

Southern England has two parallel ranges of low hills, the Cotswolds and the Chilterns. Both are in areas renowned for their beautiful countryside.

Most of England's rivers have their sources in the central uplands. Those flowing to the E coast include the Tweed, the Tyne, the Humber and the Thames. Those flowing W include the Mersey, the Severn and the

Below: Edinburgh Castle, a complex of buildings of several periods, stands on a hill 445 ft (136 m) above the centre of Scotland's capital city.

Avon. The longest rivers are the Severn and the Thames.

Scotland has an area of 30,414 sq miles (78,772 sq km). It is renowned for its wild and haunting mountains and lochs. The land falls naturally into three regions: the Highlands, the Central Lowlands and the Southern Uplands.

The Highlands consist of two mountain ranges, the North-West Highlands and the Grampian Mountains. Each runs roughly NE to SW. Between them, extending from coast to coast, is the great valley, Glen More. A series of lochs lying in the glen includes Loch Ness. The lochs are linked to form the Caledonian Canal. Just S of the glen, in the Grampians, is Ben Nevis (4,406 ft; 1,343 m), the highest peak in the British Isles.

The Central Lowlands consist chiefly of the valleys of the Tay, Forth and Clyde. The Tay (118 miles; 190 km)—which, like the Forth, has its firth on the North Sea—is the longest river in Scotland. The Clyde, however, is of far greater industrial importance. The Clyde flows to the North Channel. The Lowlands have Scotland's best agricultural land, and also its largest cities.

The Southern Uplands is a region of hilly moorland. Much of it consists of grazing land, but there are also pockets of intensive cultivation.

Wales has an area of 8,016 sq miles (20,761 sq km). About two-thirds of it is occupied by the Cambrian Mountains, a series of grassy plateaux interspersed with mountain ranges and peaks. The highest peaks are in the N and include Snowdon (Eryri) which rises to 3,560 ft (1,085 m). There are rich agricultural areas in the coastal plains and in the river valleys, such as the valley of the Wye.

Northern Ireland, in the NE of Ireland, has an area of 5,452 sq miles (14,121 sq km). It consists of six of the nine counties of Ulster. Lough Neagh (153 sq miles; 396 sq km) in the E-centre of the province is the largest lake in the British Isles. There are three mountain groups: the Sperrin Mountains in the NW, which rise to 2,240 ft (683 m) in the Sawel Mountain; the Antrim Plateau in the NE, with its beautiful valleys, the *Glens of Antrim*; and the granitic range of the Mourne Mountains in the SE, where Slieve Donard rises to 2,796 ft (852 m).

Agriculture and Industry

The U.K. has one of the world's most productive agricultural industries, largely because of the use of scientific farming methods. About half of the total land area is put to some agricultural use. The largest areas of arable land are in eastern England. There, farmers grow cereals—wheat, barley and oats—sugar-beet and vegetables. But there are rich farms in many other parts of the U.K., too. For example, some of the best farming land is in the Central Lowlands of Scotland. Fruit farming is of major importance in Kent and in Hereford and Worcester. The Fens are famous for their potato crops.

Dairy farming is common in all regions. Many famous breeds of cattle—both beef and dairy—originated in the British Isles. Pig-breeding is widespread. In some places, sheep farming is predominant.

The Guild Chapel in Stratford-upon-Avon, built by the Guild of the Holy Cross in 1269 and rebuilt in the 1400s. William Shakespeare, who was born in Stratford-upon-Avon in 1564, attended the grammar school beside the chapel.

Above: Punting on the river, a pastime known to generations of undergraduates at Cambridge University.

Right: Blenheim Palace in Oxfordshire, built by the nation in the early 1700s for the first Duke of Marlborough.

Below: Stonehenge, a group of huge stone blocks on Salisbury Plain, is believed to date from about 2000 B.C.

Britain's industrial importance was once based on its resources of coal and iron. Their declining value may be replaced in the 1980s by oil from the North Sea, where several fields are already in operation. The Sullom Voe oil terminal was opened in 1981. Two years later, production of coal started at Selby in Yorkshire—the largest drift mine complex in Europe.

British industries are extremely varied. To a large extent, they depend on imported raw materials. Ranking high among them is the manufacture of steel. Much of this steel goes to other industries in the U.K. Many of these are engineering industries with long traditions of inventiveness and skilled workmanship: shipbuilding; and the manufacture of motor vehicles, rolling stock, engines of all kinds, farm machinery, industrial and construction machinery, machine tools and electrical equipment.

Other important branches of industry are concerned with chemicals, plastics, glass, paper, cutlery and ceramics. The making of textiles, one of the oldest industries, is still important. Today, much of it is based on man-made fibres.

Cities and Communications

The largest cities are London (the capital), Birmingham, Glasgow, Leeds, Liverpool, Sheffield and Manchester. Communications are highly developed in all parts of the U.K. except in very thinly-populated areas. Although a number of unprofitable rail routes have been closed since the 1960s, much of the remaining track and rolling stock has been modernized and fast inter-city services have been developed.

The road system has been greatly improved by the building of more than 1,000 miles (1,600 km) of motorways on the major through routes. Work on the M25 orbital motorway for London continues. In 1981 the Humber Bridge was opened. There are several international airports.

The People

The U.K. is one of the world's most densely-populated countries. The greater part of the population lives in urban areas, the chief concentrations being in the London area, the Midlands, the Central Lowlands of Scotland and the Belfast area. Since the end of World War II more than a million people have emigrated to the U.K. from the West Indies, Africa, India and Pakistan. Although almost everybody speaks English, some other languages are spoken, too, usually as second languages—Welsh in Wales, Gaelic in Scotland and Northern Ireland, and Manx in the Isle of Man.

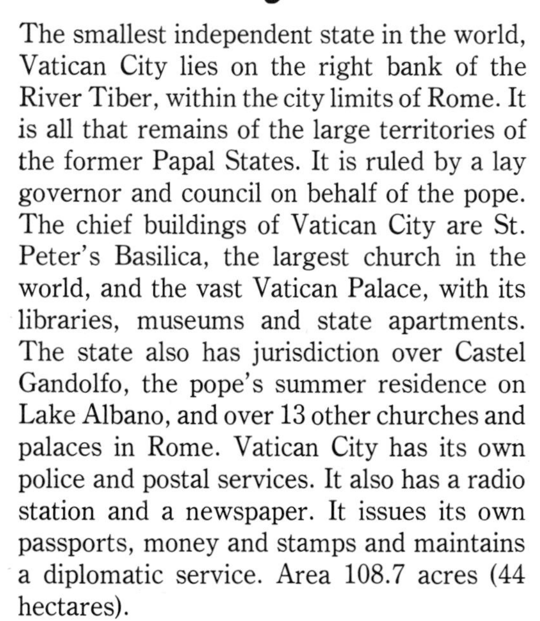

Vatican City

The smallest independent state in the world, Vatican City lies on the right bank of the River Tiber, within the city limits of Rome. It is all that remains of the large territories of the former Papal States. It is ruled by a lay governor and council on behalf of the pope. The chief buildings of Vatican City are St. Peter's Basilica, the largest church in the world, and the vast Vatican Palace, with its libraries, museums and state apartments. The state also has jurisdiction over Castel Gandolfo, the pope's summer residence on Lake Albano, and over 13 other churches and palaces in Rome. Vatican City has its own police and postal services. It also has a radio station and a newspaper. It issues its own passports, money and stamps and maintains a diplomatic service. Area 108.7 acres (44 hectares).

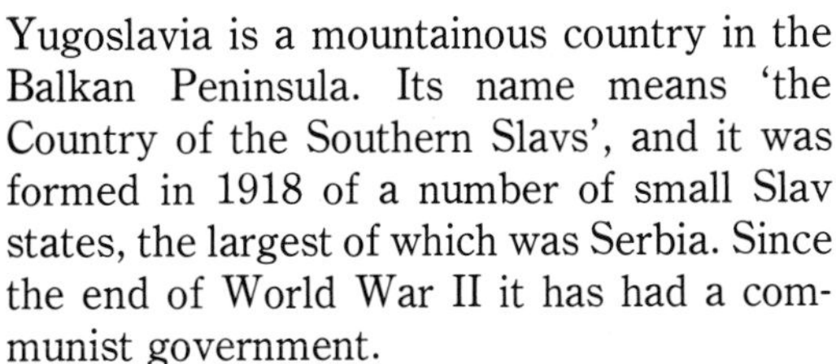

Yugoslavia

Yugoslavia is a mountainous country in the Balkan Peninsula. Its name means 'the Country of the Southern Slavs', and it was formed in 1918 of a number of small Slav states, the largest of which was Serbia. Since the end of World War II it has had a communist government.

Yugoslavia lies in the W of the Balkans. It is bordered by Austria and Hungary in the N, Romania and Bulgaria in the E and Greece and Albania in the S. In the W it has a long coastline on the Adriatic Sea. Off the coast there are many islands. The coastal plain is narrow for most of its length. Behind it rises the *Karst*, a high limestone region that includes some mountain ranges, principally the Velebit Mountains and the Dinaric Alps. Farther inland is a vast, broken plateau, rising here and there to some 8,000 ft (2,400 m) above sea-level. In the NE is the wide Pannonian lowland, a fertile plain in the basin of the River Danube. The Danube has four large tributaries in this region, the Sava, Drava, Tisza and Morava. The Vardar, a river whose outlet is on the Aegean Sea, flows in a broad valley in the SE of the country.

Mostar, in the east of Yugoslavia, is the chief city of Hercegovina. Its stone bridge across the River Neretha was built in 1566 by the Turks.

Yugoslavia has many lakes. The largest are Scutari and Ohrid on the Albanian border and Prespa on the border with Albania and Greece. The climate in the Danube basin is hot in summer and cold in winter. In the mountains the climate also tends to be extreme. But coastal regions are mild and warm.

Yugoslavia has rich mineral deposits, principally lead, antimony, aluminium, coal and iron. Copper, chromium, mercury and other metals are also found. The best farmlands are in the Danube basin. There, farmers grow wheat, maize, rye, hemp, sugar-beet and potatoes. There are many vineyards. Other crops include tobacco, olives and plums. The raising of livestock is extremely important.

A number of new manufacturing industries have been developed in recent years, but the country also has some industries that are well established. Chief among these are shipbuilding and the making of machinery. Yugoslavia has three principal languages, Serbo-Croatian, Slovene and Macedonian. They have a common origin and many points of similarity. The chief cities include Belgrade (the cap.), Zagreb, Skopje, Sarajevo and Ljubljana. Area 98,750 sq miles (255,760 sq km).

Asia

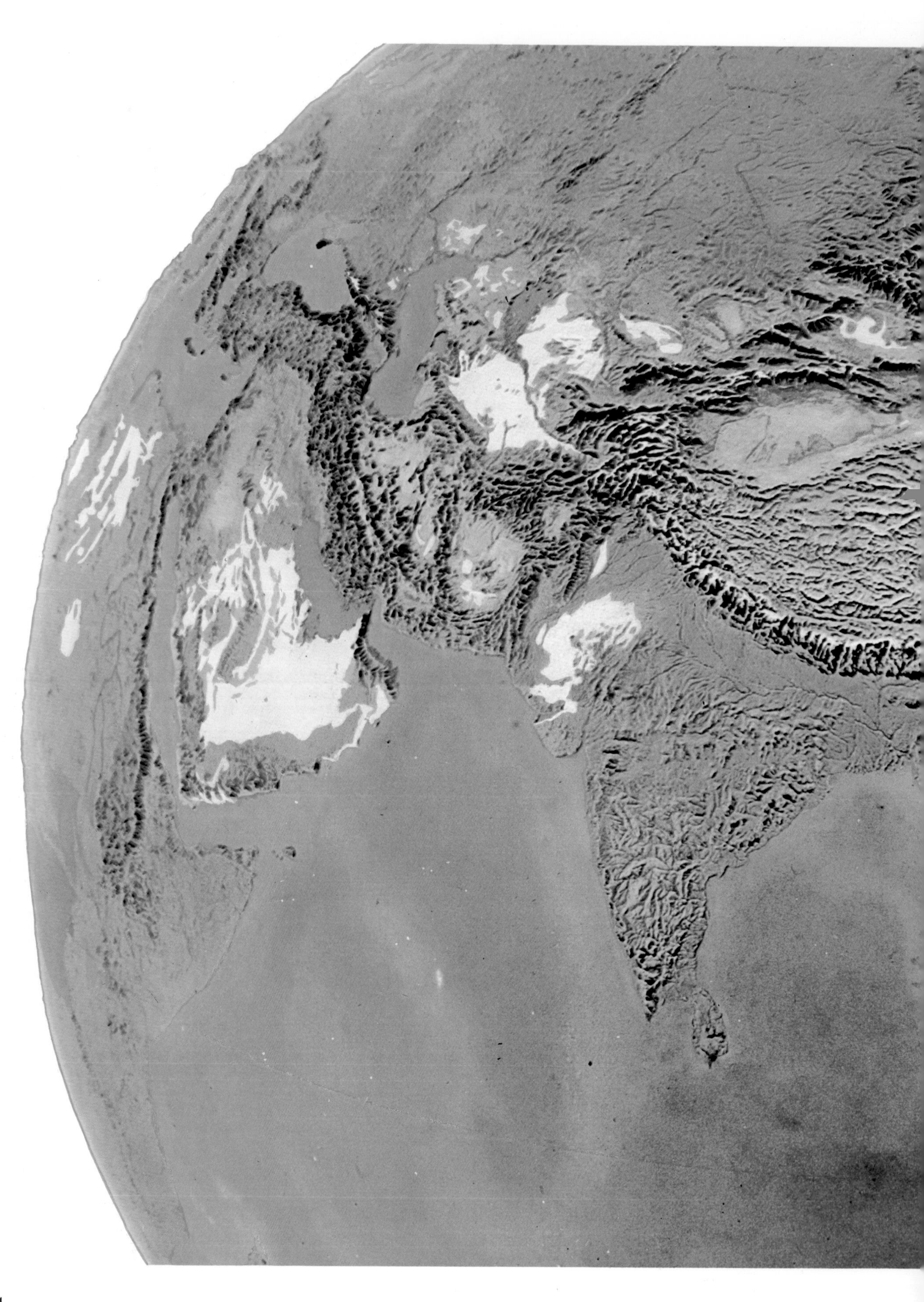

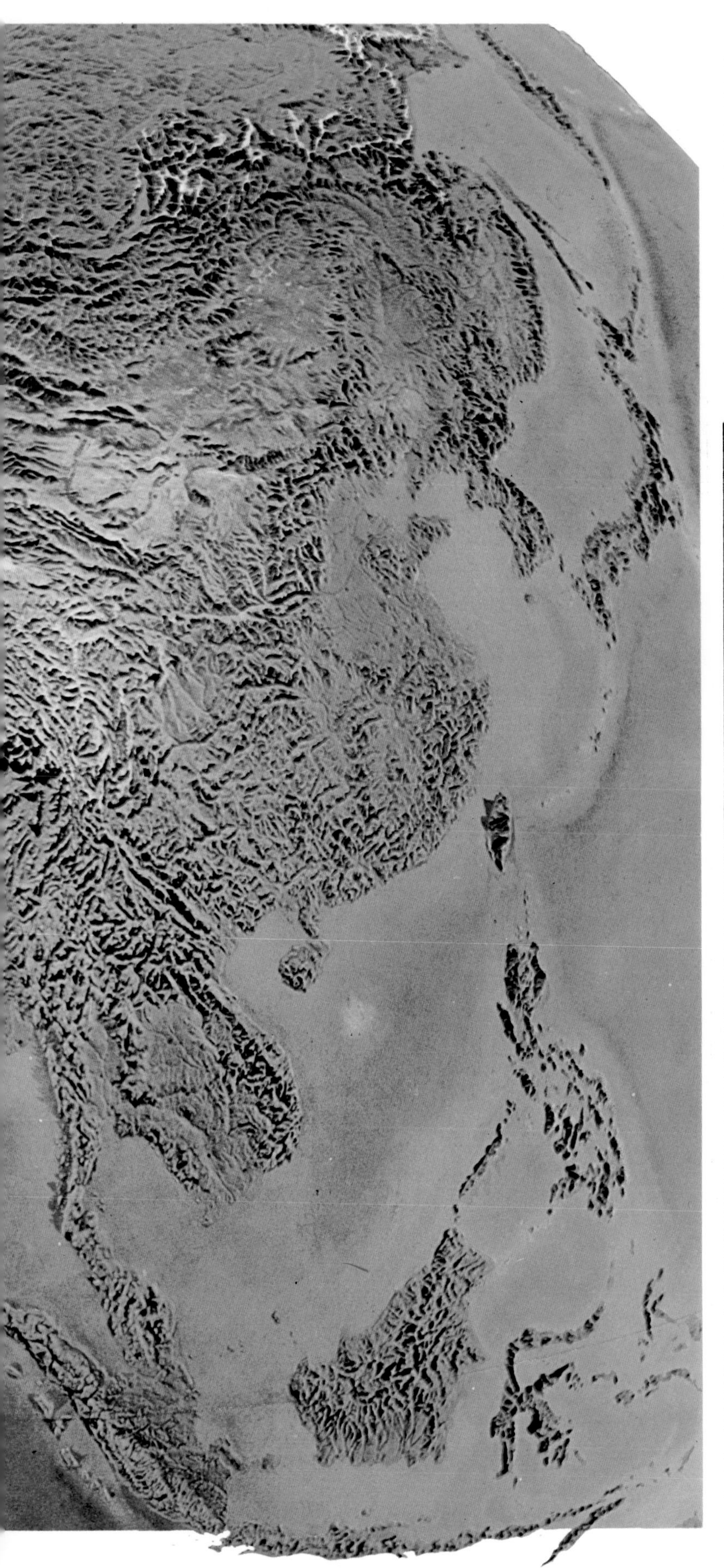

Asia, the largest of the continents, covers almost a third of the Earth's total land area. It is also the most varied of the continents. It has rich green forests; hot, arid deserts; fertile river valleys; vast dry windswept plains; and regions of permanent snow and ice. In it are the world's highest peak, Mount Everest, and the world's lowest land region, the deep valley around the Dead Sea. It also contains some of the richest countries and some of the poorest.

Some countries of Asia, such as Kuwait, are small and newly independent. China, on the other hand, has more people than any

COUNTRIES OF ASIA
POPULATIONS AND MONETARY UNITS

Country	*Population*	*Money*
Afghanistan	16,363,000	Afghani
Bahrain	359,000	Dinar
Bangladesh	89,655,000	Taka
Bhutan	1,325,000	Rupee
Brunei	240,000	Dollar
Burma	33,640,000	Kyat
China	1,008,175,000	Yuan
Cyprus	637,000	Pound
Hong Kong	5,154,000	Dollar
India	683,810,000	Rupee
Indonesia	147,490,000	Rupiah
Iran	39,320,000	Rial
Iraq	13,527,000	Dinar
Israel	3,954,000	Shekel
Japan	117,645,000	Yen
Jordan	2,779,000	Dinar
Kampuchea	6,828,000	Riel
Korea, North	18,317,000	Won
Korea, South	38,723,000	Won
Kuwait	1,464,000	Dinar
Laos	3,811,000	Kip
Lebanon	2,688,000	Pound
Macau	271,000	Pataca
Malaysia	14,415,000	Dollar
Maldives	157,000	Rupee
Mongolia	1,732,000	Tugrik
Nepal	15,020,000	Rupee
Oman	919,000	Rial
Pakistan	84,579,000	Rupee
Philippines	49,530,000	Peso
Qatar	248,000	Riyal
Saudi Arabia	9,319,000	Riyal
Singapore	2,443,000	Dollar
Sri Lanka	14,988,000	Rupee
Syria	9,314,000	Pound
Taiwan	17,479,000	Dollar
Thailand	48,125,000	Baht
Turkey	41,875,000	Lira
United Arab Emirates	1,040,000	Dirham
U.S.S.R. (Asia)	56,160,000	Rouble
Vietnam	54,968,000	Dong
Yemen (Aden)	2,030,000	Dinar
Yemen (San'a)	5,940,000	Riyal

Above: The famous statue called the *Daibutsu* at Kamakura, in central Honshu, Japan. The Japanese word *Daibutsu* means 'great Buddha'. The bronze statue, cast in 1252, is 42 feet (13 m) high. It depicts the Buddha in an attitude of serene contemplation.

other country, and a history that goes back more than 3,500 years.

Asia is bounded on the N by the Arctic Ocean. Asian lands extend for hundreds of miles beyond the Arctic Circle. To the S lies the Indian Ocean, with the Arabian Sea, the Bay of Bengal, and the South China Sea. On the E is the Pacific Ocean. The south-eastern part of Asia is made up of many islands, the southernmost of which lie close to the continent of Australia. On the W, Asia is joined to Europe: the Ural Mountains and the Caspian Sea form the boundary between the two continents. Some geographers consider that Europe and Asia are not really separate continents and that together they form one large continent, which is given the name *Eurasia.* But, historically, Europe and Asia are always treated as different continents. In the SW, Asia is bounded by the Mediterranean Sea and the Red Sea. It is linked to Africa by the narrow Isthmus of Suez, artifically cut by the Suez Canal.

The total area of Asia is 17,336,000 sq miles (44,900,000 sq km). The greatest distance N to S is about 5,400 miles (8,700 km), and W to E about 6,000 miles (9,700 km). The total coastline measures just over 80,200 miles (129,000 km).

Physical Features and Climate

People in Europe and North America sometimes describe some parts of Asia as the Near East, Middle East and Far East. These are not true geographical divisions and lead to much confusion as Near and Middle East are not clearly defined. The *Far East* comprises China, Japan and the south-eastern part of the continent.

Geographically, Asia has six well-defined land divisions marked off by mountain ranges. Each of these regions differs from the others in character and climate.

Northern Asia stretches from the Ural Mountains in the W to the Pacific. It is the vast region known as Siberia, part of the U.S.S.R. The northern part of Siberia lies inside the Arctic Circle. It is a land of treeless plains, with long cold winters and short, warm summers. In summer the temperature may rise to 50°F (10°C), but even then the ground, except for a few inches near the surface, remains frozen. This permanently-frozen subsoil is called *permafrost,* and the great Arctic plains are often known as *tundra.*

The more southerly part of the region includes wide expanses of *steppe* (grasslands), and some forests. The southern edge of the region is marked by the Tien Shan, Sayan and Yablonovyy ranges, and by Lake Baykal, the world's deepest lake, which has depths of some 5,700 ft (1,700 m). The three biggest rivers of northern Asia, the Ob, Yenisey, and Lena, flow to the Arctic Ocean, while the Amur flows eastwards to the Pacific. All four rivers are frozen for part of the year.

Central Asia is one of the least-explored parts of the world. The Tien Shan and Sayan ranges form its northern boundary, and it extends to the Great Himalaya range in the S. It includes Mongolia and the Western Chinese provinces of Sinkiang (Chinese Turkestan) and Tibet. Mongolia and Sinkiang are areas of high, barren plateaux, with some grassland but also the Takla Makan and Gobi deserts. The Tibetan plateau is extremely high. It is an area of cold deserts, stark mountains, and deep valleys. Near the Tien Shan mountains in Sinkiang is the Turfan Depression, about 500 ft (150 m) below sea-level. It is partly occupied by salt lakes.

Southern Asia consists of the countries of the Indian sub-continent—Pakistan, India and Bangladesh in the S, and Bhutan and Nepal in the N. The island country of Sri

Lanka (formerly Ceylon) also forms part of this region. The north-western areas are bordered by the Hindu Kush Mountains, which run N and E to a tangle of peaks and valleys called the *Pamir Knot*. Eastwards from the Pamir Knot stretch the Karakoram Mountains and the Himalayas, the world's highest and most rugged mountain range. Mount Everest, on the frontier between Nepal and Tibet, rises to 29,028 ft (8,848 m) above sea-level. For many years it defied attempts of mountaineers to reach its summit. But in 1953 it was at last climbed. The Himalaya Range includes many other lofty peaks, too, of which several are more than 25,000 ft (7,620 m) high and among the world's highest mountains.

The rest of Southern Asia consists of two main areas: the first is made up of a series of wide plains watered by rivers flowing from the Himalayas: the second is the Deccan, a triangular plateau that occupies most of central and southern India.

Southern Asia is strongly influenced by the *monsoons*, winds that change their direction from one season to another. From about June to September the SW monsoon blows in from the Indian Ocean bringing torrential rains to large areas. The heaviest rains are in the Western Ghats and in the NE. There, they often cause severe floods, particularly in the Ganges delta in Bangladesh. From October to March, the winter monsoon, a much lighter wind, blows from the N over the Himalayas, bringing cool, dry weather.

South-Eastern Asia consists of Burma and the Indochina Peninsula (including the Malay Peninsula), lying to the E of India and the S of China, and a large number of large and small islands that were once collectively known as the *East Indies*. The mainland part of this region is, broadly, a mixture of wooded mountains in the N with wide river plains in the S. In places, the plains are well-drained and fertile; but in some places they are marshy, and have dank swamp forests. The islands tend to have similar terrain, with mountainous interiors covered by almost impenetrable woodland. The largest of these islands are Borneo, Sumatra, Java and Celebes (Sulawesi) to the SE. N of them lies the chain of islands that forms the Philippines.

The two main rivers of mainland South-Eastern Asia are the Irrawaddy in Burma and the Mekong in Indochina.

The eastern islands, in the Pacific, lie in an area of volcanoes and frequent earthquakes. Earthquakes also occur in Anatolia and in several other parts of South-Western Asia.

Eastern Asia lies between the uplands of Central Asia and the Pacific. It is a region of highlands, deep valleys, and fertile plains. Its main rivers are the Yangtze-Kiang and the Hwang (Yellow) Ho. Along the eastern coast

Left: The taiga, the immense forested plain that stretches across northern Asia and Europe. It makes up a quarter of all the world's forest. Until recently it was practically uninhabited, but new roads—and air communications—are now opening up some of it to settlement and industry.

Below: Mount Everest (29,028 ft; 8,848 m), the world's highest peak. It rises in the central Himalayas, on the border between Nepal and Tibet. It is named after Sir George Everest, who was surveyor-general of India in the 1830s.

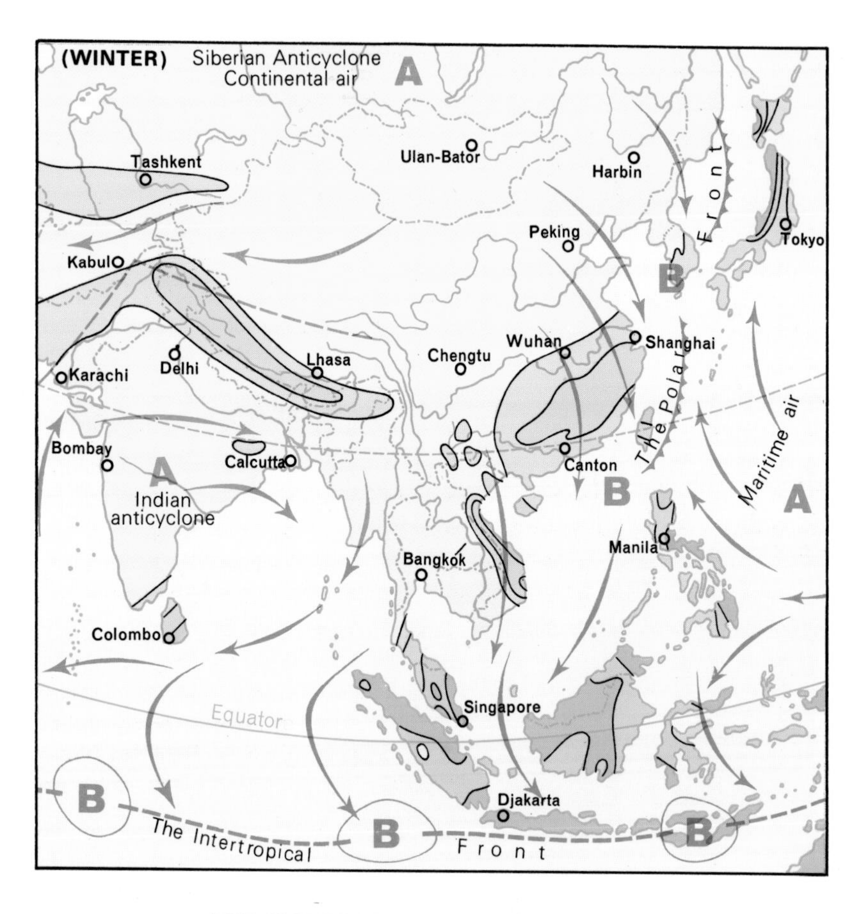

THE MONSOONS IN SOUTH EAST ASIA

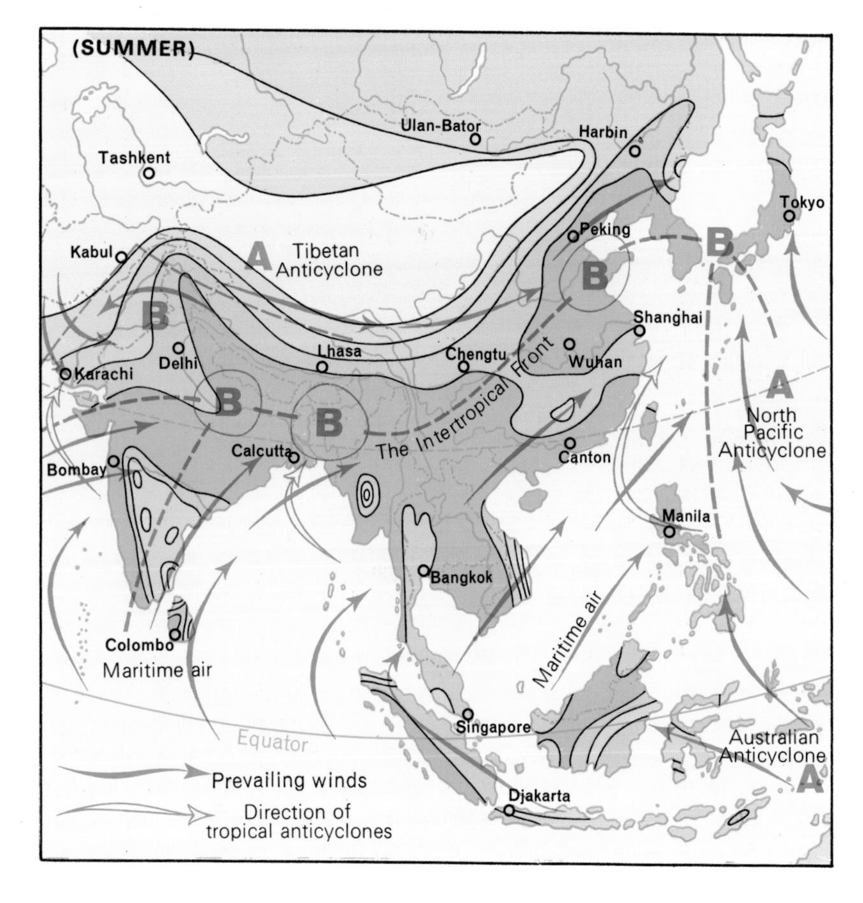

of Asia are the Kamchatka Peninsula, really part of Northern Asia and a series of islands. Four of these islands—Hokkaido, Honshu, Shikoku and Kyushu—form the country of Japan, while farther S is the smaller island of Taiwan (Formosa). The Korean Peninsula lies opposite Japan.

South-Western Asia includes most of the land commonly called the Near East or Middle East, and consists mainly of two large peninsulas: Anatolia and the Arabian Peninsula. Anatolia runs westwards between the Black Sea and the Mediterranean, and forms the larger part of the country of Turkey. To its S lies the island country of Cyprus. The Arabian Peninsula extends between the Red Sea, the Persian Gulf and the Arabian Sea.

Anatolia, which is often called *Asia Minor*, is a rugged plateau whose many rivers and lakes are dry for much of the year. Most of the Arabian Peninsula is hot and dry, and it includes great areas of sandy desert. The land bordering the eastern Mediterranean is a mixture of hot, arid deserts and fertile valleys. The River Jordan flows southwards through this area in a deep cleft in the Earth's surface, ending in the Dead Sea. The Dead Sea, which is 1,268 ft (391 m) below sea-level at its surface, is the world's saltiest body of water. The Dead Sea and the Jordan Valley, together with the Red Sea and the Great Rift Valley of Africa, are sections of a natural fault in the Earth's crust.

The Caspian Sea, to the NE of the region, lies below sea-level, but only by some 90 ft (27 m). Its waters contain more salt than is usual in sea water, but are not as salty as those of the Dead Sea. The Caspian is the world's largest inland body of water.

E of it, on the northern borders of Western Asia, lies the Aral Sea, another sea of high salinity, fed by two large rivers of central Asia, the Amu Darya and the Syr Darya.

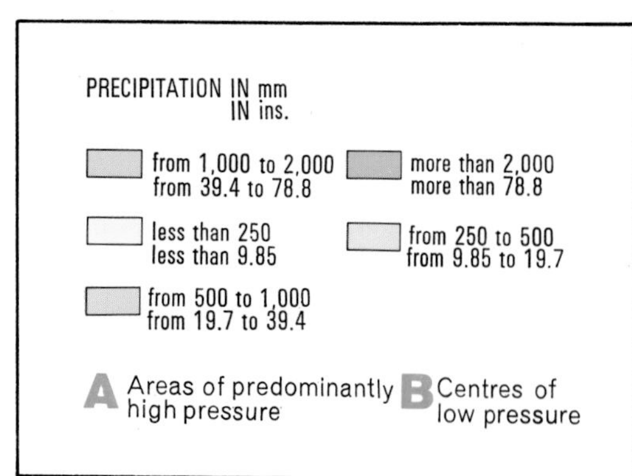

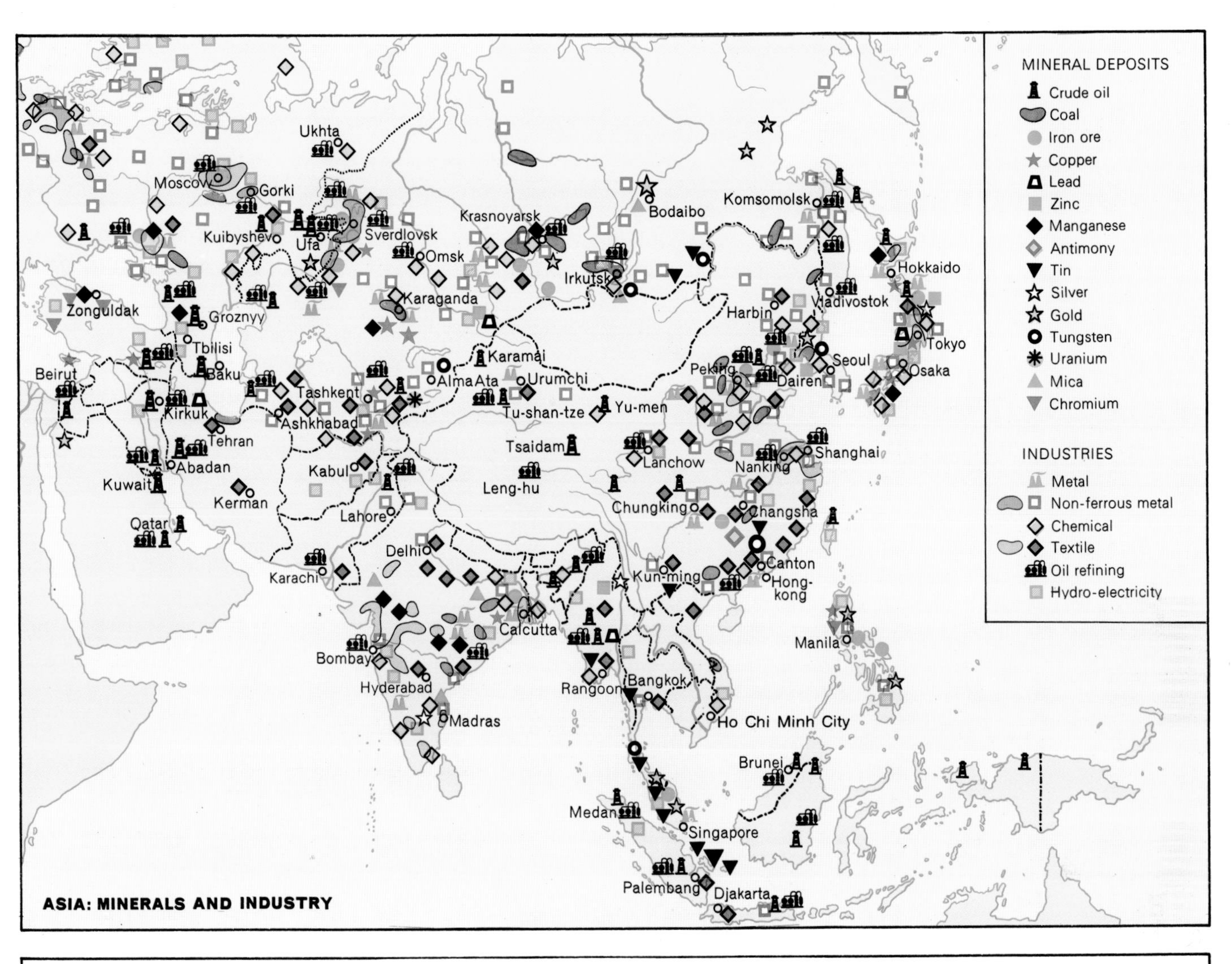

IMPORTANT DATES IN THE HISTORY OF ASIA

1500s B.C. The Aryans from central Asia invaded the Indus Valley.

1200s B.C. Moses led the Israelites out of Egypt.

c. 563 B.C. The Buddha born in India.

336–323 B.C. Alexander the Great created an empire from south-west Asia to the Indus Valley.

300s B.C. Building of the Great Wall of China began.

269–232 B.C. Asoka ruled in India—a Golden Age.

c. 6 or 4 B.C. Birth of Jesus Christ in Palestine.

300s A.D. Beginnings of the Byzantine Empire in SW Asia.

300s A.D. The Huns from Mongolia established a kingdom in northern China.

570 Muhammad born in Arabia.

600s Muslim warriors conquered much of south-west Asia.

1206–1227 Genghis Khan and his Mongols conquered China, Persia and part of India.

1264–1294 Kublai Khan ruled the Mongol Empire.

1368 Rule of the Ming dynasty began in China.

1526 Babar, the ruler of Kabul, established the Mughal Empire in India.

1500s The Ottoman Empire in south-west Asia flourished.

1500s European soldiers and traders into Asia.

1644 Rule of the Manchu dynasty began in China.

1757 The British East India Company ruled Bengal after Robert Clive's victory at Plassey.

1774 Warren Hastings became first British governor-general of India.

1842 China, under duress, opened five ports to British trade.

1854 An American naval force obliged the Japanese to open their ports to American trade.

1857–1858 The Indian Mutiny. After its suppression, the Mughal emperor was deposed.

1876 The British parliament declared Queen Victoria Empress of India.

1904–1905 The Russo-Japanese War. Russia defeated.

1911 Revolution in China. End of the Chinese Empire.

1931 The Japanese seized Manchuria.

1937 Full-scale Japanese attack on China.

1939–1945 World War II. The Japanese occupied large areas of Asia, but were forced to relinquish their gains.

1940s–1960s End of colonial regimes in many countries and regions.

1950–1953 The Korean War.

1959–1973 The Vietnamese War.

1967 Six-day Arab–Israeli War.

1971 Indo–Pakistan War.

1973 Arab–Israeli War.

1979 Israel–Egypt peace treaty.

1980 USSR invasion of Afghanistan; war between Iran and Iraq.

1982 Israeli invasion of Lebanon.

The eastern part of South-Western Asia, comprising Iran and Afghanistan, is semi-arid and sparsely vegetated with cool winters, hot summers and only light rainfall.

Resources and Industry

Because Asia is so immense, it has almost every kind of industrial and agricultural resource in one area or another. But many of Asia's mineral resources have not yet been fully exploited, and probably there are enormous mineral riches the locations of which are still unknown.

Northern Asia—that is, the Asiatic part of the U.S.S.R.—has major coal deposits both in the E and in the W, and experts believe that there are vast untapped resources in the permafrost regions of the N. Oilfields are actively exploited to the N of the Caspian Sea and also on the long, slender island of Sakhalin, off the Pacific coast to the N of Japan. Other important minerals which the Russians are extracting include asbestos, chrome, cobalt, copper, gold, iron, mercury, manganese, nickel, silver, tin, uranium, and vanadium. In addition, the major rivers have been harnessed as a source of energy, and several industrial regions are dependent on the hydro-electric power that they provide.

Central Asia is very largely unexplored, but gold, iron and other metals are mined there, together with coal and petroleum.

Eastern Asia is rich in coal, and is known to have small deposits of a great many metal ores, particularly those of antimony and tungsten. One of the most important resources in this region is iron ore, but surveys have so far indicated only small amounts of petroleum.

South-Western Asia is rich in petroleum and coal. The area around the Persian Gulf is one of the world's chief sources of oil, and there is coal in Iran and Turkey.

Southern Asia is rich in iron ore and also has valuable supplies of coal, the chief concentrations being found in India. The region also has some petroleum, and Pakistan and Bangladesh both possess natural gas deposits. But other minerals are scarce, though much of Southern Asia has yet to be surveyed.

South-Eastern Asia has good supplies of petroleum, mostly located in the islands, and there are also coalfields and beds of iron ore. Other minerals relatively plentiful in the region include bauxite (aluminium ore), gold, lead, nickel, tin and zinc.

Japan, despite having to import nearly all of its raw materials, has become one of the world's leading industrial nations. Its factories produce a wide range of goods for export, many of them products requiring high technical knowledge and skill. The U.S.S.R., too, has many highly-developed manufacturing centres. Most other Asian countries have, by comparison, few and small manufacturing industries, and have to import many of the goods they need.

Below right: Work on a communal farm in China. Despite increasing industrialization, agriculture is still of primary importance all over Asia. The introduction of machinery and improved methods of farming have helped to increase productivity in some regions, but in most of the continent work is still carried out slowly and methodically by hand.

Below left: Hong Kong is one of Asia's most important ports and a major trading centre. It is also one of the world's most overcrowded cities. More than 100,000 of its inhabitants have no homes except the small boats called sampans.

Agriculture, Fisheries and Land Use

Two Asians out of every three live by farming. Many have their own smallholdings where they grow just enough food for themselves and their families; in bad years they often go hungry.

The best agricultural lands are in southern Asia, south-eastern Asia, China and the south-western parts of Siberia. The two chief subsistence crops are rice and wheat. Rice needs warmth and plenty of rain; consequently, it does best in regions that are affected by the monsoon winds. The largest crops of rice are produced in China, India and Pakistan. Wheat is grown in Siberia, China and the drier parts of India. Although improved agricultural methods and the use of better seeds have led to larger crop yields in recent years, many parts of Asia do not manage to grow enough food to support their inhabitants if the weather is unfavourable and have to import wheat from North America.

A number of valuable crops are grown for export. India, Sri Lanka and China produce most of the world's tea. Malaysia and Indonesia are major producers of rubber. Rubber trees were introduced into these countries from Brazil in the 1870s. Other important cash crops include cotton, citrus fruits, jute and tobacco. The islands of the East Indies produce most of the world's spices. In the past they were usually known as the 'Spice Islands'.

Japanese farm workers on the island of Honshu thresh the newly-gathered grain. Behind them rises the symmetrical cone of Fujiyama, the sacred mountain. Japan has relatively little farmland, and what there is is intensively cultivated. The main crop is rice.

Most of central and south-western Asia is unsuitable for crop-growing. But on the scanty pasture-lands of these regions, nomadic (wandering) herdsmen tend goats, sheep, and cattle. They also raise camels, which are an important means of transport. In southern Asia cattle are little used for food; in some countries they are protected for religious reasons, and in other places they are protected because they represent wealth. But they are widely employed for ploughing and for drawing carts. In Tibet the yak, a hardy animal which can stand up to cold weather, is used both as a working animal and for food. Many wild animals still roam the forests of Asia, but some, such as the tiger, are becoming rare and are in danger of becoming extinct. In some countries wild animals are now protected and there are severe penalties for killing them without licence. Elephants live wild in southern Asia and Burma, but many are domesticated and are used as work animals. Reindeer roam the tundra lands of the N; they are herded, are used for work, and are a valuable source of food in regions where there are few crops.

Forests still cover large parts of Asia, including the remote areas of Siberia. The Russians extract a great deal of softwood timber from those parts of the forests that lie near rivers or railways. In south-eastern Asia there are thick teak forests that are exploited for timber, but most of the forests of the islands, including the Indonesian islands, are not fully exploited.

Fish are important as food in the coastal lands and islands of eastern Asia. Both China and Japan possess large fishing fleets which roam the waters of the northern Pacific in search of catches. Southern Asia, Bangladesh, Sri Lanka and Burma all rely heavily on

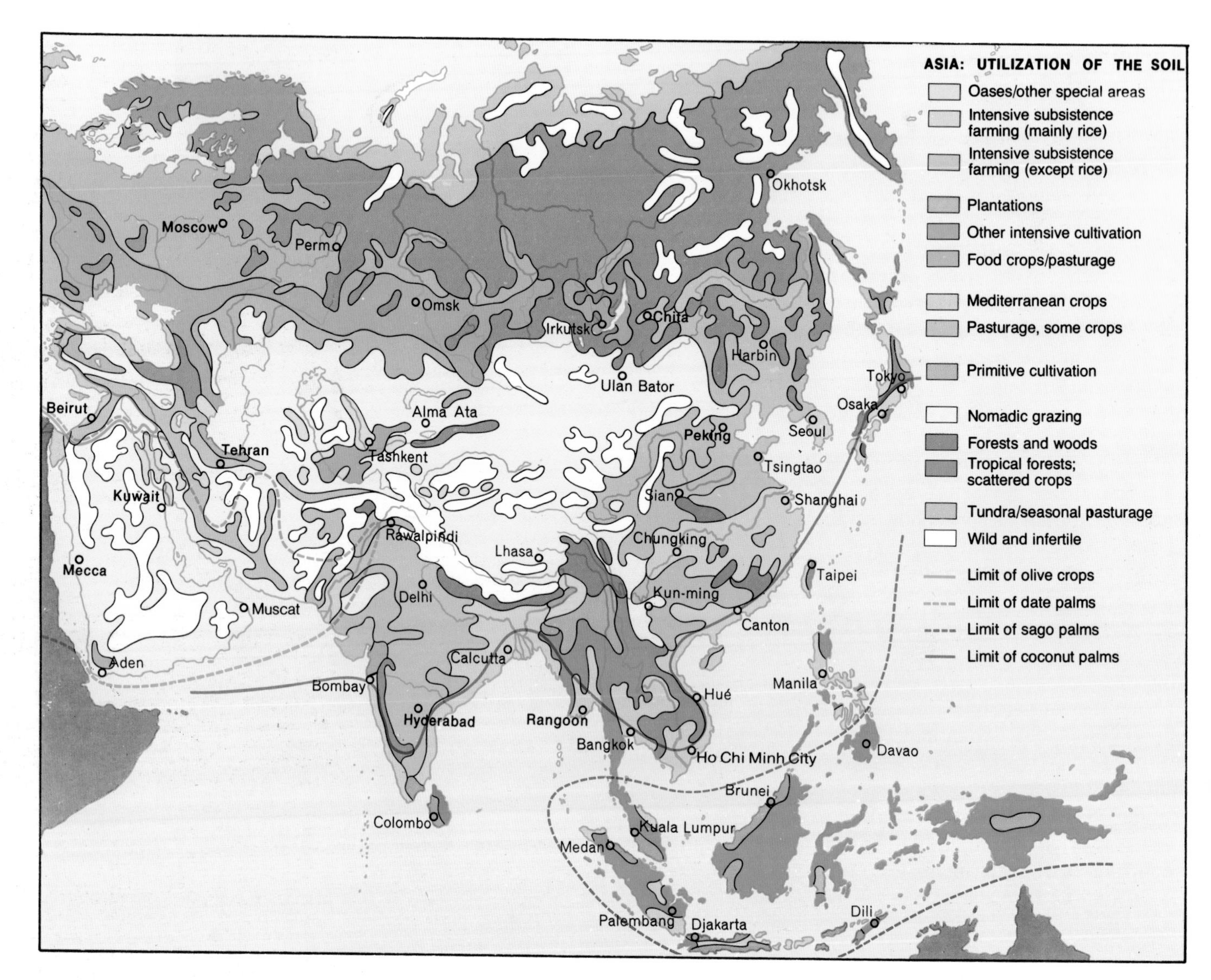

Opposite page: Demon figures guard the royal library in the Grand Palace in Bangkok. The Grand Palace is really the walled royal city. It is about one square mile (2.5 sq km) in extent.

fish as an item of food. River and lake fishing are important in many parts of south-eastern Asia, particularly in the deltas of the great rivers, such as the Mekong. China, too, has lakes and rivers in some areas that are well stocked with freshwater fish.

Japan possesses a major whaling fleet, which hunts not only in home waters but also in the Antarctic Ocean. Edible seaweeds are collected by the Japanese, and some of this valuable foodstuff is exported. In Japan and the Persian Gulf, pearl fishing is important.

Land use is very uneven in Asia. Half the world's people live in six Asian countries—China, India, Indonesia, Japan, Pakistan and Bangladesh. But huge areas of central Asia and the Arabian Peninsula are almost uninhabited, because they are desert or semi-desert. Over most of the continent there are fewer than 25 people to each sq mile of land (less than 10 to each sq km).

Communications

In general, communications are poorly developed in Asia, and vast tracts of the continent are almost inaccessible to travellers. However, Japan, India and Pakistan have first-rate rail systems, and the Trans-Siberian Railway spans northern Asia. Israel, Japan, and the countries of southern Asia have good roads and highways; other countries are improving their roads gradually. The fastest progress in developing communications is being made by the wealthy oil-producing countries of the Persian Gulf area. Rivers and canals are still major links in many parts of Asia, and there is an increasing use of air transport. Aircraft have been particularly useful in opening up northern Asia. Most places are within reach of radio broadcasting, but there is little television, and newspapers reach only a comparatively few people.

ANCIENT CIVILIZATIONS OF ASIA

Indus Valley

Excavations at Mohenjo-daro and Harappa in the north-west of the Indian peninsula (today in Pakistan) have shown that an advanced civilization existed in the Indus Valley in about 3000 B.C., and reached its peak between about 2500 and 1500 B.C. It is thought that the people of the city of Mohenjo-daro traded with the people of Sumer. They wrote in a hieroglyphic script, and had a distinctive art. Their craftsmen produced fine work in bronze. The religion of Mohenjo-daro had something in common with the later religion of Hinduism. Little is known of the reasons for the decline—apparently, a rapid decline—of Mohenjo-daro, but it may have been overrun by the Aryans.

The Aryans, a fair-skinned people from central Asia, invaded the Indus Valley in the 1500s B.C. The Sanskrit word *Aryan* means 'lord of the land'. The Aryans gradually extended their rule over most of India, and developed the Vedic civilization that underlies Hindu life and culture.

Central China

The earliest developed civilization in China was probably that of the Shang dynasty between 1500 and 1000 B.C. The people of this age were accomplished builders and had a simple, pictorial form of writing. The Shang dynasty was succeeded by the Chou dynasty, from the valley of the Yangtze Kiang. The Chou emperors built up a system of feudalism, and controlled much of China until 249 B.C. During this period art and literature were encouraged, numerous important discoveries were made in science, and there were many inventions. In the teachings of philosophers, a basis was found for attitudes of mind that the world looks upon as peculiarly Chinese. Among the great thinkers of the period were Lao Tse (c. 605 B.C.) and Confucius (551?-479 B.C.).

Mesopotamia

The Greeks gave the name Mesopotamia to the land between the Tigris and Euphrates rivers in south-western Asia. In Greek, *Mesopotamia* means 'between the rivers'.

Some of mankind's earliest civilizations developed in and around Mesopotamia. In the 3000s B.C., the Sumerians were building cities and organizing small states. They had a system of writing using a cuneiform (wedge-shaped) script. Clay tablets bearing cuneiform writing have been preserved, and tell us much about the life of the Sumerians. They were acquainted with many sciences, and were skilful farmers, architects, weavers, potters, and jewellers.

Near Sumer, another civilization developed—that of the Babylonians. They borrowed ideas and techniques from the Sumerians, but also contributed much that was original. Their language was a dialect of a Semitic language known as *Akkadian*.

Babylonia was closely linked with Assyria. The Assyrians also spoke an Akkadian dialect, and attained their greatest power and prosperity in the 800s to 600s B.C. For part of this period, they controlled Babylonia.

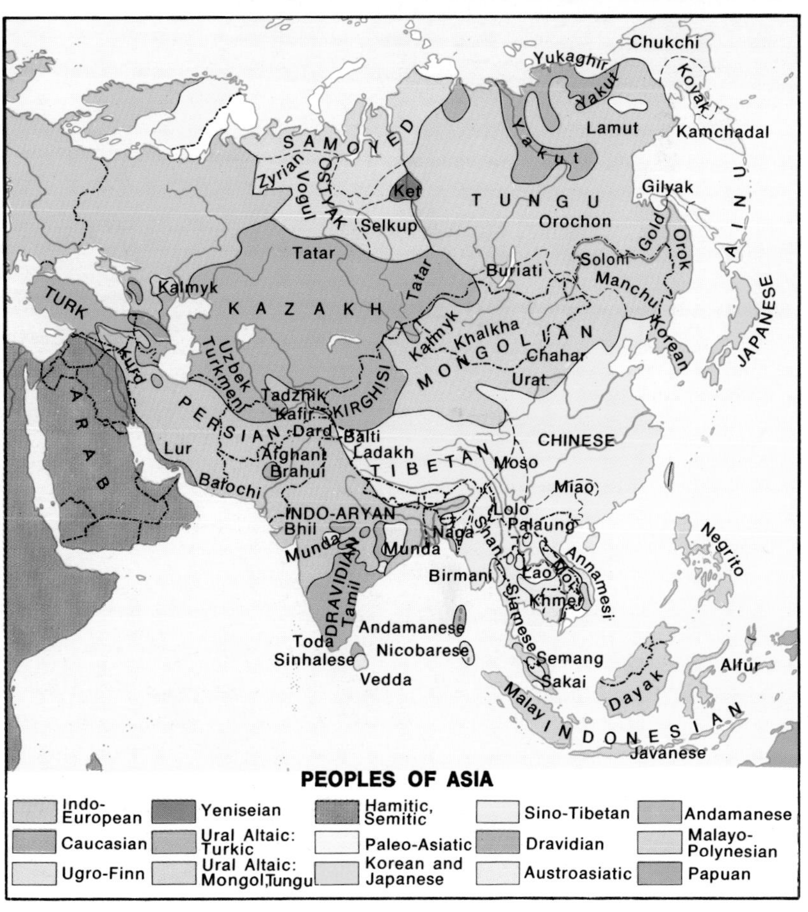

Far right: Refugees look for a new life—anywhere. With them they carry the few possessions they were able to salvage. In this century, wars and natural disasters have uprooted countless people from their homes.

Top: A huge torii in the Inland Sea at the island of Miyajima, Japan. A *torii* is a gateway built at the approach to a Shinto temple. Miyajima has a temple celebrated for its elusive beauty.

Centre: A giant statue, one of 24 depicting men and animals that line the road to the mausoleums of the Ming emperors in Peking, in China. The Ming dynasty ruled China from 1368 to 1644.

Below: A family caravan moves along an ancient road in Afghanistan. With it travel possessions that represent considerable wealth.

People and Ways of Life

Asia is not only the largest continent but also much the most populous. It has more people than the whole of the rest of the world, and its population is increasing faster than that of any other part of the Earth. Asia's peoples include members of all the main racial groups of Man and their ways of life are as varied as the places in which they live.

Mongoloid people live in eastern Asia and in most of central Asia. Some have emigrated to the countries of south-eastern Asia, too. Typically, they have straight, dark hair and yellowish skins. They are related to the American Indians, whose ancestors are believed to have crossed the Bering Strait into North America about 20,000 years ago.

Caucasoid people, related to the people of Europe, live in southern Asia, south-western Asia and Siberia. Most of the Caucasoids of Europe originally came from Asia.

Negroid people, similar to the peoples of Africa S of the Sahara, live in many of the islands of south-eastern Asia. Pygmies, called *Negritos*, live in Malaya and several of the larger islands.

In most countries there are representatives of more than one racial stock. Though people of mixed race are found all over Asia, they are most numerous in the SE.

The world's principal religions began in Asia; some of them now have more members in other continents. Hinduism, the main religion of India, began nearly 4,500 years ago. From it developed Buddhism about 500 B.C. This religion spread eastwards to China and Japan. Confucianism, China's other great religion, and Shintoism, the traditional Japanese religion, began around the same time. In the W, Judaism evolved among the ancient Israelites more than 3,000 years ago. From it have come two other great religions, Christianity, founded nearly 2,000 years ago, and Islam, which dates from the A.D. 600s. Today, Islam is the main religion of south-western Asia and parts of the S and SE; India is largely Hindu; the Chinese are Buddhists, Confucianists, or a mixture of religions; and many of the people of Siberia are nominally Christians.

Religion plays a very large part in the daily lives of most Hindus and Muslims (followers of Islam). In the U.S.S.R., and still more in China, the place of religion has been largely taken over by Communism, which serves as a philosophy of life as well as a political belief.

The way people live in the various regions of Asia depends not only on tradition and religion but also on the physical characteristics of their homelands. In the belt of dry, semi-arid land that runs through south-western Asia to central Asia, many of the people are nomadic herdsmen, tending herds of such animals as goats, sheep, cattle, yaks and camels. However, in the oil-rich regions around the Persian Gulf, the people are rapidly adopting a Westernized way of life.

The most Westernized and industrial countries in the continent are Japan, which depends for its wealth largely on industry, and Israel, which is peopled largely by Jewish settlers from Europe and North America. The former European colonial territories of southern and south-eastern Asia have ways of life that reconcile tradition with Western ideas. One of the chief effects of Western influence has been ever-accelerating urbanization.

The rapid increase in Asia's population in this century has been largely confined to the Indian peninsula, China and Indonesia. The principal cause of the increase has been a reduction in the death rate. Until comparatively recently, famine, floods and disease took a fearful toll of human life. Some Asian countries have agonizing difficulty in finding enough food for their peoples. Millions of Asians live at subsistence level—that is, they are able to grow barely enough food to feed themselves. In Bangladesh, which is one of the world's poorest countries, countless people never have enough to eat. And every few years, the fertile delta of the Ganges—which is where most of the people live because it has soil suitable for crop-growing —is visited by devastating floods in which many die.

Over large parts of Asia, people's ways of life are centred on the village, whether it is in a forest clearing, a river valley, or an upland plain. In most villages people work together to gain a living. In China this type of co-operative effort is controlled and led by the government, which also provides technical advice and help. One of the big problems for most Asian countries is education. In a world increasingly dominated by technology, real progress is not possible until a majority of the people have mastered the arts of reading, writing and simple mathematics.

THE COUNTRIES OF ASIA

Afghanistan

Mountainous country of southern Asia, bordered on the N by the U.S.S.R., on the NE by China, on the S by Pakistan and on the W by Iran. Following internal political disturbances it was invaded in 1979 by the U.S.S.R.

The Hindu Kush crosses the country from NE to SW. Some of its peaks rise to about 25,000 ft (7,600 m). In the SW is an arid plain, but the N has wide, fertile valleys. Afghanistan has a dry climate, with hot summers and cold winters. Farmers grow wheat, rice, fruit and cotton, and keep goats and sheep. There are many nomads. The country is famous for its lambskins, carpets and lapis lazuli. The chief towns are Kabul (the cap.) and Kandahar. Area 250,000 sq miles (647,000 sq km). For population, see Table on page 105.

Bahrain

Tiny but oil-rich state consisting of several islands in the W central Persian Gulf, about 15 miles (24 km) E of the Saudi Arabian coast. The largest island is Bahrain; others include Muharraq, Sitra, Umm Nasan and the Hawar group. The islands are predominantly desert with daytime temperatures exceeding 100°F (38°C).

There are large natural gas deposits as well as oilfields. The main industries are oil production and oil refining. A refinery on Sitra processes local oil as well as oil from Saudi Arabia. Bahrain also has an aluminium smelter using imported ore. A petrochemical plant and a huge dry dock opened in 1977.

Crops include dates, rice and vegetables. Pearl fishing was once important but is now declining.

The main language is Arabic. The chief towns are Manama (the cap.) and Muharraq. Area 420 sq miles (1,080 sq km). For population, see Table on page 105.

A high valley in the bleak mountains of north-east Afghanistan.

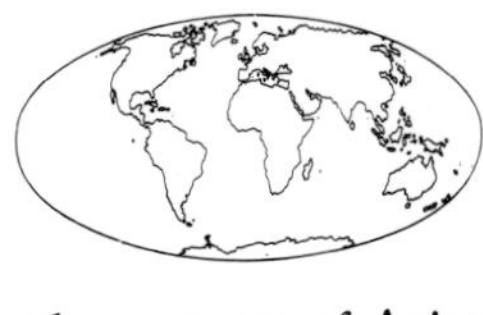

Bangladesh

The newest and among the poorest of Asian nations. Bangladesh was formerly East Pakistan. It became an independent republic in 1972 after a civil war. Its territory consists of the old province of East Bengal and part of Assam.

The country has a 440-mile (710-km) coastline in the S on the Bay of Bengal. In the SE it has a short boundary with Burma. On all other sides it is bordered by India. The land is mainly alluvial plain, with many river valleys and river mouths, tropical jungles and swamps and a tropical monsoon climate. The chief rivers are the Ganges and the Jamuna, the main stream of the Brahmaputra before its junction with the Ganges. The mouths of the Ganges, the *Sundarbans*, form the largest delta in the world.

There are reserves of natural gas and some coal deposits. The country's principal crops are jute, rice, tea and tobacco. Jute milling is the main industry.

With more than 1,500 people per sq mile, Bangladesh has one of the world's highest population densities. Most of the people are Muslim Bengalis, with a few tribal people in the SE hills. The chief cities are Dhākā (the cap.) and Chittagong. Area 55,126 sq miles (142,776 sq km).

Bhutan

Kingdom in the Himalayas, bordered on the N by China, and on other sides by India. High mountain peaks along the N border give way to densely populated, fertile valleys in central Bhutan and forested plains in the S. There are several rivers. Most Bhutanese are farmers working their own land. The main crops are rice, wheat and barley. The official language is Dzongka. The cap. and chief town is Thimphu. Area 18,000 sq miles (46,000 sq km). For population, see Table on page 105.

Brunei

The sultanate of Brunei, on the N coast of Borneo, consists of two small enclaves within the territory of Sarawak. Brunei has a coastline on the South China Sea. Much of it is forested.

Its chief sources of revenue are oil, rubber and timber. The capital is Bandar Seri Begawan. Area 2,226 sq miles (5,765 sq km). For population, see Table on page 105.

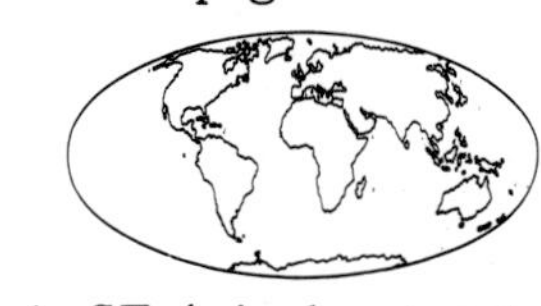

Burma

The Union of Burma, in SE Asia, has coastlines on the Bay of Bengal and the Andaman Sea. It is bordered on the NE by China, on the E by Laos and Thailand, and on the W by Bangladesh and India. The Burmese climate is monsoonal. Coastal districts are hot and wet, but hilly areas are often cool. The country has rich mineral resources, of which petroleum is the most important. There are now several refineries. Lead, tungsten, silver, zinc, wolfram and tin are found too. Manufacturing industries produce building materials, chemicals, textiles and other goods.

Most people live by farming. The chief crop is rice. Other crops are wheat, maize, millet, ground-nuts, tea, tobacco, rubber and sugar-cane. About three-quarters of the people are Burmans. Area 262,000 sq miles (679,000 sq km). For population, see Table on page 105.

Left: The port of Chittagong in Bangladesh, near the mouth of the Karnaphuli River on the Bay of Bengal. It is an important commercial and industrial town.

Below: The Shwe Dagon pagoda in Rangoon, Burma, built over a shrine believed to contain eight of Buddha's hairs. The pagoda stands on a hill and is 368 ft (112 m) high. It is built of brick, covered in gold. By tradition, it was founded some 2,500 years ago, but in its present form it dates from the 1500s.

China

More than one-fifth of all the people in the world live in China, though in area China is only the third largest of the world's countries, after the U.S.S.R. and Canada.

China has one of the world's oldest and most highly-developed civilizations, with writings, philosophies and traditions going back for more than 3,500 years. It has contributed greatly to the culture of neighbouring countries, including Japan, and its influence in thought, arts and invention has been felt in every continent. In the 1900s, China has undergone violent change, brought about by revolution and war. Today it is a leading communist nation, but its relations with the U.S.S.R. are strained over differences in belief and boundary disputes.

Most of the people of China live in small villages and work on the land. China has many of the world's largest cities. Most of them are in the eastern third of the country.

The capital of China is Peking, which is in the NE. The area of the country is estimated at about 3,700,000 sq miles (about 9,600,000 sq km).

Land and Climate

China, occupying more than one-fifth of Asia is cut off from the rest of the continent by natural barriers—high mountain ranges that fringe the country on the N, S and W. The Great Plain of eastern China is the largest lowland and the part of the country in which

Below: The Great Wall winds across China in an east-west direction from the Gulf of Khihli, on the Yellow Sea, to a point some 1,500 miles (2,400 km) inland. It was built in the 300s and 200s B.C., and has been reconstructed many times since. Its purpose was to protect China against attacks by Mongols from the north. The total length of the wall is probably about 3,000 miles (4,800 km), because it is not constructed in a straight line but follows the natural contours of the land.

most people live. It has many waterways. Tibet, in the SW, is a high plateau or tableland lying between the world's greatest mountain range, the Himalayas, in the S, and the Kunlun Mountains in the N.

The flat, low plains of Manchuria occupy NE China. On the country's N-central border in Inner Mongolia (the Chinese part of Mongolia) is the vast Gobi Desert.

Eastern China is 'China proper'—that part of the country that is enclosed by the Great Wall of China. It is divided into northern and southern parts by the Chin Ling or Central Mountains, which stretch from W to E and separate the great river plains of the Hwang Ho, or Yellow River, to the N, and the Yangtze-Kiang to the S. Both of these rivers rise in the mountains of the W and flow in a generally eastern direction to the sea. Each of them is liable to frequent flooding, caused by melting mountain snows. Over thousands of years, these floods have levelled the lower Yangtze and Hwang Ho valleys into broad plains. The Hwang Ho gets its name from the large amount of yellow mud it carries.

The Si or West River is SE China's largest river. It empties into the South China Sea near Macao and Hong Kong.

Because of China's great size, climatic conditions vary considerably. In the Tibetan Plateau and Inner Mongolia, the climate is very dry and marked by great extremes of heat and coldness.

China's two great river basins, being nearer to the ocean, have a wetter, more equable climate. In this part of China the dry season lasts from November to February. The rest of the year, particularly May, is extremely wet. In the southern coastal regions, the climate is sub-tropical, with intense summer heat and intermittent typhoons, which cause much destruction.

Agriculture and Industry

The fertile soil of the lowland regions is one of China's principal natural resources. The richest farmland is found in the Yangtze-Kiang plain. Farther N, in the Hwang Ho plain and in Manchuria, the soil is also fertile. But cold winters and irregular rainfall limit agriculture particularly in the most northerly areas. Rich, red soil is found in S-central China. This soil is particularly suitable for grain crops. NW China is covered with windblown *loess*, the fine yellow particles of which give the Hwang Ho its colour.

A cheerful family meal in a busy city street. The food is being eaten with kwaitsze—chopsticks. The word *kwaitsze* means 'quick ones'.

Farming is the basis of the Chinese economy and most of the people live on the land. Every usable patch of soil is planted with crops. Especially in the S, hillsides are terraced far up their slopes to prevent erosion of the soil. Fertilizers, both natural and artificial, are widely used to increase fertility.

The most important food crops are rice and wheat. Rice needs a hot climate and wet soil requiring a plentiful supply of water for irrigation. These conditions are found in southern China, where most of the country's rice is grown in paddy fields. The paddies must be kept flooded for several weeks after planting.

Other crops cultivated in the warm S include tea (formerly the most important export crop) and sugar, cotton, sweet potatoes, fruit and vegetables. These crops are planted mainly on the terraced hillsides.

Wheat, which needs far less water than rice, is the main crop in northern China. Other crops grown in this region include millet, maize, sorghum, soya beans, fruit and vegetables.

Water buffaloes are raised as draught animals. Nearly every farmer keeps chickens, ducks and pigs. Most of China's sheep, horses and cattle are raised on the grasslands of Manchuria and Inner Mongolia.

China has considerable mineral wealth, particularly in the S, where there are deposits of copper, tin, antimony, lead and zinc. In the N there are quite large coal deposits and iron is found in many places. Petroleum has been found in northern, central and western China.

Each of China's major port cities is a centre of light industries, including the production of cotton and silk textiles, and food processing. The main heavy industries include the manufacture of iron and steel, heavy machinery, motor vehicles and farm machinery. Other industries include the production of chemicals and fertilizers.

Cities and Communications

China's largest city and chief port, Shanghai, is situated near the mouth of the Yangtze-Kiang. Some cities on the Yangtze, which is navigable by large ships, are also major ports. They include Nanking and Wuhan (really three cities—Hankow, Wuchang and Hanyang—in one).

N of Shanghai, at the northern end of the Hwang Ho lowland, is the port city of Tientsin, located at the mouth of the Pei River. It is the main port for the Hwang Ho plain. Manchuria's chief port is Lü-Ta (Dairen). S China's leading port is Canton, on the bank of the Pearl River (Chu Kiang) and in the delta of the Si-Kiang.

Other large cities important as political or industrial and commercial centres include Harbin and Mukden in Manchuria, and Peking, the national capital, located in the northern part of the Hwang Ho plain.

Transport chiefly depends on rivers or roads, though the railway system is being

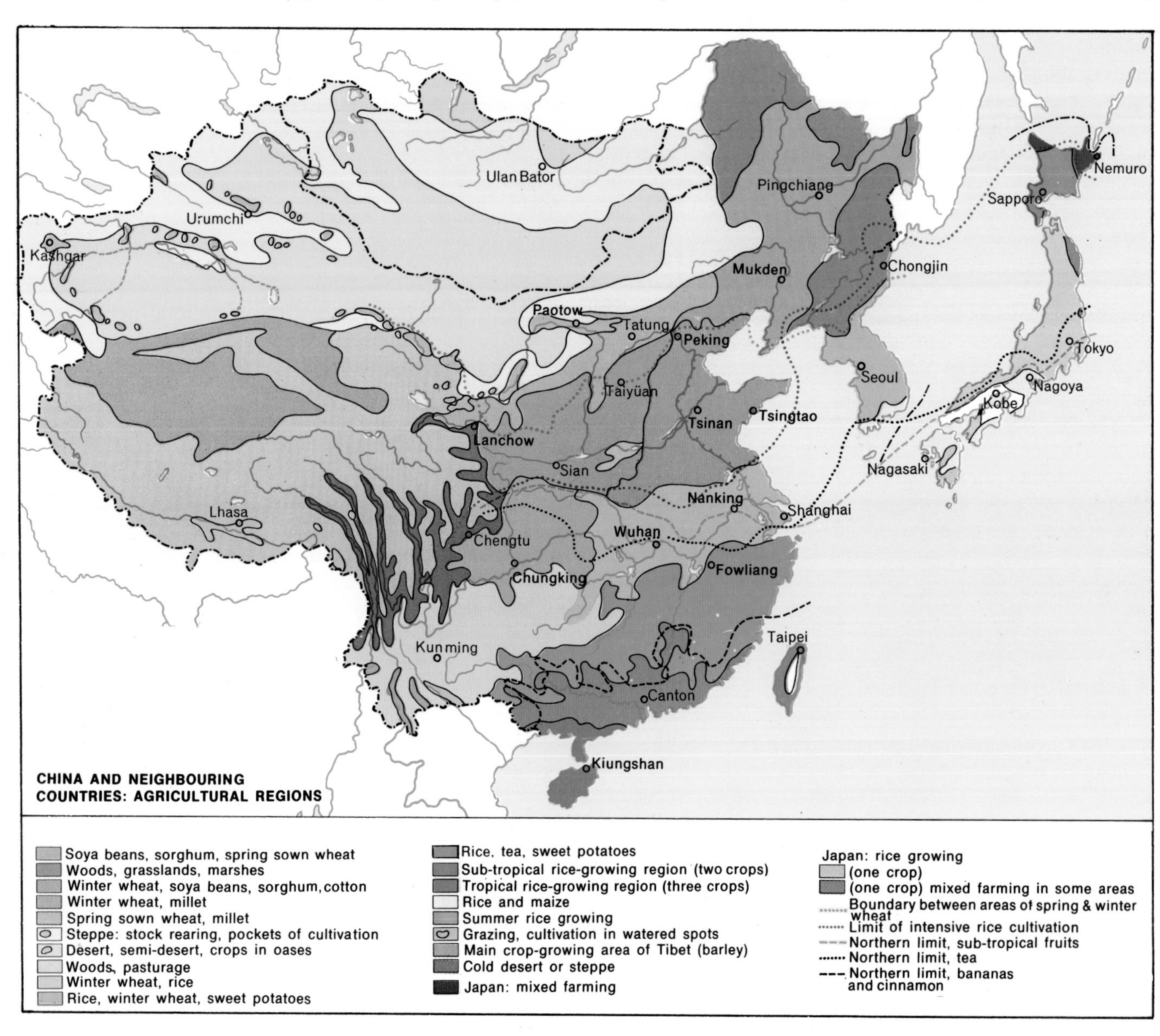

CHINA AND NEIGHBOURING COUNTRIES: AGRICULTURAL REGIONS

extended, especially in the N and NW. There, centres of heavy industry are being built up on the basis of oil and iron deposits. The port of Lü-Ta is the eastern terminus of the Trans-Siberian Railway. Coastal shipping is important along the China and Yellow Seas. Many port cities depend almost entirely on this kind of transport. In the villages, goods are carried in carts or wheelbarrows, or on pack animals or carrying-poles.

China has a total of about 300,000 miles (480,000 km) of surfaced roads and about 24,000 miles (38,000 km) of railways.

The People

With almost a quarter of the world's population, China has a massive problem of population growth. The bulk of the people are crowded into about a third of the country's area, mainly in the farms and cities of the great river plains.

The Chinese language is spoken in many different regional dialects, but is written in a uniform script.

The Forbidden City, in Peking, within the Inner City. It was once the residence of the emperor and only members of the imperial court could enter it. Today, the Forbidden City, with its palace, temples, gardens and priceless works of art, is the site of several museums and is open to the public.

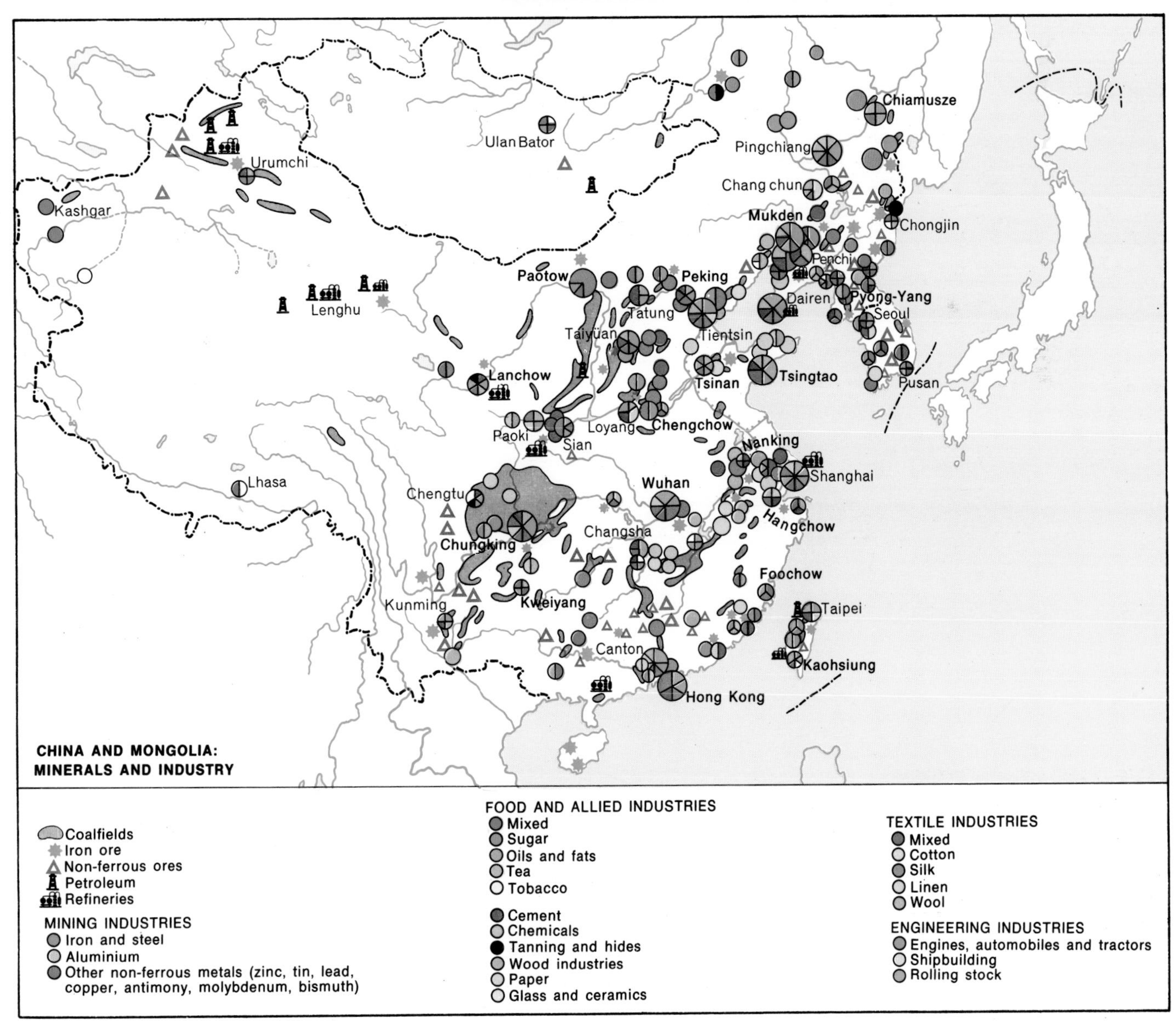

Cyprus

In 1974, troops from Turkey invaded Cyprus and forcibly partitioned the island, expelling many Greeks from the NE sector, which was declared unilaterally independent in 1983.

Minerals account for a third of Cyprus's exports; mineral deposits include copper, asbestos, iron pyrites and chromite. The chief agricultural products are wine, raisins, cattle, cotton and barley. Nearly four-fifths of the people are of Greek descent and one-fifth of Turkish. Area 3,572 sq miles (9,251 sq km).

Hong Kong

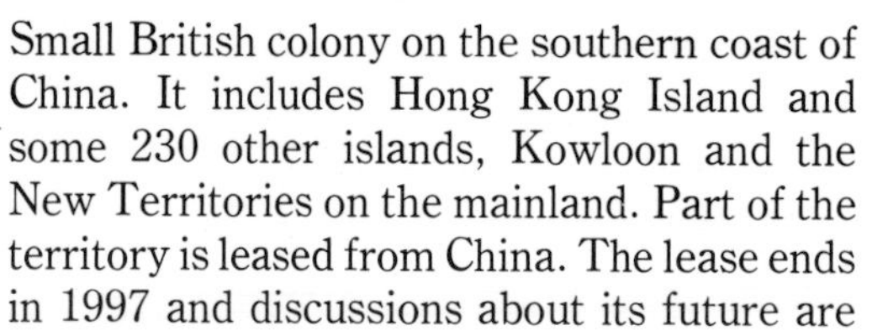

Small British colony on the southern coast of China. It includes Hong Kong Island and some 230 other islands, Kowloon and the New Territories on the mainland. Part of the territory is leased from China. The lease ends in 1997 and discussions about its future are now taking place. The climate is sub-tropical. Hong Kong is a major shipping, banking, commercial and industrial centre, and an important entrepot for China. Hong Kong Island is linked to the mainland by rail and road tunnel. Farming and fishing are intensive, but much food is imported from China, who also supply water. Most of the people are Chinese. Chinese is generally spoken, but the official language is English.

Right: The splendid ritual of a service in an Orthodox church in Cyprus.

Below: Hong Kong, one of the world's finest harbours. The name *Hong Kong* means 'Fragrant Harbour'.

The capital is Victoria, on Hong Kong Island. Area 398 sq miles (1,031 sq km).

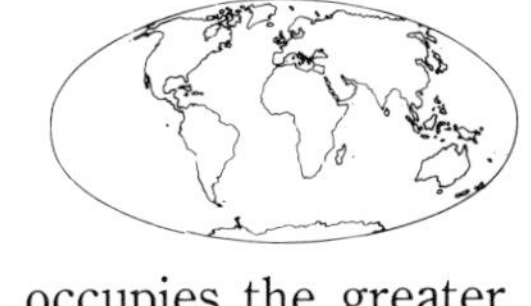

India

The republic of India occupies the greater part of the Indian sub-continent, the large triangular peninsula in Asia that projects southwards into the Indian Ocean. The rest of the peninsula is occupied by Pakistan and Bangladesh. On the N, India is bordered by China (Tibet), Nepal and Bhutan, and on the E by Burma. At its southern tip, it is separated from the island of Sri Lanka (Ceylon) by the Palk Strait.

India has one of the world's longest histories of civilization. The ruins of the cities of Mohenjo-daro and Harappa (now in Pakistan) testify to an Indian civilization that dates back to 2500 B.C. In the India of today, ancient traditions and habits exist side-by-side with modern technology.

In landscape, India is extremely varied. It is often spoken of as a land of vast plains, high mountains and great rivers. But it has regions with almost park-like scenery, and it also has deserts, swamps and jungles.

Its people are as varied as its countryside. They belong, racially, to several different stocks and speak scores of different languages. Some have great wealth, some have not enough to eat. Because of the great pressure of population—India has more people than any other country except China, and its population is increasing rapidly—one of the government's most difficult problems is providing enough food. Attempts to control population growth have so far had little success. The use of modern farming methods is encouraged in order to increase the productivity of the soil. But when India's two worst enemies, drought and floods, ruin the harvest, many people still die of starvation.

India's capital is New Delhi, in Delhi Territory in the N. The country has an area of 1,262,000 sq miles (3,268,400 sq km). Its greatest length from N to S is about 1,900 miles (3,060 km) and its greatest width E to W about 1,700 miles (2,740 km).

Land and Climate

The land is divided naturally into three main regions: the triangular Deccan plateau, which comprises the greater part of southern and central India; the fertile northern plains; and the Himalaya Mountains.

13th-century Kesava temple at Somnathpur, Mysore, in India. It is typical of the Vesara style of Hindu architecture: star-shaped and having a raised platform, pyramidal towers and lavish ornament. The style influenced architecture in other countries apart from India, notably Cambodia (the Khmer Republic).

The Deccan Plateau is the largest single region. It is bounded on the N by a number of mountain ranges, of which the most important is the Vindhya Range, which extends ENE across the country as far as the valley of the Ganges River. The highest peaks in this range reach about 5,000 ft (1,500 m). The plateau is highest in the S and W, and is tilted downwards towards the E. But on both its western and its eastern edges it rises to ranges of mountains called the Ghats, which meet near Cape Comorin at the southern tip of the peninsula. The Western Ghats rise to about 5,000 ft (1,500 m) and extend for about 800 miles (1,300 km) as far N as the Tapti River. Between the Western Ghats and the sea is a narrow coastal plain on which is situated the city of Bombay.

The principal rivers of the Deccan follow the slope of the plateau. They rise in the Western Ghats and flow eastwards to empty into the Bay of Bengal.

The Northern Plains stretch right across the sub-continent, from sea to sea. They are crossed by some of the world's greatest rivers and they comprise the largest area of alluvial lowland in the world. Their width is some 1,500 miles (2,400 km) and they have an average depth of about 200 miles (320 km).

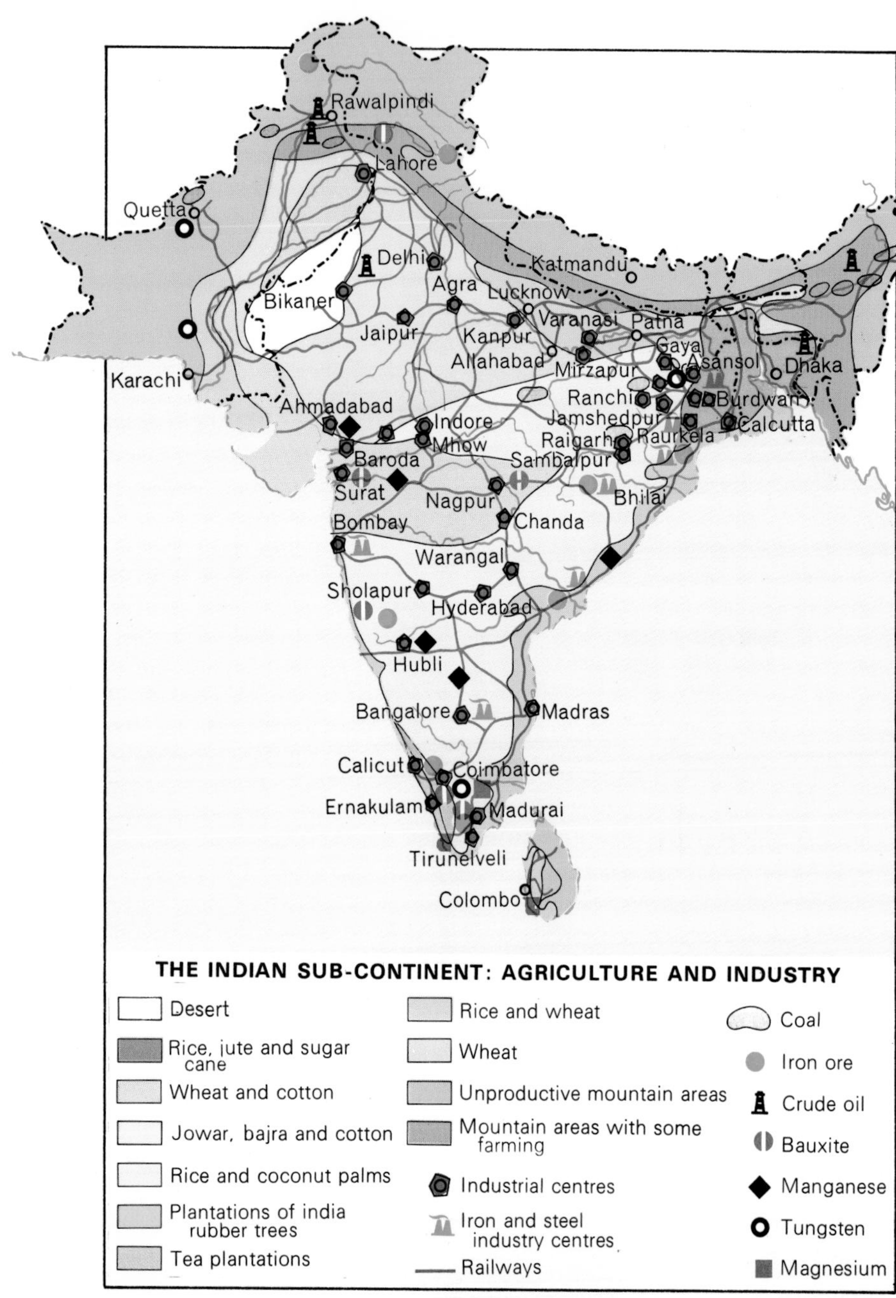

The Hindu temple of Chindambaram with its tank, the reservoir of water that is so important a feature of village life. In areas that may have to go for several months without rainfall, water is a precious commodity that must be carefully gathered and protected.

The soil of this region is extremely fertile and it produces a major part of the country's crops. It is the most densely-populated part of India.

The chief rivers of the northern plains are the Indus in the W, and the Ganges and Brahmaputra in the E. The Indus, some 1,800 miles (2,900 km) long, rises in Tibet on the northern slopes of the Himalayas and flows through NW India before entering Pakistan and swinging SW to its mouths on the Arabian Sea. It was on its banks that the civilization of Mohenjo-daro and Harappa developed. The Ganges, the holy river of the Hindus, rises on the southern slopes of the Himalayas and flows eastwards into Bangladesh, and thence to the Bay of Bengal. The Brahmaputra also rises in the Himalayas, on the northern slopes.

The Himalaya Mountains, the world's greatest mountain range, tower over some 1,500 miles (2,400 km) of India's northern border. However, for the greater part of the way only the foothills extend into India. The highest peak within Indian territory is Nanda Devi, which is 25,645 ft (7,817 m) high. Several other peaks rise to more than 20,000 ft (6,100 m).

Because of India's enormous size and the diversity of its land features, its climate varies considerably from place to place. The Himalayas have a major effect on climate, shielding the country from the influence of the great land mass of central Asia.

The hot season lasts from March to June. The highest temperatures are in the N; though places in the mountains remain cool and the so-called 'hill stations' or 'hill towns' owe their popularity to this fact. The rainy season begins in about the middle of June and goes on until the end of September. This is the period of the SW *monsoon*, when rain-bearing winds blowing in from the Indian Ocean bring torrential rainstorms to much of the country. The rains are heaviest in the Western Ghats and, particularly, in the NE, where they often cause disastrous flooding—though the farmers in many areas depend upon them for the growth of their crops. In October, the rains ease off, temperatures rise, and the air becomes extremely humid. The third climatic season, the cool season, lasts from October to February. In most areas the weather is mild, generally somewhat warmer in the S than in the N. But the mountains of the N experience extreme cold at this time, temperatures dropping lower and lower as altitude increases.

Agriculture and Industry

Most of the people of India live by farming, many of them at subsistence or near-subsistence level. Many farmers rely on age-old methods of tilling the land, with the result that the yield is low. But government plans to increase productivity have had some success. Farmers have been taught to use modern tools, have been supplied with fertilizers and high-yield seeds and have been given grants to buy machinery. Unfortunately it is often only the more prosperous who can afford to make changes. The poor continue in poverty. A wide range of crops is grown. Rice is the most important cereal and is the staple food of southern and eastern India. In the drier parts of northern India, wheat is the principal grain crop. Millet, maize and barley are also grown. Sugar-cane is grown in the NW and in the Ganges valley. Oilseed crops, including ground-nuts, sesame, rapeseed and linseed, are an important source of income in many parts. Cotton thrives on the soils of the W-central Deccan and is the basis of a major industry. It grows also in NW India, where the plantations are artificially irrigated. Tea is the chief cash crop in the NE, in the foothills of the Himalayas and in the hills of Assam. There are also extensive tea plantations in the Nilgiri Hills of southern India. Jute is the main crop of the Ganges delta region. The forests produce teak and other valuable woods.

India has millions of cattle, including buffalo, but most of them are kept as draught animals, not for their meat, because to Hindus the cow is a sacred animal. The ox-drawn cart or plough is one of the country's commonest sights. Sheep and goats are kept for food, both on the plains and in the hilly regions, and there are vast numbers of poultry. The ox is not the only animal raised as a beast of burden. Elephants, camels, mules and horses are used too.

In a country subject to such extremes and uncertainties of rainfall, irrigation and drainage are very important. Canals, tanks (ponds) and wells provide irrigation.

India has enormous mineral resources, though they are not yet fully exploited. It has some of the world's richest deposits of iron ore, much of it of high grade. Some iron ore is exported, and some, together with home-produced coal (of which there are huge deposits), forms the basis of the country's steel industry, which is centred at Jamshedpur, near Calcutta. Other important minerals include manganese, bauxite, copper, chromite, barytes, antimony and gold. The country produces about nine-tenths of the world's supply of mica.

The Taj Mahal at Agra, often described as the world's most beautiful building. It was built in 1632–43 by the Mughal Emperor Shah Jahan as a mausoleum for his wife Mumtaz-i-Mahal.

The chief industry is the manufacture of textiles, principally cotton, silk and wool materials. The leading centres are Bombay, Nagpur, Jabalpur, Amritsar and Varanasi (Banares). Other major industries are the manufacture of cement, paper, burlap products (from jute), chemicals, plastics and leather goods. The engineering industry is being progressively developed, and the manufacture of machine tools and rolling stock is important. In towns and villages many people work in cottage industries, producing goods in their own homes.

Cities and Communications

Less than 20 per cent. of India's enormous population lives in cities, the largest being the ports of Bombay on the W coast, Calcutta on the NE coast, and Madras on the E.

India has about 200,000 miles (320,000 km) of surfaced roads and some 38,000 miles (60,000 km) of railway.

The People

India has people of many different races. The two major groups are the Indo-Aryans, who form the majority and who live chiefly in the northern part of the country, and the Dravidians, who are most numerous in the S. Generally, the Indo-Aryans are taller and lighter skinned than the Dravidians.

The official language of the country is Hindi, which is spoken by about 35 per cent. of the population. English is also widely spoken. In all, there are about 60 languages and dialects, including 14 national languages recognized by the Indian Constitution.

More than four-fifths of the people are Hindus and the *caste* (class) system has a strong influence on their lives. A person remains for life in the caste into which he was born. The four main castes are the *Brahmins* (priests and scholars), *Kshatriyas* (warriors), *Vaishyas* (merchants) and *Sudras* (those whose function is to perform menial tasks). Some people, the 'untouchables' (called *harijans*, 'children of God', by Mahatma Gandhi), are traditionally outside the caste system altogether. Since 1950, any form of discrimination based on caste has been illegal in India.

India also has a large Muslim community, living mainly in the North. Smaller

Left: The Victoria Memorial in Calcutta, built to commemorate the reign of Queen Victoria, was opened in 1921. It houses a museum of art and history.

religious minorities include Christians, Sikhs, Buddhists, Jains and Parsees (Zoroastrians).

Indonesia

Indonesia, the largest country of South-East Asia, consists of about 3,000 islands lying between the Pacific and Indian Oceans. The more northerly islands are crossed by the equator. Malaysia and the Philippines are N of Indonesia and Australia is to the SE. The capital is Djakarta, on the island of Java. Area 735,000 sq miles (1,904,000 sq km).

Land and Climate

The three largest wholly Indonesian islands are Sumatra (Sumatera), Celebes (Sulawesi) and Java (Djawa). In addition, two large islands are shared between Indonesia and other countries: Borneo with Malaysia and New Guinea with Papua New Guinea. The Indonesian part of Borneo is called *Kalimantan*, and the Indonesian part of New Guinea, *West Irian*. Between Celebes and West Irian are the Moluccas (Maluku), once famous as the *Spice Islands*.

Java, the most densely populated island, has a fertile alluvial plain in the N-centre, heavily cultivated and supporting most of the island's people. A belt of volcanic mountains runs the length of Java.

Sumatra is similar, with a volcanic mountain range in the W, and a swampy alluvial plain inland from the coast. Kalimantan is mostly covered with primeval forest and swamps, with mountains in the centre.

Celebes has a deeply indented coast and a volcanic, mountainous and thickly forested interior. West Irian is mostly forested, with vast mountain ranges. The large islands have many rivers.

The climate is generally hot and wet. The southernmost islands, such as Java, are less hot than equatorial Indonesia, and have alternate wet and dry seasons. All the islands are subjected to the monsoons.

Agriculture and Industry

About 70 per cent. of the people live by farming. Much of the farming is at a subsistence level, but there are also many plantations. Crops include rice, maize, tapioca, sago, bananas, beans, cassava and spices. The chief plantation crops are rubber,

Above: The 15th-century Varadarajaswami Vaishnavite temple at Kanchipuram, one of the seven Hindu sacred cities. It is built in the Dravidian style.
Below: The great temple of Tiruvannamalai, some 50 miles (80 km) from Vellore, dedicated to the god Siva. In modern times, a famed holy man, Sir Ramana Maharshi, lived nearby.

Right: A temple dancer of Bali, one of the Lesser Sunda Islands. Balinese culture is highly developed, and the music, dance and drama of the island are famous. Most islanders are Hindus.

coffee, tea, sugar and sisal. Forest products include palm oil, hardwoods and bamboo.

Indonesia has large deposits of oil, tin, bauxite and other minerals. Oil and tin, together with rubber, are the main exports. The main industries are oil refining and food processing. Java is well-known for *batik*—fabric elaborately printed by means of a wax process.

Cities and Communications

More than 60 per cent. of the people live on Java. The chief cities are Djakarta, Surabaja, Semarang, Bandung and Surakarta.

Indonesia has about 50,000 miles (80,500 km) of surfaced roads and 3,500 miles (5,600 km) of railways.

The People

The two main population groups are Malays and Papuans. Chinese form the largest minority.

Below: Rice growing in terraced *paddies* or *cuts*, small fields surrounded by dikes. Because the young plants need a great amount of water, the paddies are kept flooded. Towards harvest time, the water is allowed to drain away.

Iran

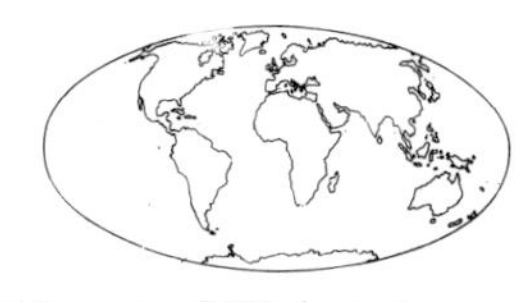

The former kingdom of Iran in SW Asia has a history that goes back to the Persian Empire of 2,500 years ago. As a result of strong opposition the Shah left Iran in 1979 and the country was declared an Islamic republic.

The country has a coastline in the S on the Persian Gulf and the Arabian Sea, and in the N on the Caspian Sea. Its land borders are with the U.S.S.R. in the N, Afghanistan and Pakistan in the E, and Iraq and Turkey in the W. Most of the country consists of a plateau averaging some 4,000 ft (1,200 m) above sea-level. In the N the plateau is rimmed by the Elburz Mountains, the highest peak of which is Mount Damavand (18,933 ft; 5,771 m). Another mountain system rims the plateau on the SW. These mountains, the Zagros, rise in several parallel ranges, their highest point, Zardeh Kuh, reaching 14,921 ft (4,548 m). To their E is another series of ranges, the Central Mountains. The central plateau consists of two great deserts: the Dasht-i-Kavir (Great Salt Desert) in the N; and the Dasht-i-Lut (Great Sand Desert) in the centre and E. In the cool season small rivers flow into the Dasht-i-Kavir from the mountains, forming shallow salt-water lakes and marshes. A low, fertile strip lies between the Elburz Mountains and the Caspian Sea. This is one of the country's few agricultural regions. Another is the Khuzestan plain at the head of the Persian Gulf, which is really part of the Mesopotamian lowlands—mainly in Iraq. Khuzestan is the fount of much of the country's wealth because of its oil-fields. In the interior, the climate is one of very hot summers and very cold winters. Towards the S, the climate becomes more temperate, but the autumn period is still extremely hot.

Iran is one of the world's leading petroleum producers. Other mineral resources of importance include iron, copper, coal, chrome, lead and zinc.

Despite the small proportion of cultivable land, more than half the people live by farming. Crops include wheat, barley, rice, fruits and cotton. The most widely-used language is Farsi. The chief cities are Tehran (the cap.), Tabriz, Esfahan, Mashhad and Shiraz. Area 628,000 sq miles (1,626,500 sq km).

Top: The tomb of Cyrus the Great at Pasargadae. Cyrus, who founded the ancient Persian Empire, reigned from 559 to 529 B.C. He invaded and conquered many lands, including Babylonia, and called himself *King of the World*.

Centre: The mosque of Shaykh Luft Allah at Esfahan, begun in 1601, was built by Shah Abbas as a private oratory. Like the imperial mosque it stands on the city's great maidan, scene of splendid parades.

Below: The exquisite imperial mosque at Esfahan, one of the world's greatest architectural treasures. Its construction was begun in 1612, during the reign of Shah Abbas I. The dome and minarets are faced with turquoise-coloured majolica and much of the building is decorated with intricate and subtle mosaics.

Iraq

Arab republic of SW Asia at the head of the Persian Gulf. Sparsely-wooded highlands in the NE and the arid Syrian Desert in the W are separated by the broad lowlands and plateaux crossed by the Tigris and Euphrates rivers. The NE highland region is part of *Kurdistan*—the land of the Kurds. In 1980 a war with Iran developed over control of the Shatt al Arab waterway. The climate is dry and hot, especially in the desert region. The country is an important producer of petroleum—the chief source of its wealth. Other minerals include sulphur, phosphates, salt and gypsum. More than half the people live by farming. Their crops include dates, wheat, barley, rice, tobacco and cotton. Livestock provide wool and hides. Industries include food processing and the manufacture of cement, soap and textiles. Efforts are being made to develop industry and agriculture by hydro-electric and irrigation works.

Most of the people are Arabs and the rest are mostly Kurds. Arabic is the official language. The people are nearly all Muslims, divided two to one between the Shi'a and Sunni sects. There is a small Christian minority. The chief cities are Baghdad (the cap.), Basra and Mosul. Area 172,000 sq miles (445,500 sq km). For population, see Table on page 105.

The giant hospital known as 'Medical City' is built on the banks of the Tigris in Baghdad, Iraq.

Below: Tel Aviv, Israel's largest city. It lies in the Plain of Sharon, on the Mediterranean coast. The city was founded in 1909 and is laid out in a regular pattern, with modern buildings arranged in rectangular blocks. It has a university and is an important centre of industry and communications.

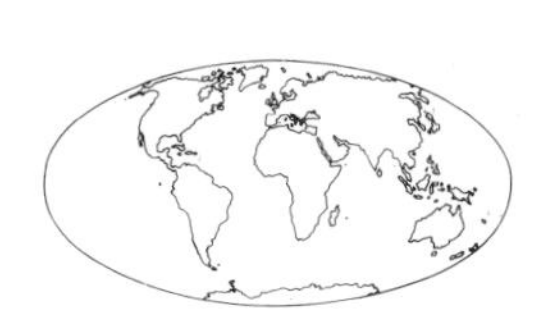

Israel

Located in SW Asia, Israel has a long Mediterranean coastline. About half of Israel consists of the Negev desert in the S. The rest includes a hilly region in the N and a narrow coastal strip in the centre. The climate is hot and dry. The chief rivers are the Jordan and Kishon. The country's mineral resources include potash, phosphates, copper, iron and gypsum. Through irrigation and careful cultivation, much formerly arid land has been made productive and Israel is a major exporter of citrus fruits and vegetables. Industries include food processing and the manufacture of machinery, textiles and chemicals.

Almost all the people are Jews, with the rest mostly Arabs. Hebrew is the official language but English is widely used in business. The chief cities are Tel Aviv-Yafo, Jerusalem, Haifa (the main port), and Ramat Gan. The Israeli parliament declared Jerusalem the cap. in 1950, but this is not recognized by the United Nations. Area 7,992 sq miles (20,700 sq km). For population, see Table on page 105.

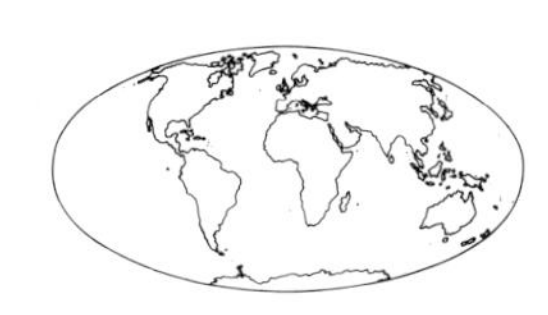

Japan

The island nation of Japan is a long archipelago off Asia's eastern coast. The Sea of Japan separates it from the Asian mainland and to its E is the Pacific Ocean. Japan's capital is Tokyo, on the island of Honshu. The country's area is 142,810 sq miles (369,880 sq km). For population, see Table on page 105.

Land and Climate

Japan consists of four main islands and more than 3,000 smaller ones. The four largest, in order of size, are Honshu (area 87,800 sq miles; 227,400 sq km); Hokkaido, the northernmost major island (30,140 sq miles; 78,060 sq km); Kyushu, the southernmost major island (14,110 sq miles; 36,447 sq km); and Shikoku (7,050 sq miles; 18,260 sq km). The chief groups of smaller islands are the Ryukyus and Bonins.

All the larger islands are extremely mountainous. The highest mountains are the Japanese Alps in central Honshu. Japan has nearly 200 volcanoes, some of them active. The famous landmark of Mount Fuji, or Fujiyama, on Honshu, is Japan's highest peak, rising 12,388 ft (3,776 m). It is an inactive volcano.

Japan's many rapid mountain rivers are an important source of hydro-electric power. The main rivers are the Ishikari (Hokkaido); and the Shinano, Tone and Kitakami (Honshu). The islands have many beautiful lakes, including Biwa and Inawashiro (Honshu) and Shokotsu (Hokkaido).

Japan's climate ranges from cool temperate in the N to warm temperate in the S. The variation is caused by the influence of the warm Japan Current in the S and the cold Oyeshio Current in the N. All Japan has plentiful rain. Many of its mountain peaks are snow-covered in the winter.

Agriculture and Industry

Only about 15 per cent. of Japan's land can be farmed. As a result, the available farmland has to be intensively cultivated. The main crops are rice, wheat, barley, fruit, and potatoes and other vegetables. There being so little lowland, many of the hills and mountainsides are cultivated by means of terracing.

Most of Japan's mountains are covered with forests, providing valuable timber and charcoal, which is widely used as a domestic fuel. Japan is among the world's leading fishing nations, catching about 8.5 million tons of fish every year.

The country is poor in natural resources. Only coal is mined in large amounts and most of it is low grade. There are small amounts of copper, zinc, bauxite, lead and nickel. Despite the shortage of resources, Japan is one of the world's great industrial countries, importing almost all the raw materials it needs. It is one of the leading steel producers and leads the world in shipbuilding.

Other heavy industries include the construction of locomotives and industrial

The sacred mountain of Fujiyama or Fujisan, on Honshu. Its snow-capped peak rises to 12,388 ft (3,776 m). Fujiyama, about 70 miles (112 km) from Tokyo, is in Fuji-Hakone National Park, and its beauty has provided inspiration for many Japanese artists and poets.

Below right: Cherry-blossom time at the temple of Dai-goji in the ancient city of Kyoto. Right: A garden without trees or flowers. This peaceful spot is a place of contemplation for Zen Buddhists. Below left: The temple of Heian-Jingu in Kyoto. The city has hundreds of temples and shrines.

machinery. The country has also earned a high reputation for the manufacture of textiles, motor vehicles, optical instruments, electronic equipment, chemicals and precision tools.

Cities and Communications

More than two-thirds of the people live in cities or other urban communities. The chief cities include Tokyo, Osaka, Yokohama, Kyoto, Nagoya and Kobe, on Honshu; Fukuoka on Kyushu and Sapporo on Hokkaido.

Japan has about 80,000 miles (130,000 km) of surfaced roads and a rail network of 15,000 miles (24,000 km) covering the four main islands.

The major ports are Tokyo, Yokohama, Nagoya, Kobe and Osaka.

The People

The Japanese are a Mongoloid people, closely related to other groups of eastern Asia, but with some mixture of Malaysian and Caucasoid strains. Japan has solved a serious problem of population growth through a widespread birth control programme.

Japanese is the official and universal language. Its structure is similar to Korean, and both languages resemble those of the Ural-Altaic family, which also includes Turkish and Finnish. Written Japanese is based on Chinese characters, some of them in simplified form.

Buddhism and Shintoism are the main religions. Almost all Japanese follow one or the other religion and many people practise both.

Jordan

Arab kingdom of SW Asia. It lies near the eastern coast of the Mediterranean, but its only outlet to the sea is at the port of Aqaba on the Gulf of Aqaba leading to the Red Sea. It is bordered on the N by Syria, on the NE by Iraq, on the E and S by Saudi Arabia, and on the W by Israel.

In the W the River Jordan flows N-S through the country. At its S end is the Dead Sea, a salt lake, which is the earth's lowest surface point—1,286 ft (391 m) below sea-level. E of the river, arid, stony uplands slope away to the Syrian Desert in NE Jordan. The climate is hot and dry.

Exploited minerals include phosphates, potash and marble; there are also deposits of manganese, iron, sulphur and copper.

About a third of the people live by farming. The chief crops are wheat, barley, citrus fruits, olives and vegetables. Industries include oil refining, tanning and the manufacture of electrical goods, cement and soap.

Almost all the people are Arabs, divided between Bedouin to the E of the Jordan and Palestinians to the W. The largest cities are Amman (the cap.), Irbid and Nablus. The Jordanian sector of Jerusalem was occupied by Israel in 1967. Area 37,700 sq miles (97,600 sq km). For population, see Table on page 105.

Khmer Republic

Country of SE Asia, in the Indochina Peninsula. Formerly the kingdom of Cambodia and later the Khmer Republic. Since 1970 development has been hindered by civil war and intervention by foreign powers.

It is bordered on the N by Thailand and Laos, and on the E and SE by Vietnam. In the SW it has a coastline on the South China Sea. The E of the country lies in the alluvial basin of the lower Mekong River, with extensive jungles and the large Tonle Sap ('Great Lake') in the W part. The climate is tropical and affected by the monsoon.

The country's mineral resources are little exploited; they include phosphates, iron and limestone. Agriculture and fishing are the two main occupations. Many farmers live at a subsistence level, growing, chiefly, rice. Other crops include rubber, tobacco, maize and sugar.

Nearly 90 per cent. of the people are Khmers. Khmer (or Cambodian) is the official language. The chief cities are Phnom Penh (the cap.), Battambang, Kompong Cham and Kampot. Area 70,000 sq miles (181,000 sq km). For population, see Table on page 105.

Above: Angkor Wat, a ruined Hindu temple in the north of the country, believed to be the largest religious structure in the world.

Below: Cultivating rice in a paddy field, a farming technique that is probably 5,000 years old.

Korea, North

The Democratic People's Republic of Korea occupies the N part of the Korean Peninsula on the NE coast of China. It is bordered on the N by China, on the extreme NE by the U.S.S.R. and on the S by South Korea. It has coastlines on the Yellow Sea in the W and the Sea of Japan in the E. North Korea is a mountainous land with thousands of peaks. The highest mountains are in the N, where some rise to more than 8,000 ft (2,440 m). Another impressive mountain range extends along the E coast. The only flat and low-lying land is in the W. The chief rivers are the Tumen, Yalu, Ch'ongch'on and Taedong. The climate is subject to the monsoon, but is more temperate and somewhat cooler and drier than in South Korea. The country is rich in minerals. There are large deposits of coal, iron, lead, zinc, copper, gold, graphite, silver and tungsten. More than half the people work in mining and industry. The main industrial products are steel, machinery and textiles. Only a fifth of the country's area is suitable for farming and less than one-tenth of the people live on the land. The chief crops are rice, maize, barley, wheat and cotton. The people of North Korea are related to the Mongolians with some Chinese mixture. Korean is the official language. The chief cities are Pyongyang, Chongjin, Hungnam, Hamhung and Kaesong. Area 46,800 sq miles (121,200 sq km). See also *South Korea.*

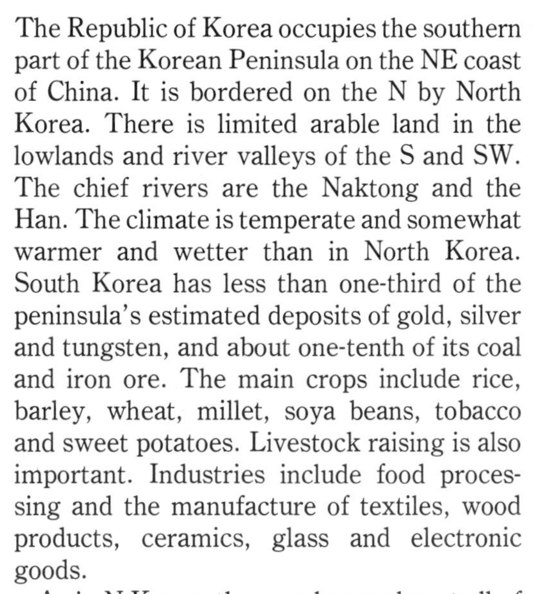

Right: The harbour at Pusan, South Korea's largest port. The city lies on the Korea Strait and is an important industrial centre.

Below: The Eastern Gate of the city of Seoul, one of the three surviving gates in the ancient city walls. As well as being South Korea's capital, Seoul is the cultural centre and has large industries.

Korea, South

The Republic of Korea occupies the southern part of the Korean Peninsula on the NE coast of China. It is bordered on the N by North Korea. There is limited arable land in the lowlands and river valleys of the S and SW. The chief rivers are the Naktong and the Han. The climate is temperate and somewhat warmer and wetter than in North Korea. South Korea has less than one-third of the peninsula's estimated deposits of gold, silver and tungsten, and about one-tenth of its coal and iron ore. The main crops include rice, barley, wheat, millet, soya beans, tobacco and sweet potatoes. Livestock raising is also important. Industries include food processing and the manufacture of textiles, wood products, ceramics, glass and electronic goods.

As in N Korea, the people are almost all of Korean stock. Traditional religions include Buddhism and Confucianism. The chief cities are Seoul (the cap.), Pusan, Taegu and Inchon. Area 38,400 sq miles (99,460 sq km). Annexed by Japan in 1910, Korea was divided into Russian and American occupation zones in 1945 after Japan's defeat in World War II. This division became the basis of two states, communist North Korea and non-communist South Korea.

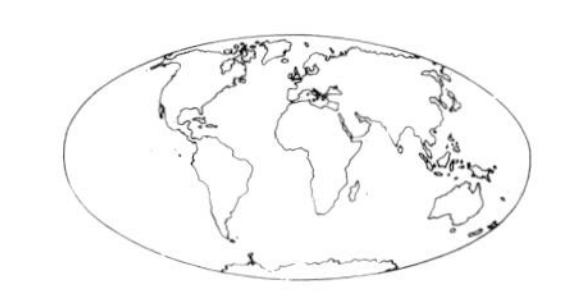

Kuwait

Located at the NE corner of the Arabian Peninsula, at the head of the Persian Gulf, the tiny, oil-rich state of Kuwait is bordered on the N and W by Iraq and on the S by Saudi Arabia. The country is almost all flat desert, with a few fertile oases. Only a very small part of the land is cultivated, producing fruit and vegetables.

The economy is almost totally based on the country's oil deposits, amounting to nearly 20 per cent. of the world's known resources. Natural gas is also produced. Kuwait's oil revenues give it almost the world's highest per capita income.

The population is mainly Arab, but less than half are Kuwaitis. Area 7,500 sq miles (19,000 sq km).

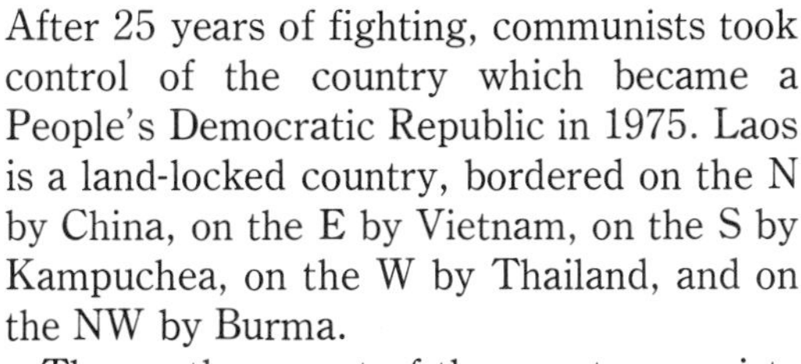

Laos

After 25 years of fighting, communists took control of the country which became a People's Democratic Republic in 1975. Laos is a land-locked country, bordered on the N by China, on the E by Vietnam, on the S by Kampuchea, on the W by Thailand, and on the NW by Burma.

The northern part of the country consists of jungle-covered mountains and plateaux, cut by narrow river valleys. In this region is Phu Bia, the highest peak in Laos, rising to 9,240 ft (2,816 m).

The S part has sparsely-forested, arid limestone terraces, descending to the valley of the River Mekong. The Mekong is the largest and most important river. Like surrounding lands, Laos has mainly a tropical monsoon climate.

The country's chief mineral resource is tin. It also has deposits of iron, gold, copper and manganese, but these are largely unexploited.

Laos is one of the least developed countries in SE Asia. Most of the people live by subsistence farming, but cash crops are grown, too. The main crops are rice, maize, coffee, cotton, tea and tobacco. The country's physical features hamper development. And the lack of an adequate road network makes communication difficult between the Mekong basin in the W, where most of the people live, and the rest of the country. About half of the people are Lao, descendants of Chinese settlers of the 1200s. Various scattered hill tribes live in the rest of the country. Lao is the official and main language. French is widely used in business.

Theravada Buddhism is the official religion. The chief cities are Vientiane (the administrative cap.), Pakse, Thakhek and Luang-Prabang (the ancient royal cap.). Area 91,400 sq miles (236,725 sq km). For population, see Table on page 105.

Bottom: Houses built on piles line a swampy creek in Vientiane, the capital of Laos. The city extends along 3 miles (5 km) of the left bank of the Mekong River.

Below: Part of El Sabaah hospital in Kuwait, one of the most modern and lavishly-equipped medical centres in the world. Much of the state's large income from oil has been spent on hospitals, schools, roads and houses.

Lebanon

The smallest country of SW Asia, Lebanon is located on the E coast of the Mediterranean. In 1975 fighting broke out between opposing factions, and in 1982 Israel invaded to overcome Palestine Liberation Army guerrillas. Lebanon's main physical features are a narrow coastal plain, behind which rise the Lebanon Mountains. Inland is the fertile Bekaa Valley, behind which rise the Anti-Lebanon Mountains, along the Syrian border. Lebanon has a hot, Mediterranean climate, but has fairly heavy rainfall and snow in the mountain regions. The country's mineral resources include deposits of bitumen, iron, limestone and salt. Its industries include food processing, oil refining and the manufacture of textiles and cement. About a quarter of the land is cultivated. The main crops are citrus fruits, vegetables and tobacco. There are also large numbers of livestock. Much of Lebanon's income is derived from commerce, finance and banking. Most Lebanese are Arabs, but, unusually for SW Asia, they are divided nearly equally between Muslims and Christians. Arabic is the official language, but French and English are widely spoken in business and commerce. Beirut, the cap. and largest city, is the financial cap. of the Middle East. Area 4,300 sq miles (11,000 sq km).

Below: The national mosque in Kuala Lumpur in Malaysia. The city has many fine and imaginative modern buildings. Bottom: A village in Sarawak, on the island of Borneo.

Malaysia

The Federation of Malaysia in South-East Asia consists of the southern part of the Malay Peninsula—known as West Malaysia —and most of the northern part of the island of Borneo—known as East Malaysia. East Malaysia consists of two territories, Sarawak and Sabah. Malaysia's capital is Kuala Lumpur, in West Malaysia. The country's area is 130,000 sq miles (337,000 sq km).

Land and Climate

Malaysia lies just N of the equator. More than two-thirds of its land area is covered by dense tropical rain forest. Forested mountains form the backbone of the Malay Peninsula. They are flanked by coastal plains, more extensive in the W than in the E.

The climate is hot and humid, with heavy rainfall, and is affected by the monsoons.

Agriculture and Industry

Malaysia's main natural resources are rubber and tin, which with timber and palm oil account for about three-quarters of the country's exports. With Indonesia, Malaysia is the world's leading natural rubber

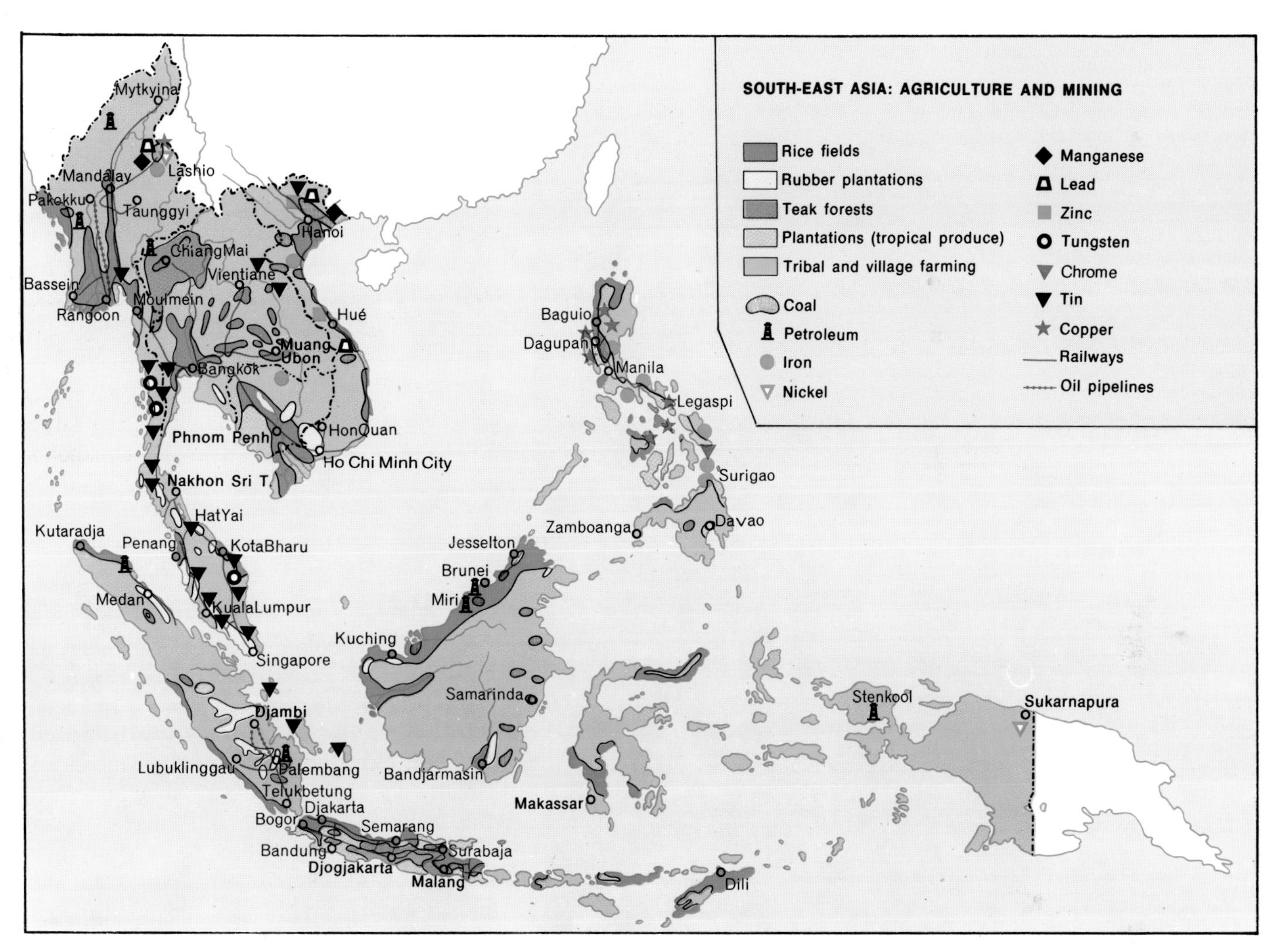

producer, its climate being ideally suited to rubber production. The rich tin deposits in the Malay Peninsula make Malaysia the world's main producer of this metal. Other mineral deposits include oil, coal and bauxite. The dense forests produce valuable hardwoods.

Most Malaysians live by farming. The main food crops are rice, palm oil, coconut oil, fruits, sago, sugar, tea and pepper.

Many industries are based on the processing of food, metals and rubber.

Cities and Communications

Nearly 40 per cent. of Malaysia's people live in cities. Apart from Kuala Lumpur, the capital, the largest cities are Penang and Ipoh, both in the western part of the Malay Peninsula. Kuching is the chief town of Sarawak, and Kota Kinabalu of Sabah. Malaysia has extensive road networks and a comprehensive rail system, and its main population centres are linked by internal air services.

The People

Malays (45 per cent.) and Chinese (35 per cent.) are the two largest population groups. About one-tenth of the people belong to non-Malay tribal groups. They include the Dyaks in Sarawak.

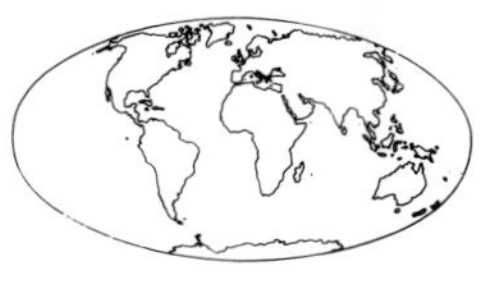

Maldives

The Maldive Islands are an archipelago of some 2,000 coral islands in the Indian Ocean, about 400 miles (640 km) SW of Sri Lanka. They became a member of the commonwealth in 1982. The islands are grouped into a dozen atolls. About 220 islands are inhabited. The climate is hot and humid. Fishing and copra and coconut oil production are sources of revenue. The people are Muslim and speak Maldivian, a Sinhalese dialect. The capital is Malé. Area 115 sq miles (298 sq km).

MACAU

The tiny Portuguese possession of Macau lies at the mouth of the Pearl River, on the SE coast of China. It is about 40 miles (64 km) W of Hong Kong. It lives by trade. Area 5 sq miles (8 sq km).

Top: A herdsman on the great plains that form the eastern part of Mongolia. Mongol horsemen have been famed for their skill and fearlessness since the 1100s. Then, under Genghis Khan, they swept over Asia and eastern Europe and established the largest empire in history.

Centre: In Nepal, onlookers mount decorated 'beasts' for a better view of a religious procession. The majority religion in Nepal is Hinduism.

Bottom: A fertile valley on the southern slopes of the Himalayas, near the city of Katmandu in Nepal.

Mongolia

Mongolia, in central Asia, is the world's most sparsely populated country. It lies between the U.S.S.R. to the N and China to the S. Its climate ranges from extremely hot in the desert to very cold in higher regions. Coal and oil are the main mineral deposits. Most of the people are Mongols, many of them herdsmen. Area 600,000 sq miles (1,550,000 sq km).

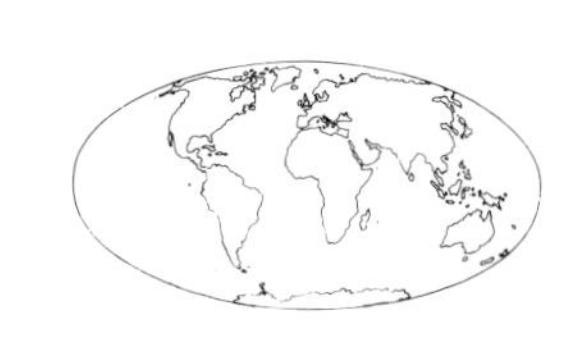

Nepal

The Hindu kingdom of Nepal is situated in the Himalayas. It is bordered on the N by Tibet (China) and on the E, S and W by India. The central part of the country is traversed by high, fertile valleys and sharp mountain ridges. The High Himalaya Range lies along the northern border, and includes Mount Everest (29,028 ft; 8,848 m), the world's highest peak. The climate, mild along the Indian border, is bitterly cold in the High Himalayas.

Nepal's mineral resources include coal and iron, but are largely unexploited. Nepal is now linked to both India and Tibet by road. Nearly all the people live by farming. The chief crops include rice,maize, wheat, millet and jute. Forestry is also important. There are numerous population groups, of which the Gurkhas (the largest) and the Sherpas are the best known. Area 54,362 sq miles (140,797 sq km).

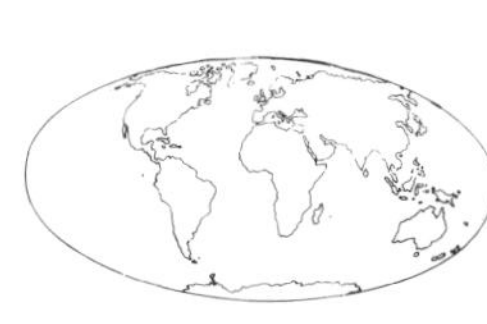

Oman

Oman, in the SE of the Arabian Peninsula, is bordered on the N by the United Arab Emirates and Saudi Arabia, and on the W by Yemen (Aden). High, barren hills in the NW are separated from the Dhofar Plateau in the SW by hundreds of miles of desert. The principal source of revenue is petroleum. The capital is Muscat. Area 82,000 sq miles (212,000 sq km).

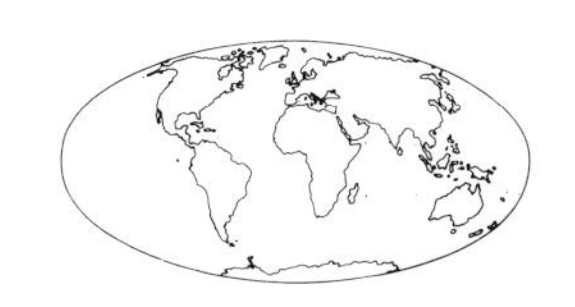

Pakistan

The Islamic Republic of Pakistan, the northwestern part of the Indian sub-continent, lies on the northern shore of the Arabian Sea. Its NE frontier is in Jammu and Kashmir, a territory whose ownership it disputes with India. Until 1972, Pakistan also included the country that is now Bangladesh. Pakistan's capital is Islamabad. The country's area is about 310,400 sq miles (804,000 sq km).

Land and Climate

Pakistan includes all or part of several historic lands: Baluchistan, North-West Frontier Province, Sind and the Punjab. It has a mountain wall in the N and NW which includes the towering peaks of the Himalaya, Hindu Kush and Karakoram Ranges. Nanga Parbat in the W Himalayas rises to 26,660 ft (8,126 m) above sea-level. On the NW frontier with Afghanistan is the famous Khyber Pass.

Pakistan's main river, the Indus, flows generally SW through the country. The Thar Desert of NW India extends into SE Pakistan.

Pakistan's climate ranges from hot and dry in the desert regions to cold and wet in mountainous areas. Rainfall varies greatly. The climate of the whole country is strongly affected by the monsoons.

Agriculture and Industry

Pakistan has deposits of a number of minerals, including coal, petroleum, iron, chromite, gypsum, sulphur and antimony. Dams on the Indus and its tributaries provide hydro-electric power and water for irrigation.

About two-thirds of the people of Pakistan live by farming. They grow wheat, maize, millet, barley and fruit. Other important crops are cotton, sugar, tobacco and tea.

Pakistan's industries include food processing and the manufacture of cotton and other textiles, chemicals, cement, paper and metal goods.

Cities and Communications

The chief cities are Karachi, Lahore, Lyallpur, Hyderabad, Multan, Rawalpindi and Islamabad (the cap.), Pakistan possesses well-developed road and rail systems, though communications are difficult in mountainous and desert regions.

The People

The chief population groups are Pathans, Punjabis, Sindhis and Baluchis. Most of the people are Muslims, and Islam is the state religion.

Left: A busy street in Rawalpindi, in the Punjab, Pakistan. The city lies in the shadow of the Himalayas and is a major industrial centre.

Below: The valley of the River Swat, on the historic North-West Frontier of Pakistan. Much of the region is stark and barren. The people hold tenaciously to their own way of life.

Philippines

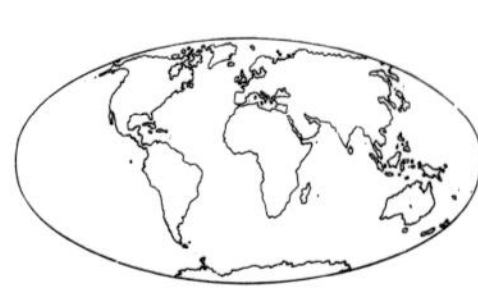

The republic of the Philippines is an archipelago of more than 7,000 islands in South-East Asia. It stretches for more than 1,000 miles (1,600 km) between the Pacific Ocean and the South China Sea. The northernmost islands are about 330 miles (530 km) from the coast of China, and the southern tip of the chain is about 30 miles (50 km) from the island of Borneo. The capital of the Philippines is Manila on Luzon Island. The area of the Philippines is 115,800 sq miles (299,900 sq km).

Right: The slow, strong *carabao*, the water-buffalo, is the most common beast of burden in the Philippines. It can work in flooded rice fields where other animals would be unable to walk.

Below: The port of Manila, the country's largest city.

Land and Climate

The three largest islands are Luzon (41,765 sq miles; 108,170 sq km) in the N. Mindanao (38,344 sq miles; 99,310 sq km) in the S, and Palawan (5,751 sq miles; 14,895 sq km) in the W.

Most of the larger islands are mountainous. All have coastal lowlands and some have fertile river valleys. The country's largest lowland area is in central Luzon. Volcanic peaks rise in many of the islands and there are numerous hot springs.

The Philippines lie just N of the equator. The climate is warm and humid, with little seasonal variation in temperature. But rainfall varies considerably. The northern islands of the Philippines are frequently hit by typhoons, and earthquakes occur from time to time, especially in Mindanao.

Agriculture and Industry

More than half of the people live by farming. They grow rice, maize, sweet potatoes, manioc and tropical fruits. Sugar, tobacco, Manila hemp (abaca), and coconuts are important cash crops. Many goats, sheep, pigs and water buffalo are raised. Forest products and fishing are also important.

The Philippines are rich in minerals. The main metal is gold. Other minerals include silver, iron, copper, chromite, manganese, limestone, coal and petroleum. Industries include oil refining, food processing and the manufacture of cement, textiles, chemicals, and rope and fibre products.

Cities and Communications

About a third of the people live in cities. The chief cities are Manila, Quezon City, Cebu City, Davao City and Isabela. The country has good roads and a railway system.

The People

The majority of the people are of Malay stock, with Indonesian and Mongoloid strains. There are about half a million Chinese and smaller minorities of Americans and Europeans.

Qatar

Qatar, on the E coast of the Arabian Peninsula, occupies a 120-mile (190-km) long peninsula extending into the Persian Gulf. It is bordered by Saudi Arabia and the United Arab Emirates. The country is a low desert plain. Its climate is hot and dry. The basis of the country's prosperity is its petroleum. Revenues from the production of petroleum have been used to develop roads, water supplies and schools and welfare services. Natural gas is also produced, and fishing is important. Vegetables are grown, and there are herds of goats and camels. Most of the population lives in and around the capital, Doha. Area 4,000 sq miles (10,300 sq km).

Saudi Arabia

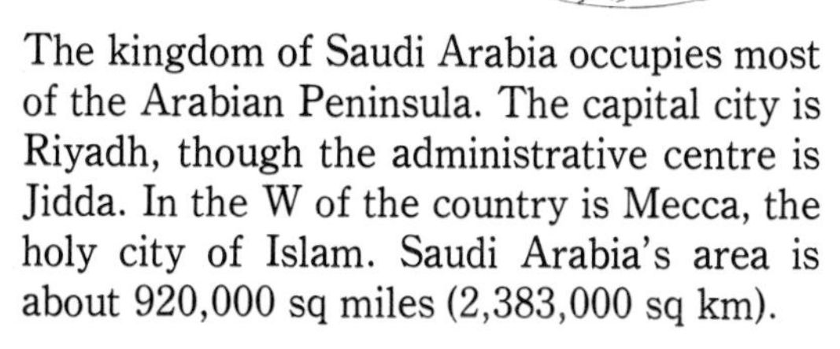

The kingdom of Saudi Arabia occupies most of the Arabian Peninsula. The capital city is Riyadh, though the administrative centre is Jidda. In the W of the country is Mecca, the holy city of Islam. Saudi Arabia's area is about 920,000 sq miles (2,383,000 sq km).

Land and Climate

Saudi Arabia is a country of rocky plains, deserts and mountains. Stretching along the Red Sea coastline is an arid, narrow plain, the Tihama. Inland is the great central plateau, the Nejd. This barren region of rocky and sandy deserts covers most of Saudi Arabia. The largest deserts are the Rub'al Khali, or Empty Quarter, in the S, and the An Nafud in the N. The eastern coastlands along the Persian Gulf form a low plain, fringed by mud flats, lagoons and sandbanks.

Saudi Arabia's climate is among the world's hottest, with summer temperatures reaching 125°F (52°C) in the deserts and coastal regions. On the other hand, winter temperatures in mountainous areas can drop to freezing. The coastal regions are humid, but the interior has almost no rain.

Agriculture and Industry

Over half the people are farmers, growing crops in desert oases, and they depend heavily on irrigation. The main crops are dates, alfalfa, barley, citrus fruits, millet and wheat. Nomadic Bedouin tend herds of camels, sheep and goats.

Oil refinery at Mina Saud, on the Persian Gulf in Saudi Arabia. Oil is the source of the country's wealth.

Saudi Arabia's immense oil deposits make it the world's fifth largest oil producer. Much of the oil is transported by pipeline to Bahrain and the Lebanon to be refined. There is a major oil refinery and deep-water terminal at Ras Tannura on the Persian Gulf coast.

Cities and Communications

About one-tenth of the population live in cities. The main cities, apart from Mecca, Jidda and Riyadh, are Alf Hufuf, Medina, Al Khobar and Dammam.

Camel and truck caravans are the main means of transport. There are about 10,000 miles (16,000 km) of roads, mostly unsurfaced. There is a 350-mile (560-km) railway link between Riyadh and Dammam, and extensions are planned.

The People

Almost all the people are Arabs, divided between nomadic Bedouin, town dwellers and farmers. The official language is Arabic, and Islam is the official religion.

Left: Barges line the banks of the Singapore River. Singapore's harbour is crowded with craft, from great liners to sampans.

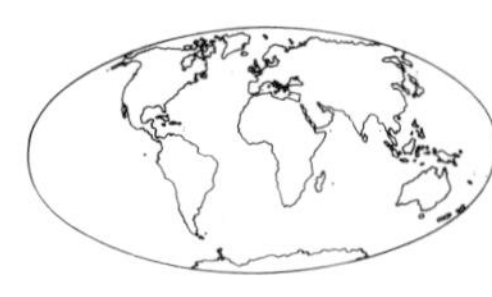

Singapore

Island republic in SE Asia, off the southern tip of the Malay Peninsula, from which it is separated by the Johore Strait. The island is linked to Malaysia by a causeway. Most of Singapore is low lying. Its highest point is Bukit Timah (581 ft; 177 m). There are several small rivers. Singapore's climate is hot and humid.

Singapore City has one of SE Asia's few natural harbours and is the world's fourth largest port. The economy is based on the international trade passing through the port and on the processing of raw materials from other countries. More than three-quarters of the people are Chinese. The cap. is Singapore City. Area 224 sq miles (580 sq km).

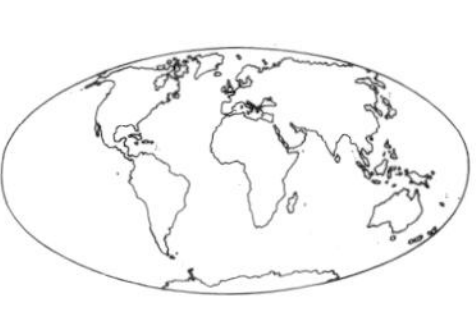

Sri Lanka

Formerly called *Ceylon*, the island republic of Sri Lanka lies in the Indian Ocean off India's southern tip, from which it is separated by the Palk Strait. The S central part of the island is mountainous and richly forested. The rest of the land is a fertile, low-lying plain. The chief river is the Mahaweli Ganga. Sri Lanka has a hot, humid, tropical monsoon climate.

More than half the people work on farms and plantations. The chief products are tea, rice, rubber and coconuts. Mineral deposits include graphite and semi-precious and precious stones. Industries include food processing and the manufacture of chemicals, cement, paper and textiles. Three-quarters of the people are Sinhalese and are Buddhist. But more than 20 per cent. are Tamils, and are Hindu. The official language is Sinhala; Tamil is used, too, and many people speak English. The chief cities are Colombo (the cap.), Jaffna, Kandy, Galle, Negombo and Trincomalee. Area 25,332 sq miles (65,610 sq km).

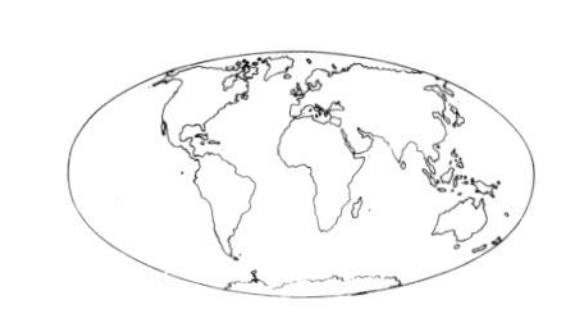

Syria

Arab republic of SW Asia on the eastern coast of the Mediterranean. N of the desert is the fertile valley of the River Euphrates, and beyond that a region of rough plains. Parallel with the coast in the W is the valley of the River Orontes. The climate is mild along the coast, with plenty of rain. Inland it ranges from very hot to very cold, with little rain. Syria is self-sufficient in food production and more than half the people live by farming. The chief crops are wheat, barley, cotton, tobacco and fruit. Cattle and sheep raising is also important. In desert areas there are many nomadic herdsmen. Industries include food processing and the manufacture of textiles and cement. A dam on the Euphrates provides hydro-electric power and aids irrigation. Oil pipelines from Saudi Arabia and Iraq cross the country to Mediterranean ports. Most Syrians are Arabs. Area 71,800 sq miles (185,960 sq km).

Syria. The construction of a pipeline across the Syrian Desert. Pipelines carry oil from Iraq for hundreds of miles across Syria to Mediterranean ports.

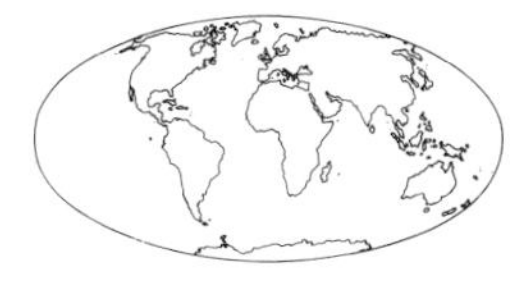

Taiwan

The island of Taiwan or *Formosa* lies about 100 miles (160 km) off the SE coast of China, from which it is separated by the Formosa Strait. Officially called the *Republic of China*, Taiwan is the seat of the Nationalist government of China which was driven from the mainland in 1949. About two-thirds of Taiwan is mountainous. The W part is a flat coastal plain that supports most of the population. Taiwan has few minerals, but industry, including textiles, chemicals, food-processing and electrical goods, is well developed. Area 13,885 sq miles (35,962 sq km).

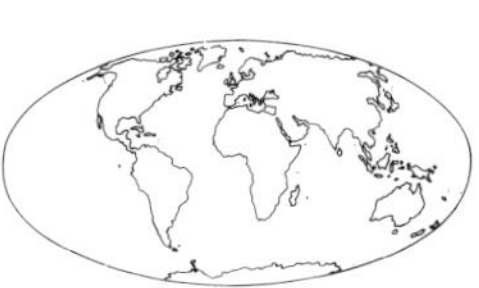

Thailand

Formerly known as *Siam*, Thailand, in SE Asia, is bordered on the NE by Laos, on the SE by Kampuchea (Cambodia) and on the S by Malaysia. It has a long coastline on the Gulf of Siam in the S, and a short one on the Andaman Sea in the SW. The northern part of the country is on the SE Asian mainland, and the southern part stretches down the Malay Peninsula. The country can be divided into four main natural regions. The mountainous northern region lies between the Salween River in the NW and the Mekong River in the NE. The eastern region is a high and arid plateau, stretching from the mountains to the border with Laos. Central Thailand consists of a low, flat alluvial plain formed by the Chao Phraya and other rivers. Southern Thailand is long, narrow and mountainous. As well as the Chao Phraya, Mekong and Salween, Thailand's many rivers include the Chi, Nan, Pattani, Ping, Wang and Yom. Most of them are navigable and they form an important means of

The Grand Palace—the walled royal city—in Bangkok, Thailand. In the centre background is the magnificent *Wat Phra Keo*, the Temple of the Emerald Buddha, which houses a jasper figure of the Buddha.

Opposite page, above: The mosque of Sultan Ahmed in Istanbul, in Turkey. It has 6 minarets instead of the usual 4.

Opposite page, below: A floating market in Bangkok.

transportation. In southern Thailand there is a large lagoon called Thale Luang. There are also important rail and road systems. A deep-water harbour is under construction at Sattahip. Bangkok is the main port. About half of the country is covered by forests that provide valuable teak, ironwood, rattan and bamboo.

Most of the people of the country live by farming. Rice, the main crop, is chiefly grown in the central region. It is the staple food and main export. Rubber is the second most important crop. Other crops include maize, sugar, pepper, coconuts, jute, tapioca and fruits. Fishing, for local consumption and for export, is important. Tin, tungsten and sapphires are mined. There are also considerable deposits of other minerals, including coal, gold and lead, but as yet these have been little exploited.

The few industries include food and rubber processing and there are traditional handcraft industries such as the making of cotton and silk cloth and lacquerware.

Most of the people are Thais, the descendants of Mongol invaders and settlers from China. A large Chinese minority, numbering about three million, dominates Thailand's retail trade.

The official language is Thai, which belongs to the Sino-Tibetan group. It is written in a script derived from Sanskrit. Most of the people are Buddhists, but there are some Christians, Confucians and Muslims. The country has nearly 20,000 Buddhist temples. Traditionally, Thai men who are Buddhists spend at least a short period of their lives as monks. Most of the people live in farming villages. Each village has a central group of buildings called a *wat*, which is used as a religious and social centre. The chief cities are Bangkok (the cap.), Chiang Mai and Lampang. Area 198,460 sq miles (513,810 sq km).

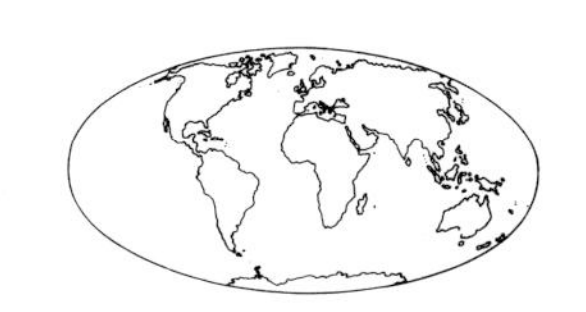

Turkey

The republic of Turkey is mainly in SW Asia ('Asia Minor'). But a small part of it is in Europe. European Turkey (called *Thrace*), about 3 per cent. of the total area, is separated from Asiatic Turkey (the greater part of which is called *Anatolia*) by the Dardanelles, the Sea of Marmara and the Bosporus. Turkey has long coastlines on the Black Sea in the N, and on the Mediterranean in the S. Part of the Mediterranean coastline is on the Aegean Sea. The European part has rolling farmland and is hilly in the NE. Asiatic Turkey has fertile coastal strips. The semi-arid Anatolian Plateau in the centre is fringed by hills and mountains, including the Pontic Mountains and the Taurus Range. The chief rivers include the Kizil, Seyhan, Menderes (Maeander), Tigris and Euphrates. The climate ranges from mild and moist along the coasts to harsh and dry on the inland plateau. Turkey has good mineral deposits, but they are mostly unexploited. More than two-thirds of the people live by farming. The chief crops are cereals, tobacco, fruit and cotton. Animal husbandry is also important. Industries include the production of iron and steel, sugar refining and the manufacture of textiles.

Most people are Turks. Minorities include about two million Kurds and small groups of Greeks, Arabs, Armenians and Jews. The chief cities are Ankara (the cap.), Istanbul (Constantinople), Izmir (Smyrna), Adana and Bursa. Area 301,300 sq miles (780,360 sq km). For population, see Table on page 105.

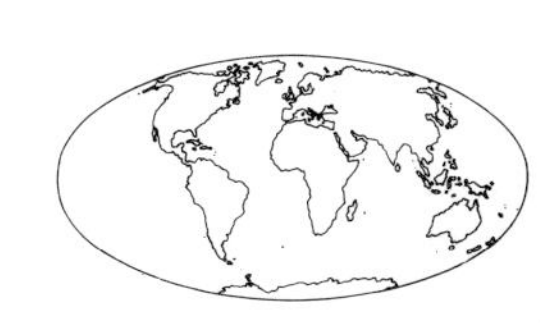

U.S.S.R.

The U.S.S.R., the world's largest country, lies mainly in Asia. About three-quarters of its territory is E of the Ural Mountains, the traditional dividing line between Europe and Asia. But the majority of its important cities —including the capital city, Moscow—are in the European section. So are more than 60 per cent. of its people. For the main article on the U.S.S.R., see page 95.

United Arab Emirates

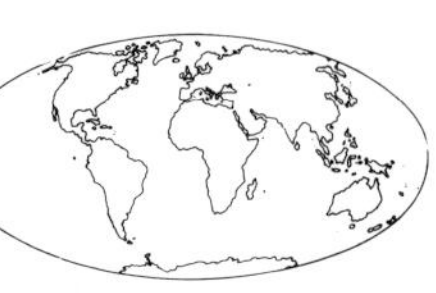

A federation of seven emirates or sheikhdoms, the United Arab Emirates is located on the Persian Gulf, in the SE 'horn' of the Arabian Peninsula. The area was previously known as the *Trucial States* and as *Trucial Oman*. The sheikhdoms are Abu Dhabi, Dubai, Sharjah, Ajman, Umm al Qaiwain, Ras al Khaimah and Fujairah. The U.A.E. is bordered by Oman, Saudi Arabia and Qatar. It consists mostly of low, flat desert land, with some coastal hills. There are vast oil deposits in Abu Dhabi, and oil has also been found in Dubai. Many people live by herding or fishing, and dates are grown. The people are mainly Arabs. The chief cities are Abu Dhabi (the cap.) and Dubai. Area 32,000 sq miles (82,900 sq km). For population, see Table on page 105.

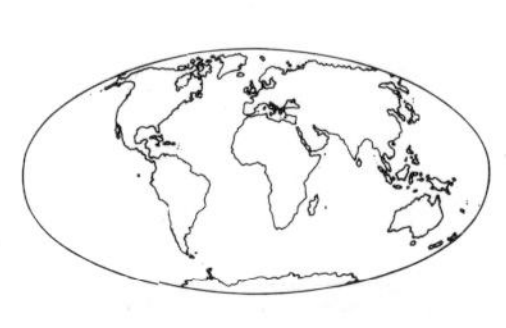

Vietnam

Located on the W coast of the South China Sea, Vietnam is bordered on the N by China, and on the W by Kampuchea (Cambodia) and Laos. On the W it also has a short coastline on the Gulf of Siam. Once part of French Indochina, from 1954 Vietnam was divided between a communist government in the N and a non-communist government in the S. In 1975, communist forces defeated the S Vietnam government and took control of the whole country. The country can be divided into three main natural regions: the fertile delta of the Red River in the N; the long, narrow coastal region of central Vietnam; and the delta of the Mekong River in the S. Nearly half the area of Vietnam is forested. The country has a humid tropical monsoon climate, with a hot rainy season in the summer and a cooler, drier season in the winter. Vietnam's mineral resources include coal, iron, manganese, phosphates, tin and zinc in the N; and coal, lead, limestone and molybdenum in the S. Important forest products include cinnamon, quinine and bamboo. Most Vietnamese are farmers. Rice, grown mainly in the delta regions, is by far the most important crop. Rubber is the main export crop. Other crops include coffee, tea, maize, manioc, ground-nuts, coconuts, sugar, tobacco and sweet potatoes. Vietnam's industries include coal mining, sugar refining, rubber processing and the manufacture of textiles and cement. Vietnamese is the official language in both parts of the country, and French is widely spoken. Only about 10 per cent. of the population live in cities. The main cities are Saigon (the south-

Right: The port of Dubai on the southern coast of the Persian Gulf. It was for long a port of call for coastal trade in the Gulf, and has today a large import trade for the seven states of the United Arab Emirates.

Opposite page: A girl measures cereals in a Vietnam street. The light and practical 'lampshade' hat is the typical headgear of the country.

ern cap., now renamed Ho Chi Minh City), Hanoi (the northern cap.), Da Nang, Gia Dinh, Hué and Haiphong. Area 128,400 sq miles (332,560 sq km).

Yemen (Aden)

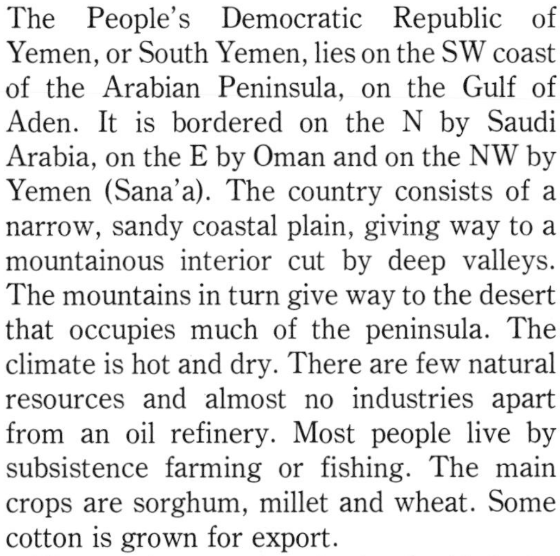

The People's Democratic Republic of Yemen, or South Yemen, lies on the SW coast of the Arabian Peninsula, on the Gulf of Aden. It is bordered on the N by Saudi Arabia, on the E by Oman and on the NW by Yemen (Sana'a). The country consists of a narrow, sandy coastal plain, giving way to a mountainous interior cut by deep valleys. The mountains in turn give way to the desert that occupies much of the peninsula. The climate is hot and dry. There are few natural resources and almost no industries apart from an oil refinery. Most people live by subsistence farming or fishing. The main crops are sorghum, millet and wheat. Some cotton is grown for export.

The people are mainly Arabs. Arabic is the official language and English is widely spoken. The capital is Aden. Area 180,000 sq miles (466,000 sq km).

Yemen (San'a)

The Yemen Arab Republic, in the SW of the Arabian Peninsula, is bordered on the N and E by Saudi Arabia and on the S by Yemen (Aden). In the W it has a coastline on the Red Sea. The country consists of two main regions: a hot, arid coastal strip along the Red Sea, and, in contrast, a mountainous interior which receives abundant rain. Little is known of Yemen's mineral resources: salt is the only mineral produced in any quantity. Most of the people live by farming or as herdsmen. The main crops are millet, maize, sorghum, fruit and dates. Coffee and qat, a narcotic, are the chief cash crops. Other exports include hides, other skins and salt.

The people are chiefly Arabs, with some African mixture along the coast. The chief cities are Sana'a (the cap.), Ta'izz, Ibb, and the coffee-exporting port of Moka. Area 75,000 sq miles (194,000 sq km).

Africa

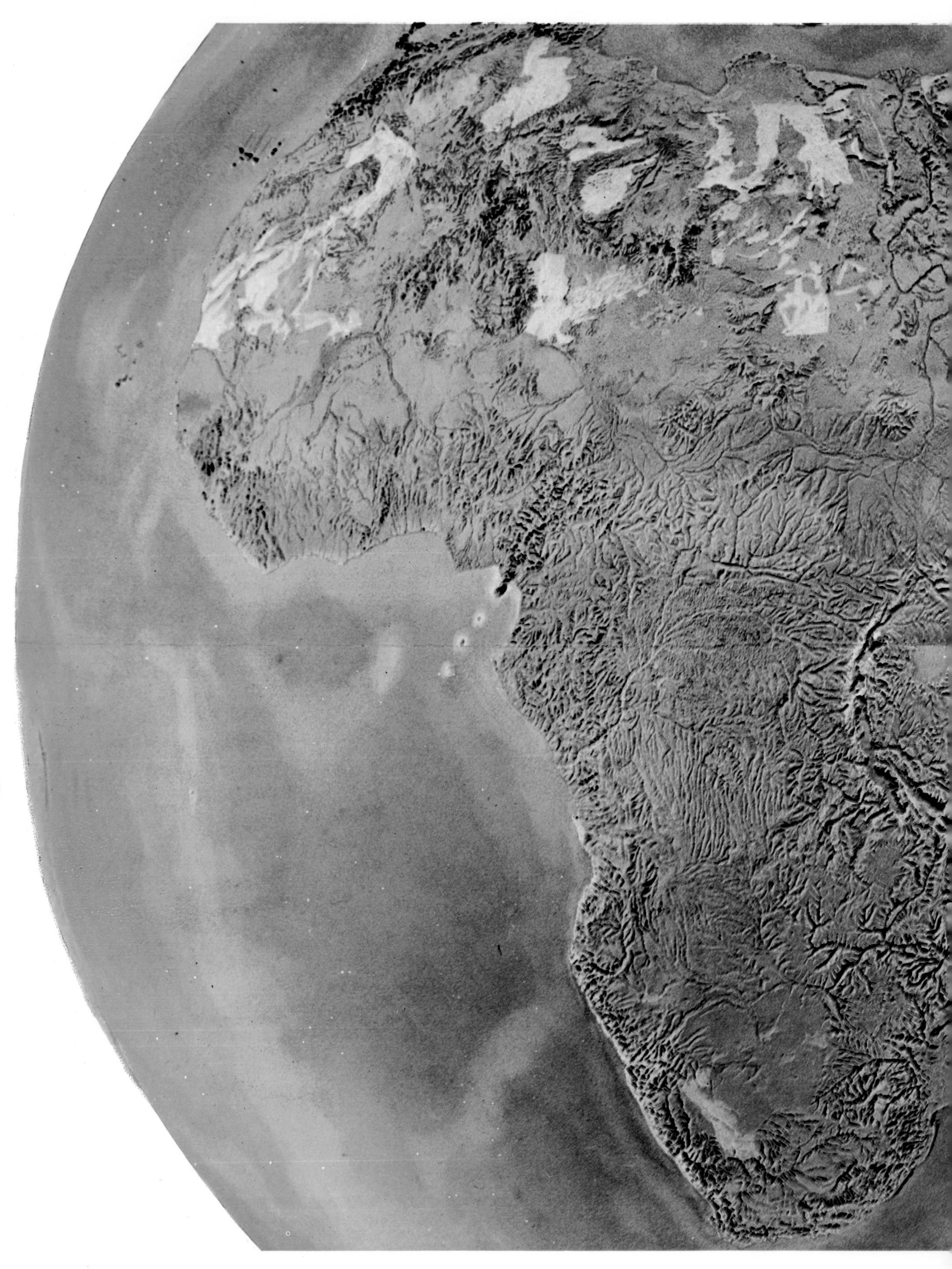

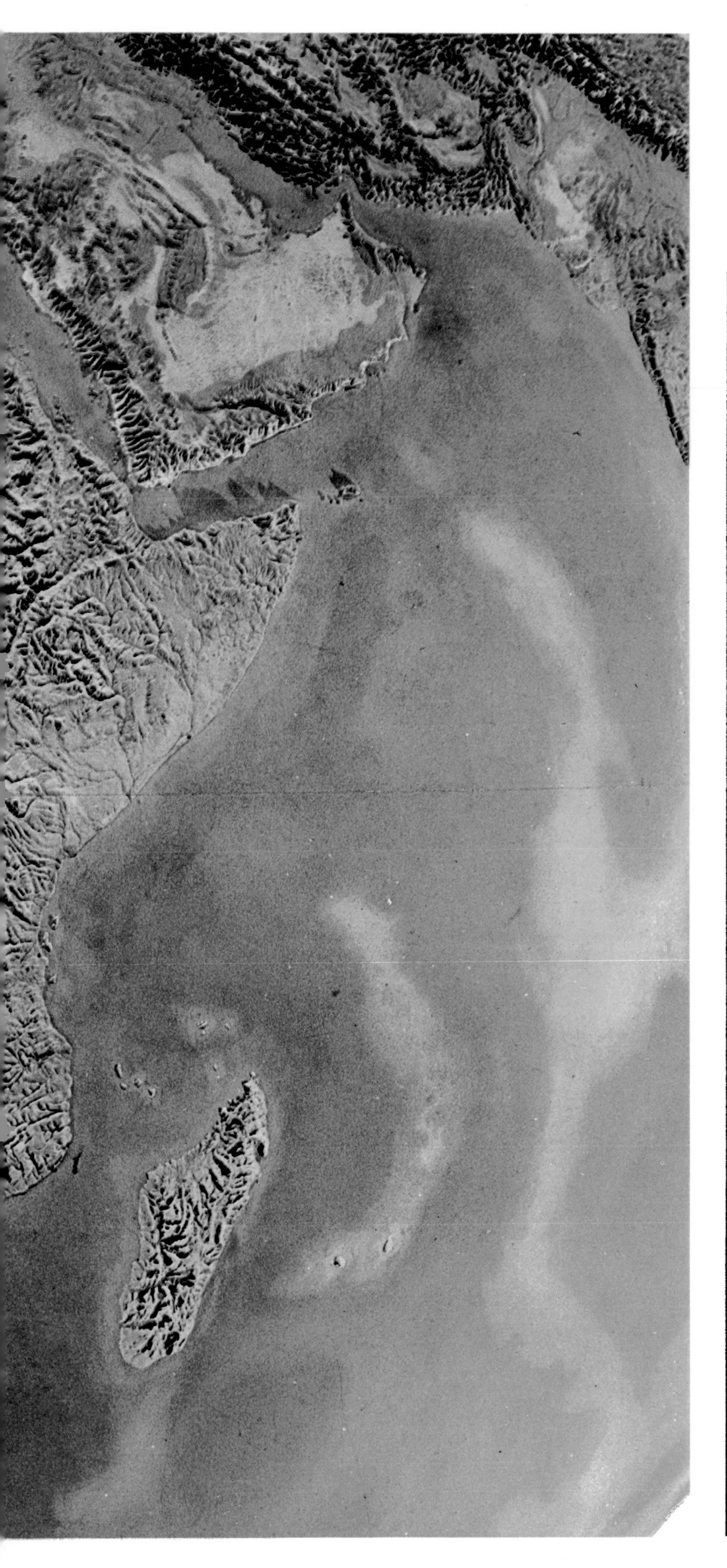

Africa, the second largest of the continents, is a land of great contrasts. It was the home of one of man's oldest civilizations, but today a few of its peoples have still scarcely progressed beyond the most primitive kind of existence.

Africa has burning deserts and luxurious tropical rain forests. In places it also has

COUNTRIES OF AFRICA
POPULATIONS AND MONETARY UNITS

Country	*Population*	*Money*
Algeria	19,590,000	Dinar
Angola	7,262,000	Kwanza
Benin	3,640,000	Franc
Botswana	937,000	Pula
Burundi	4,348,000	Franc
Cameroon	8,650,000	Franc
Cape Verde	329,000	Escudo
Central African Rep.	2,349,000	Franc
Chad	4,547,000	Franc
Comoros	370,000	Franc
Congo	1,578,000	Franc
Djibouti	323,000	Franc
Egypt	43,465,000	Pound
Equatorial Guinea	372,000	Ekuele
Ethiopia	32,158,000	Dollar
Gabon	555,000	Franc
Gambia	619,000	Dalasi
Ghana	12,063,000	Cedi
Guinea	5,147,000	Syli
Guinea-Bissau	810,000	Escudo
Ivory Coast	8,298,000	Franc
Kenya	17,148,000	Shilling
Lesotho	1,374,000	Rand
Liberia	2,034,000	Dollar
Libya	3,096,000	Dinar
Madagascar	8,955,000	Franc
Malawi	6,123,000	Kwacha
Mali	7,160,000	Franc
Mauritania	1,681,000	Ouguiya
Mauritius	971,000	Rupee
Morocco	20,646,00	Dirham
Mozambique	12,130,000	Metical
Niger	5,479,000	Franc
Nigeria	79,680,000	Naira
Rwanda	5,109,000	Franc
São Tomé and Principe	86,000	Dobra
Senegal	5,811,000	Franc
Seychelles	66,000	Rupee
Sierra Leone	3,571,000	Leone
Somalia	4,895,000	Shilling
South Africa	30,131,000	Rand
Sudan	18,901,000	Pound
Swaziland	566,00	Lilangeni
Tanzania	17,982,000	Shilling
Togo	2,705,000	Franc
Tunisia	6,513,000	Dinar
Uganda	13,225,000	Shilling
Upper Volta	7,094,000	Franc
Zaire	26,377,000	Zaire
Zambia	5,680,000	Kwacha
Zimbabwe	7,600,000	Dollar

Ice on the Equator. An elephant feeds on the tropical grasslands near Kilimanjaro, Africa's highest mountain. Although the mountain is only about 200 miles (320 km) from the equator, snow lies on its upper slopes. Kilimanjaro is an extinct volcano and has two peaks, Kibo and Mawenzi. Kibo, the higher, has a vast snow-filled crater, the source of a great glacier.

EARLY CIVILIZATIONS IN AFRICA

The civilization of ancient Egypt developed some 5,000 years ago and lasted for 2,500 years. Many great works of art from this period still exist. As well as artists, Egypt produced philosophers, scientists and mathematicians. Many of our ideas and inventions had their origin in this wonderful civilization on the banks of the Nile.

Africa S of the Sahara had its civilizations, too. One, in Nigeria, dated back to at least 3000 B.C. In the A.D. 800s, the Ghana Empire flourished. The Arabs called it 'the Land of Gold'. It was succeeded by the Mali or Mandingo Empire, whose cities were centres of trade and learning. The Songhai Empire, with its capital at Gao on the River Niger, reached the height of its power in the 1400s. Its rulers developed an extremely efficient system of administration.

some of the world's most equable climates. Its forests and plains are rich in animal life.

Politically the continent is divided into the Arab states of the N, and 'Black Africa' S of the Sahara.

Africa lies due S of Europe. The Mediterranean Sea forms its northern boundary. At the Strait of Gibraltar Africa is only 9 miles (14 km) from Europe. In the NE, the Isthmus of Suez joins Africa to SW Asia. The Suez Canal has been cut through the 72-mile (116-km) wide Isthmus to provide a passage for ships from the Indian Ocean and the Red Sea to the Mediterranean and the Atlantic Ocean.

The coastline of Africa is mostly smooth, with comparatively few bays and inlets. There are a few islands off the coast, the largest being Madagascar in the Indian Ocean. The Madeira, Canary, and Cape Verde islands are small groups off the NW coast.

The total area of Africa is about 11,700,000 sq miles (30,300,000 sq km). The greatest distance N to S is about 5,000 miles (8,000 km), and W to E 4,700 miles (7,600 km).

Since the end of World War II greater political changes have taken place in Africa than in any other continent. In 1945, all but four of Africa's countries were ruled as colonies of European countries. Now there are hardly any colonial territories left in Africa.

Physical Features and Climate

The equator crosses the middle of Africa, and most of the continent lies between the tropics of Cancer and Capricorn. The greater part of Africa is a vast plateau, lying more than 500 ft (150 m) above sea-level. The coastal plain, all round the continent, is narrow in most places, but broadens in the NE, along parts of the Indian Ocean coast, and in the NW.

The great continental plateau lies at several levels, with steep, step-like escarpments between them. On each 'step' there are wide tracts of land with few hills to break the surface.

Almost two-fifths of the continent is desert. The Sahara, the world's largest desert, stretches across the northern part of the continent, where Africa is at its widest, from the Red Sea almost to the Atlantic. It forms a barrier between the coastal lands along the Mediterranean Sea and the rest of Africa. As a result, the peoples N and S of the Sahara differ in their cultures and ways of life.

The Kalahari Desert occupies a large area of Botswana and South Africa. The Namib Desert lies along the SW coast.

In the eastern part of the continent is the Great Rift Valley, a natural fault in the Earth's crust which extends from Mozambique in the S up through the Red Sea and into Asia, where it continues as the Dead Sea and the valley of the River Jordan.

Africa's highest mountains also lie in the E. Mount Kilimanjaro, the highest peak, is almost on the equator. Its summit, 19,340 ft (5,895 m) above sea-level, is always snow covered. Some of the mountains in the Rift Valley region are volcanic, and a few of the volcanoes are still active.

Another mountain range is the Atlas Mountains, a long chain which cuts off the coastlands of Morocco, Algeria and Tunisia from the Sahara. The Drakensberg range of South Africa has several peaks of more than 10,000 ft (3,050 m) above sea-level. There are two smaller ranges of mountains in the Sahara: the Hoggar Mountains in southern Algeria, of which the highest peak is Tahat (9,557 ft; 2,913 m); and the Tibesti Massif in northern Chad, whose highest peak is Emi Koussi, an extinct volcano.

The region of the Rift Valley contains Africa's chief lakes. The largest, Lake Victoria, has an area of 26,828 sq miles (69,484 sq km); the only larger freshwater lake is Lake Superior in North America. Victoria lies between two arms of the Rift Valley. One of the other lakes in the region, Lake Tanganyika, is 420 miles (680 km) in length and nearly 5,000 ft (1,500 m) deep.

Africa has four of the world's mightiest rivers. The Nile, the world's longest river, flows 4,160 miles (6,695 km) northwards from the lakes of the Rift Valley to the Mediterranean Sea. Water from the Nile irrigates all the fertile land in Egypt and Sudan. The Aswan High Dam, near the border between the two countries, helps to

Left: Giraffes, the world's tallest animals, in a national park in Kenya. Africa has a richer variety of animal life than any other continent, but there are only one-twentieth as many animals today as there were fifty years ago.

Below: At the Victoria Falls, the waters of the Zambezi River drop into a narrow gorge, and surge through a cleft called 'the boiling pot'. The Falls, about 1 mile (1.6 km) wide, were reached by the explorer Livingstone in 1855, and named in honour of Queen Victoria.

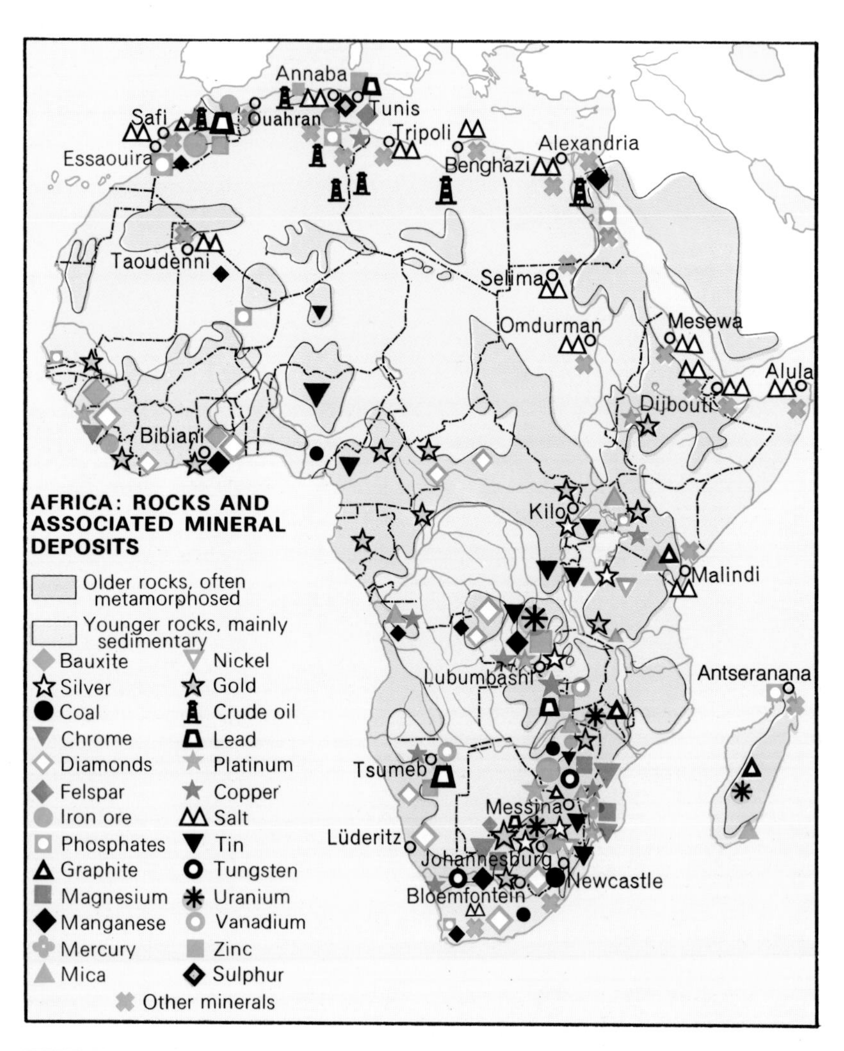

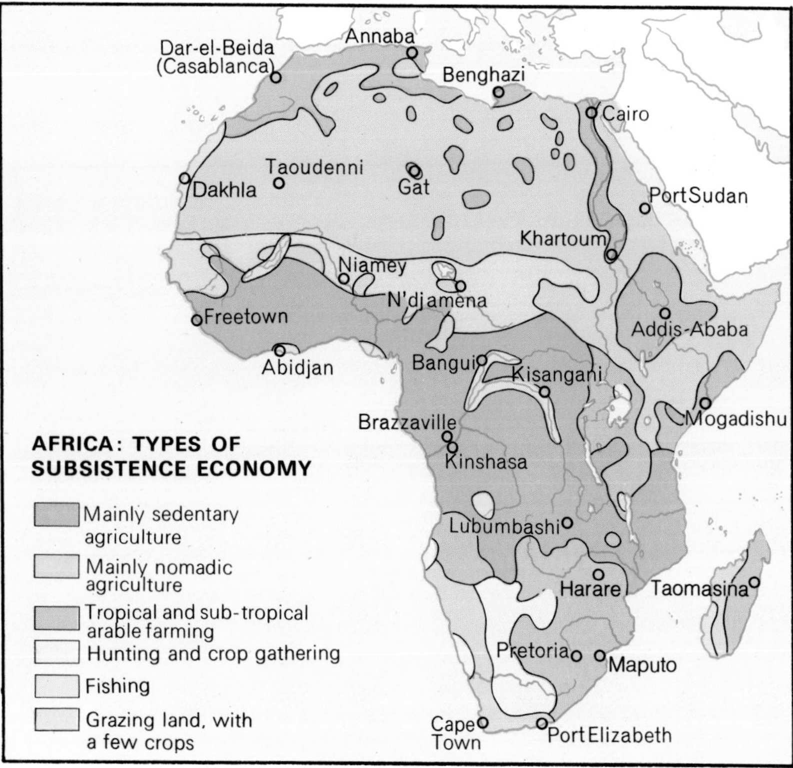

control the river's flow, and has created a huge artificial lake, Lake Nasser.

The Niger is the main river of western Africa. It rises in the hills close to the eastern boundary of Sierra Leone and flows E and N through Mali before swinging through a right-angle and coming out in a large delta to the Atlantic Ocean in the Gulf of Guinea. Its length is 2,600 miles (4,180 km). It is gradually changing course: Timbuktu, which was built on its banks 900 years ago, now lies 7 miles (11 km) from it.

The Zaire, often still called the Congo, rises in SE Zaire, and flows N and W in a wide curve to the Atlantic. At its source it is called the Lualaba. With its tributaries it drains an area almost one-eighth of the entire continent. The Zambezi rises in NW Zambia, and flows over an S-shaped course for 1,600 miles (2,575 km) to the Indian Ocean. Its many waterfalls include the Victoria Falls, which are among the world's largest.

Many parts of Africa, particularly central Africa, are thickly forested. Some forest regions have not yet been thoroughly explored.

Africa is, in general, a hot continent; the central regions are hot all the year round. The Zaire River basin has a tropical climate with heavy rainfall, but the rest of central Africa has moderate rainfall. The Sahara has little rain, and is hot nearly all the time, though some areas occasionally have frost at night. Southern Africa, outside the desert areas, has moderate or light rainfall. Most of southern Africa is covered by the *veld*, grasslands which in places are very thick and lush.

Because temperatures decrease by about 1°C for every 100 m (330 ft) of height above sea-level, many areas of Africa, even areas close to the equator, have moderate temperatures. For example, Mombasa on the coast of Kenya is hot and humid all the year round, but Nairobi, the capital, which is about 5,000 ft (1,675 m) above sea-level, is seldom more than pleasantly warm.

Resources and Industry

Africa is rich in mineral wealth, but so far only a small fraction of it has been found and exploited. The countries along the N African coast have large deposits of petroleum and some natural gas. Most of the world's diamonds come from the countries of southern Africa and industrial diamonds—those too small for use as jewellery—are found in several countries in the western part of Africa.

Gold is one of the assets of South Africa, where there are also deposits of other minerals including coal, chromium, manganese and uranium. Zimbabwe has some very important copper mines, and so has Zaire. Uranium and other valuable metals are found in South Africa and Zaire. There are important deposits of iron in Liberia and South Africa. But even with all the known mineral wealth, there are vast areas of Africa, including most of the Sahara and the forests of central Africa, that have never yet been surveyed by prospectors.

Many African countries are striving to increase their industrial output. Some have planned and built hydro-electric installations to provide power. Among these schemes are the Aswan High Dam in Egypt; the Kariba Dam on the Zambezi, between Zimbabwe and Zambia; and the Volta Dam on the River Volta in Ghana.

Communications

Communication is a major problem in Africa. Broad rivers, dry, wind-swept deserts, and thick forests make travelling difficult over large areas of the continent. Some of the rivers can be used for transport, but not over their whole length because of rapids and waterfalls. These natural barriers have slowed up development. A large part of the continent has been explored only in the past hundred years, and then with difficulty. The first E–W crossing was made in 1877 by Sir Henry Morton Stanley. It took him nearly three years.

Railways have gradually been built across Africa, but only South Africa, Morocco, Algeria and Tunisia have anything like good rail networks. Northern Egypt and the Nile Valley also have an efficient rail system. But many countries have only one or two rail lines, and some have none at all. Away from the cities there are few good roads. Most roads are dirt tracks, dusty in summer and deep in mud in winter, while floods frequently wash away large stretches. Fortunately, the development of air transport is beginning to open up the more remote areas.

Agriculture, Fisheries and Land Use

The main work of Africa's people is farming of one kind or another. In the more remote areas, the people are subsistence farmers, growing just enough food for their own needs, with perhaps a little over to barter for other goods.

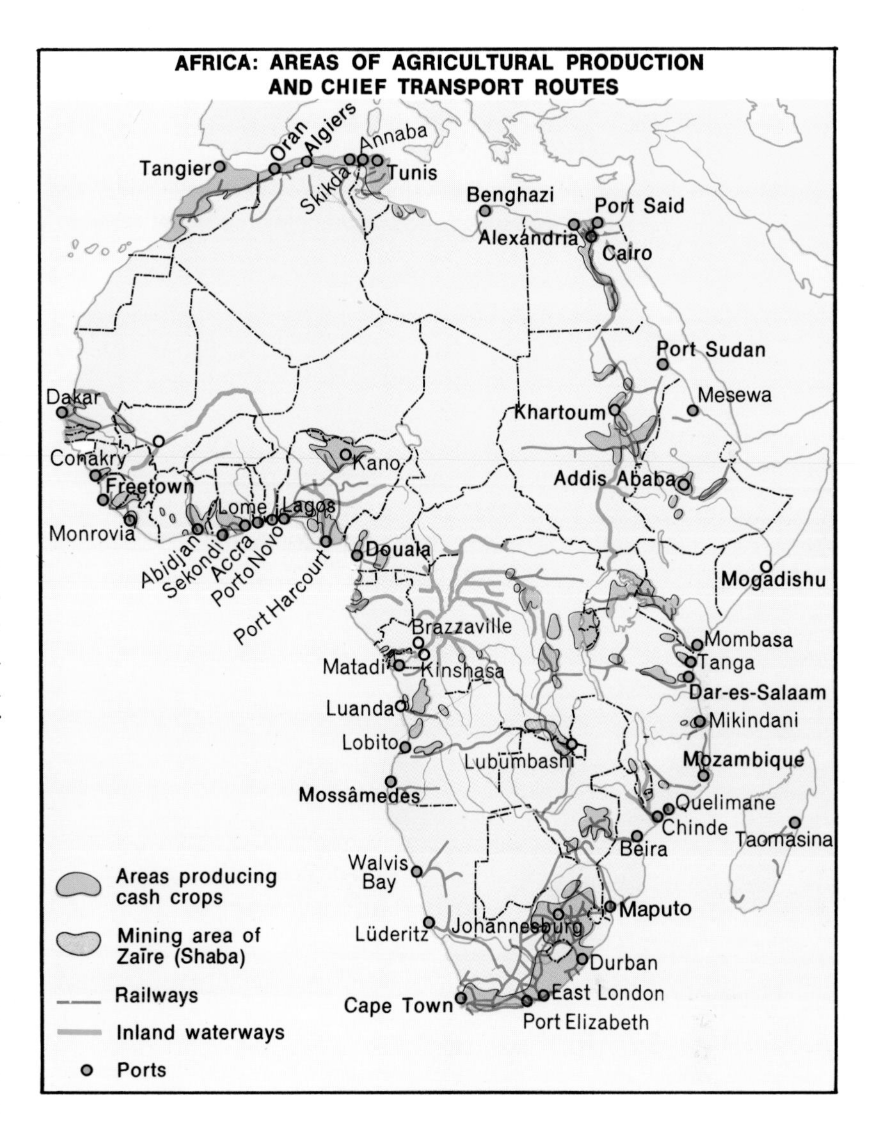

Subsistence Farming is the rule mainly in tropical Africa. Shifting cultivation is the most usual form. Tribesmen clear a part of the forest by felling trees and burning brushwood. They cultivate the land for several years until the soil is exhausted, then they move off to a fresh site and start all over again. The land has to be left for about 10 years to recover its fertility. In places where there are too many people to allow migration, farmers practise crop rotation.

Cultivation is carried out with hoes. In many regions farmers cannot keep horses or cattle because of diseases carried by the tsetse fly, so ploughing is out of the question. In any case, ploughing is not suitable in some places because the broken soil washes away in the heavy rains. Crops grown by subsistence farmers include yams, maize, rice, cassava, beans, and ground-nuts.

Tree Crops are important in the tropical forests of the Zaire River basin, on the equator. The trees cultivated include the oil palm, which yields vegetable oil; the cacao tree, which produces cocoa beans; the banana palm; the rubber tree; and the coffee tree.

Savanna Crops are those grown on the savannas or grasslands—including the veld of South Africa—that cover two-fifths of the continent. Farmers in savanna regions grow all the same crops as those who practise shifting cultivation, plus cotton, millet, and tobacco. In many areas the only cereal that will grow is maize, which forms the main food of the people there. Ground-nuts are produced as a *cash crop*—that is, one purely for sale.

Oasis Crops are grown in the oases of the Sahara. The main crop is dates from the date palm. The sap from this tree is used to make palm wine, while its leaves provide material for matting and roofing. Cotton is another important oasis crop. In addition, the people grow fruit and vegetables for their own use.

Mediterranean and Cape Crops depend on the mild winters and warm summers of these two coastal regions. They include grapes for wine-making (particularly in South Africa and Algeria), wheat and barley. Oranges, lemons and grapefruit are also grown.

River Valley Crops depend on the annual flooding of various rivers, particularly the Nile, for irrigation water. Such crops include fruit, vegetables and cereals and in the Nile valley cotton. Many flooded areas are ideal for growing rice.

Livestock cannot be reared in the areas of tropical Africa where the tsetse fly is prevalent. But livestock is an important part of farming in other regions. Sheep are reared in the Mediterranean lands and in South Africa, and cattle are grazed in the lands immediately S of the Sahara, and over large areas of southern Africa. In the Sahara, the *nomadic* (wandering) Bedouin tribes keep camels, for milk as well as for use as beasts of burden.

Fishing is carried on off the coasts of southern and south-western Africa, in the Atlantic to the NW, in the Mediterranean and in lakes and rivers.

A *casbah* or bazaar in a Moroccan city, typical of bazaars in the Arab cities of northern Africa. Jewellers and other craftsmen sell the objects they have fashioned, and traders offer fruit, vegetables, cooked dishes, cloth, leatherwork, and baskets for sale.

People and Ways of Life

The oldest known skeletons of prehistoric man have been found in Kenya and Ethiopia and may be as old as 3 million years. But scientists do not know whether Africa was the home of the first men.

Of the present-day Africans, the oldest inhabitants are the Bushmen of the Kalahari Desert, and the Pygmy Negrillos of the Zaire River basin forests. All these peoples are shorter than average.

Generally speaking, the Sahara divides the peoples of Africa into two distinct types. The Negro peoples live S of the Sahara, and this part of the continent is often referred to as 'Black Africa'. N of the Sahara are the Arabs and other related peoples.

The Negroes fall into two main groups. The Sudanese, also called the W African Negroes, live in the region of western and central Africa that lies just S of the Sahara. They include the Yoruba and Ibo of Nigeria, the Mandinka of Guinea and the Wolof of Senegal. Some of their tribes are very small.

In the northern part of this region are a number of tribes that are a mixture of Negro peoples with peoples from N of the Sahara. This intermixture has resulted from centuries of trading across the desert.

The other main group of Negroes are the Bantus of the southern half of the continent. The Bantu Negroes all speak languages that are of the same general type. They, too, are thought to be the result of intermarriage between pure-blooded Negroes and peoples from N of the Sahara. The Bantu peoples moved into the areas they now occupy during the past 2,000 years. This colonization process was still taking place 200 years ago.

The Caucasian Peoples are those living mainly N of the Sahara. They are basically of the same stock as the peoples of India and Europe. The two main groups are the Hamites and the Semites. They are both swarthy in complexion, with black hair.

The Hamites include the fellaheen of Egypt, and the nomadic Tuareg and Fulani tribes of the Sahara. The Fulani of today are found in Niger and northern Nigeria. In the NW are the Berbers, a much lighter-skinned group similar to the people of Spain and Italy in appearance. Morocco, Algeria and Tunisia are often called the *Barbary States* after the Berbers. Most of the Hamites are what are loosely called 'Arabs' today.

The Semites, also called 'Arabs', live in the lands E of the Barbary States, and in the lands in the eastern 'horn' of Africa. Like the Hamites they are Muslims.

Polynesian Peoples live in the island of Madagascar. They apparently arrived from south-eastern Asia, and speak a language similar to that spoken by the Malays and by the Polynesians of the Pacific islands.

Later Settlers include the white people from Europe who colonized Africa in the 1800s. They are mostly found in South Africa and Zimbabwe. Many of those in other countries, such as Kenya and Uganda, have left or been ejected since these countries became independent. There are also a number of settlers from India and China. But many have left or have been forced to leave in the recent past.

Top: A beautiful arched gateway in the ancient walls of the city of Fès in Morocco.

Centre: The Great Sphinx, near Gizeh in Egypt, has looked out across the Nile valley for nearly 5,000 years, seeing several civilizations come and go. It has the head of a pharaoh—one of the kings of ancient Egypt—and a lion's body. Near it is the Great Pyramid.

Below: Roman Remains. The ruins of the basilica at Apollonia (Marsa Susa) in Libya, dating from the A.D. 400s. Hundreds of such ruined buildings in northern Africa testify to the years of Greek and Roman rule.

THE COUNTRIES OF AFRICA

Algeria

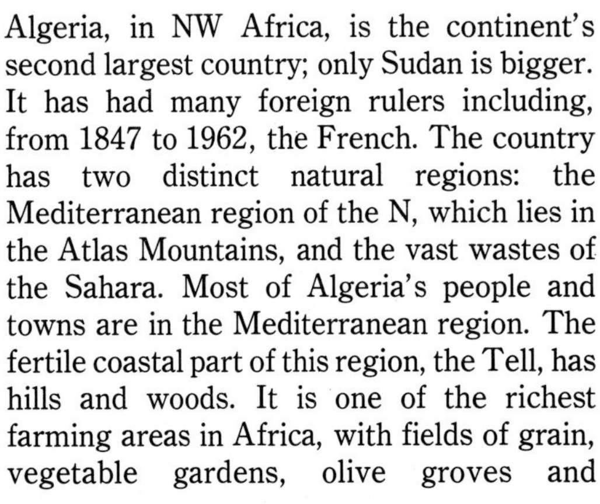

Algeria, in NW Africa, is the continent's second largest country; only Sudan is bigger. It has had many foreign rulers including, from 1847 to 1962, the French. The country has two distinct natural regions: the Mediterranean region of the N, which lies in the Atlas Mountains, and the vast wastes of the Sahara. Most of Algeria's people and towns are in the Mediterranean region. The fertile coastal part of this region, the Tell, has hills and woods. It is one of the richest farming areas in Africa, with fields of grain, vegetable gardens, olive groves and vineyards. To the S of the Tell are high plateau grasslands, bounded on their southern side by the peaks of the Saharan Atlas. Beyond, to the S, lies the Sahara, occupying nine-tenths of the country. The desert has almost no surface water except in the oases. There, villagers grow maize and dates. Elsewhere in the desert, the sparse population consists of nomadic herdsmen. In the S of the country are the spectacular Hoggar Mountains. Northern Algeria has a Mediterranean climate; on the coast, winters are mild, but in the mountains they are cold. The desert is extremely hot, and in places has no rain for many years. Algeria is rich in petroleum and natural gas. It also has iron and phosphates. There are chemical, steel and food-processing industries. But the majority of the people live by farming. The chief towns are the capital Algiers (El Djezair), Oran (Ouahran), Constantine, Annaba and Blida. Area 918,000 sq miles (2,378,000 sq km).

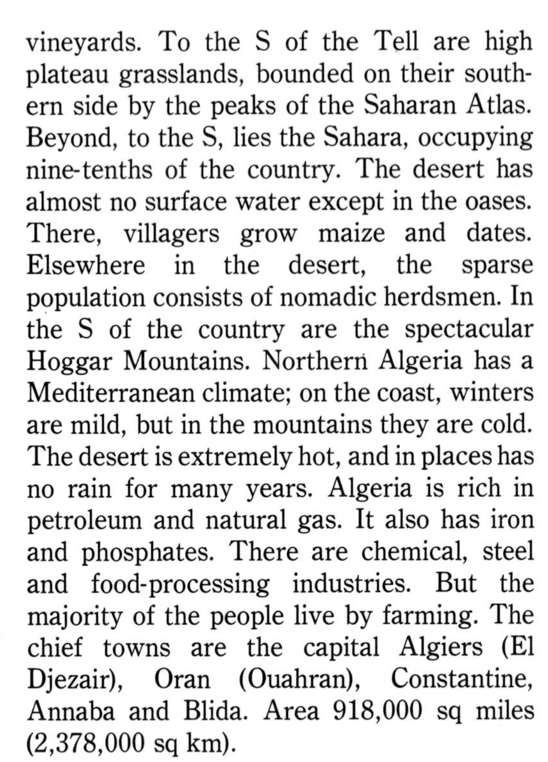

The rich, productive land of parts of northern Algeria contrasts with the barren desert wastes that make up most of the country. In the fertile areas near the Mediterranean coast, vineyards lie beside olive groves, broad fields of cereals and plantations of fruit and vegetables. One of the farmers' most difficult problems is soil erosion. Government plans to deal with it include the planting of more tree crops and bush crops.

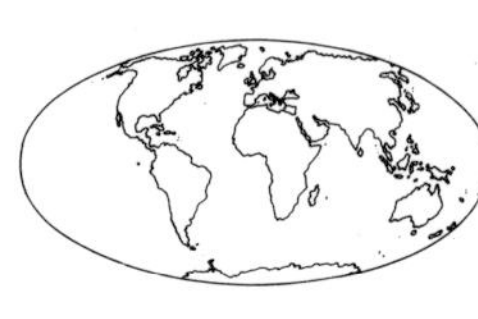

Angola

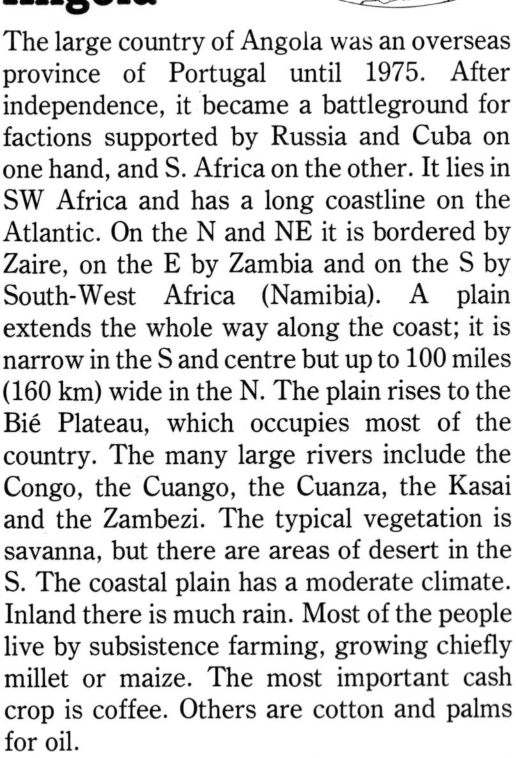

The large country of Angola was an overseas province of Portugal until 1975. After independence, it became a battleground for factions supported by Russia and Cuba on one hand, and S. Africa on the other. It lies in SW Africa and has a long coastline on the Atlantic. On the N and NE it is bordered by Zaire, on the E by Zambia and on the S by South-West Africa (Namibia). A plain extends the whole way along the coast; it is narrow in the S and centre but up to 100 miles (160 km) wide in the N. The plain rises to the Bié Plateau, which occupies most of the country. The many large rivers include the Congo, the Cuango, the Cuanza, the Kasai and the Zambezi. The typical vegetation is savanna, but there are areas of desert in the S. The coastal plain has a moderate climate. Inland there is much rain. Most of the people live by subsistence farming, growing chiefly millet or maize. The most important cash crop is coffee. Others are cotton and palms for oil.

There are some mineral resources, and timber is exported. The chief towns are Luanda (the cap.) and Lobito. Area 481,350 sq miles (1,246,700 sq km).

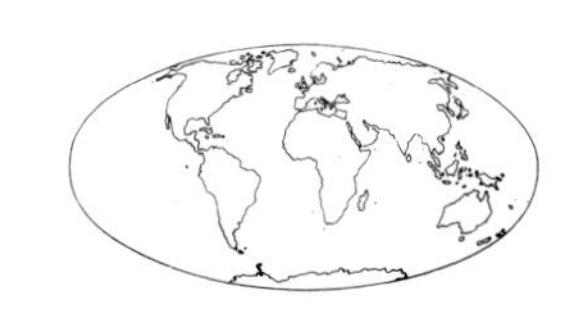

Benin

Small country of W Africa with a short sandy coast in the S on the Gulf of Guinea. The climate of the S is hot and wet. Northwards, the rainfall decreases. Benin has some mineral deposits, but its principal source of revenue is palm oil from the palm plantations of the S, where most of the people live. Before the country became independent in 1958, it was part of French West Africa. Area 44,696 sq miles (115,762 sq km).

The Bushmen of Botswana, most of whom live in the Kalahari Desert, are among the oldest of the peoples of Africa. Once, they were spread all over the southern part of the continent, but gradually they were driven into regions where few other people could survive. These tiny people, less than 5 ft (1.5 m) tall, eat roots and such animals as they can kill with their bows and arrows. Their way of life has remained unchanged for hundreds of years.

Botswana

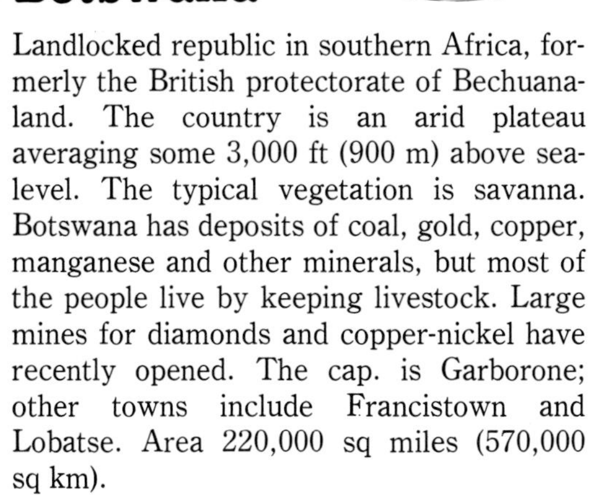

Landlocked republic in southern Africa, formerly the British protectorate of Bechuanaland. The country is an arid plateau averaging some 3,000 ft (900 m) above sea-level. The typical vegetation is savanna. Botswana has deposits of coal, gold, copper, manganese and other minerals, but most of the people live by keeping livestock. Large mines for diamonds and copper-nickel have recently opened. The cap. is Garborone; other towns include Francistown and Lobatse. Area 220,000 sq miles (570,000 sq km).

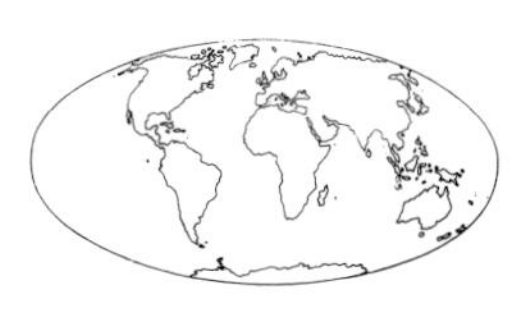

Burundi

County of E-central Africa, formerly part of the U.N. trust territory of Ruanda-Urundi. It is small but densely populated. In the western half of the country, a range of mountains extending N to S marks the edge of the Great African Rift Valley (in which Lake Tanganyika lies). The rest of the country consists of broken plateaux, with many rivers in steep-sided valleys. The higher regions are warm and often extremely wet. At lower levels there are higher temperatures, but less rain. Most of the people live by subsistence farming, growing rice, cassava, barley and wheat. The main cash crops are coffee and cotton. Area 10,747 sq miles (27,835 sq km).

Top: A busy Cameroon market place, with traders buying and selling vegetables, fruit and other local produce.

Centre: Most Cameroon villagers spend the whole of the daytime in the open air. There, the daily chores are carried out, food is cooked and families meet to eat. At night, they find shelter in huts of dried mud. Sometimes, a number of huts, the property of one family, are grouped together behind a wall.

Bottom: Village gathering house at Foumbot, in the west of Cameroon. The building has a raised floor, walls constructed of poles and a thatched roof.

Cameroon

Country of western Africa, on the Gulf of Guinea, formed by the union of two former trust territories, one French and one British. It is bordered on the E by Chad and the Central African Republic, on the S by Congo, Gabon and Equatorial Guinea, and on the NW by Nigeria.

The narrow coastal plain is swampy. Inland it rises to a plateau that occupies the greater part of the country. The southern region of the plateau is covered by thick rain forest. The central region, called the Adamawa Highlands, is forested, too. In the N there is savanna. The most northerly point of the country lies in the Lake Chad basin. Mountains rise along the western border: the Mandara Mountains in the NW, the Ladamaoua Mountains in the centre, and the Bambuto Mountains in the SW. The highest peak in the country is Mount Cameroon (13,353 ft; 4,070 m), near the coast. The coast is hot and humid. The plateau is cooler and drier.

Cameroon has a few mineral resources and a little industry including an aluminium smelter. But the bulk of the people live by farming. For most it is subsistence farming. The most important cash crops are cocoa and coffee.

Other exports include cotton, ground-nuts, rubber, palm oil, timber and bananas. Some aluminium is exported. The chief towns are Yaoundé (the cap.) and Douala. Area 183,570 sq miles (475,444 sq km).

Cape Verde

Archipelago in the Atlantic, off the W coast of Africa. It consists of two island groups, the Windwards (Santo Antão, São Vicente, Santa Luzia, São Nicolau, Boa Vista and Sal) and Leewards (Maio, São Tiago, Fogo and Brava). They were colonized by Portugal in about 1460 and became independent in 1975. A proposed federation with Guinea-Bissau was dropped in 1980 following a coup in that country. The capital is Praia. Area 1,516 sq miles (3,926 sq km).

Central African Republic

Country lying just N of the equator, formerly part of French Equatorial Africa. The northern part of the country is savanna, the southern part forested. River transport is important because lack of modern communications is one of the country's chief economic problems. All of the country is hot. The S has heavy seasonal rains. Most of the people live by subsistence farming. Exports include diamonds, gold, cotton and coffee. The chief town is Bangui (the cap.). Area 234,000 sq miles (606,000 sq km).

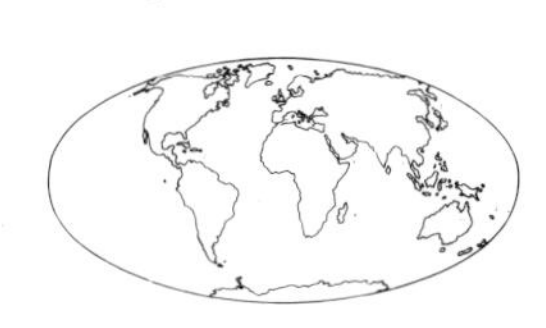

Chad

Country in N-central Africa, formerly a French colony. It is named after the large, marshy Lake Chad on its western boundary. Chad has no coastline. On the N it is bordered by Libya, on the E by the Sudan, on the S by the Central African Republic, and on the W by Cameroon, Nigeria and Niger. The northern part of the country is in the Sahara. The Tibesti Mountains in this region rise to 11,204 ft (3,415 m) in Emi Koussi. To the S of the mountains is a vast low-lying area called the Bodélé Depression. In the E are the high Ennedi Plateau and the Quaddai Massif. The typical vegetation in the southern part of the country is savanna. All of Chad is hot. The S has moderate rainfall. The northern desert region is sparsely populated, and most of the Arab and Hamitic peoples there are nomadic herdsmen. Many of the Negro peoples of the S live by subsistence farming, but important cash crops are grown too. The chief are cotton, ground-nuts and sugar. Area 495,750 sq miles (1,284,000 sq km).

Comoros

Archipelago in the Indian Ocean, between the N tip of Madagascar and the E coast of Africa. Its larger islands are mountainous. Area 838 sq miles (2,170 sq km).

A gathering of tribal chiefs and other notabilities at N'djamena, in Chad.

Congo

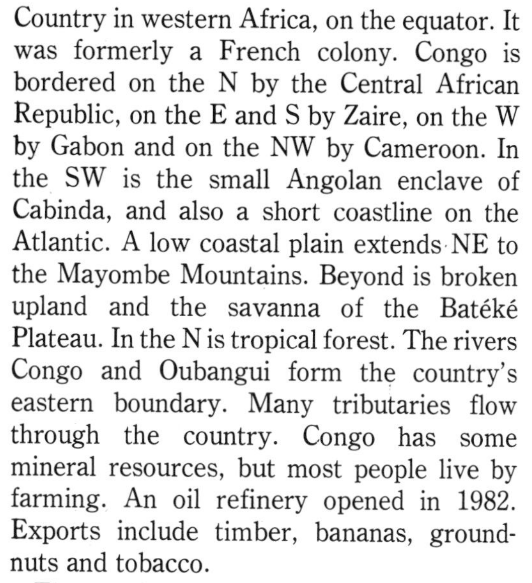

Country in western Africa, on the equator. It was formerly a French colony. Congo is bordered on the N by the Central African Republic, on the E and S by Zaire, on the W by Gabon and on the NW by Cameroon. In the SW is the small Angolan enclave of Cabinda, and also a short coastline on the Atlantic. A low coastal plain extends NE to the Mayombe Mountains. Beyond is broken upland and the savanna of the Batéké Plateau. In the N is tropical forest. The rivers Congo and Oubangui form the country's eastern boundary. Many tributaries flow through the country. Congo has some mineral resources, but most people live by farming. An oil refinery opened in 1982. Exports include timber, bananas, groundnuts and tobacco.

The chief towns are Brazzaville (the cap.) and Pointe Noire. Area 129,000 sq miles (334,000 sq km).

Right: The golden mask that covered the face of the mummy of Tutankhamun, the boy-king who reigned over Egypt in about 1350 B.C. It was among the treasures discovered when his tomb in the Valley of the Tombs of the Kings was opened in 1922. Tutankhamun succeeded to the throne when he was eight or nine years old, but died less than ten years later.

Below: A fishing village in Dahomey, on the Gulf of Guinea. Several similar villages, built on stilts, lie in the lagoons behind tongues of land on Benin's coast.

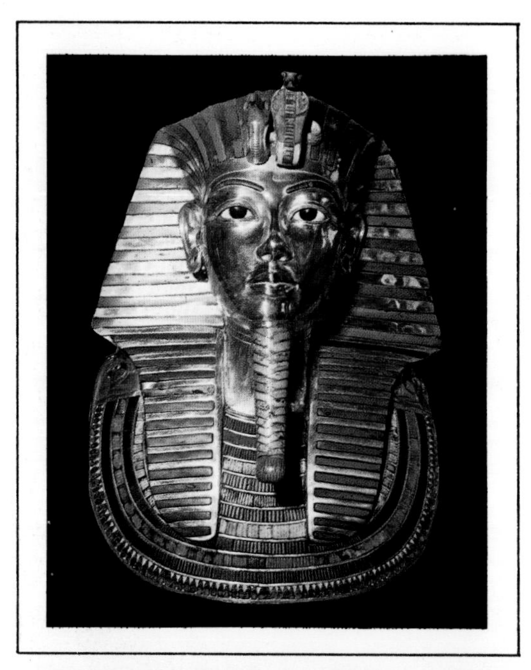

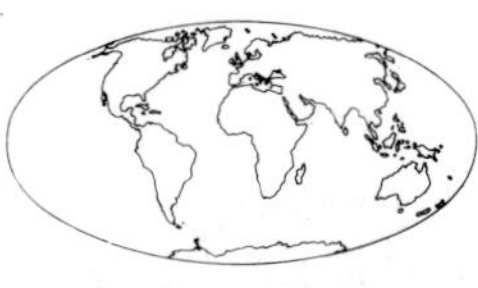

Djibouti

Small country of Eastern Africa, lying between Ethiopia and Somalia. It has a coastline on the Gulf of Aden, at the entrance to the Red Sea. Until 1967 it was called French Somaliland; it gained complete independence in 1977. Most of the country is stony desert. Crops grow only where irrigation is possible, but large numbers of sheep, goats and camels are tended by nomads. Djibouti, the cap., is a busy port, much of its trade being with Ethiopia. Area 9,000 sq miles (23,300 sq km).

Egypt, Arab Republic of

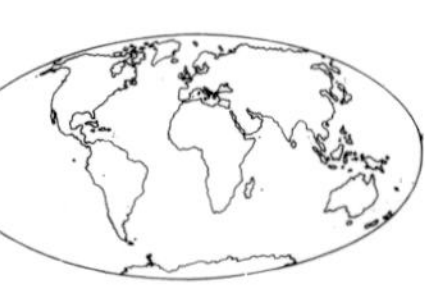

The Arab Republic of Egypt, the leading Arab country, is the home of one of the world's oldest civilizations. The signing of a peace treaty between Egypt and Israel has brought the Middle East closer to political stability. The Suez Canal cuts through Egypt's territory, joining the Mediterranean with the Red Sea and the Indian Ocean, and providing a valuable trade route for shipping between Europe and Asia. Desert covers most of the country. The Nile is the longest river in the world. The southern part of the valley has been turned into a great lake, Lake Nasser, by the building of the Aswan High Dam for irrigation and the production of hydro-electric power. Just below Cairo, the Nile enters its delta, where it forms several branches. The delta, a triangular plain, is built up of silt carried by the river. From Cairo to the sea it measures some 100 miles (160 km), and at its widest extent on the coast some 150 miles (240 km). Most of Egypt's people live in the delta and the river valley. The Nile is bounded by desert on both sides. The Western Desert is part of the Libyan Desert (in turn, part of the Sahara). Its southern end is extremely bare and windswept. In the N is the vast Qattara Depression, which sinks to 440 ft (134 m) below sea-level. The Eastern Desert or Arabian Desert, between the river and the Red Sea, rises to the Red Sea Mountains along the coast. The triangular Sinai Peninsula lies on the Asian side of the Gulf of Suez. It is the highest part of the country: the Gebel Katherina in the S rises to 8,651 ft (2,636 m). Egypt is a hot and dry country, but in Mediterranean areas the weather is often moderate and pleasant. The country has considerable mineral resources, including petroleum, phosphates and iron. The government is working to expand industry. Manufactures include textiles, chemicals, cement and processed foods. But Egypt is predominantly agricultural. The main cash crops are cotton, rice, citrus fruits, onions, sugar-cane and potatoes.

The chief cities are Cairo (the cap.), Alexandria, Ismailia, Port Said, Asyut, El Faiyum and Suez. Area 386,000 sq miles (999,700 sq km).

Equatorial Guinea

Small country in western Africa, formerly a Spanish possession. It consists of the mainland territory of Rio Muni (lying between Cameroon and Gabon), the offshore Corisco Islands, and the mountainous island of Bioko (formerly Fernando Póo) in the Gulf of Guinea. Exports include timber, coffee and cocoa. The chief towns are Malabo (Santa Isabel), on Bioko, and Bata, in Rio Muni. Area 11,000 sq miles (28,000 sq km).

Ethiopia

Ethiopia, formerly *Abyssinia*, is one of the oldest independent countries in the world. Until 1974 it was ruled by an emperor, but in that year he was deposed. In 1977 Somalia invaded Ethiopia (in support of the claim to the Ogaden region) but was repulsed. There is an armed independence movement in Eritrea. The country, which is in north-eastern Africa, had no coastline until 1952, when the former Italian colony of Eritrea, on the shore of the Red Sea, became part of it. It is a country of mountainous plateaux, separated into two sections by the Great African Rift Valley. Tana is the source of the Blue Nile. The low-lying regions in the N and SE are mostly desert. Ethiopia has some mineral resources, such as salt, potash, manganese and gold. Nearly all the people live by farming, often at a precarious subsistence level and in constant danger of famine. The main cash crop is coffee. Other crops include cotton, sugar-cane, wheat, maize, barley, corn, tobacco and fruit. In most inhabited areas, there are cattle, sheep and goats. Area 450,000 sq miles (1,165,000 sq km).

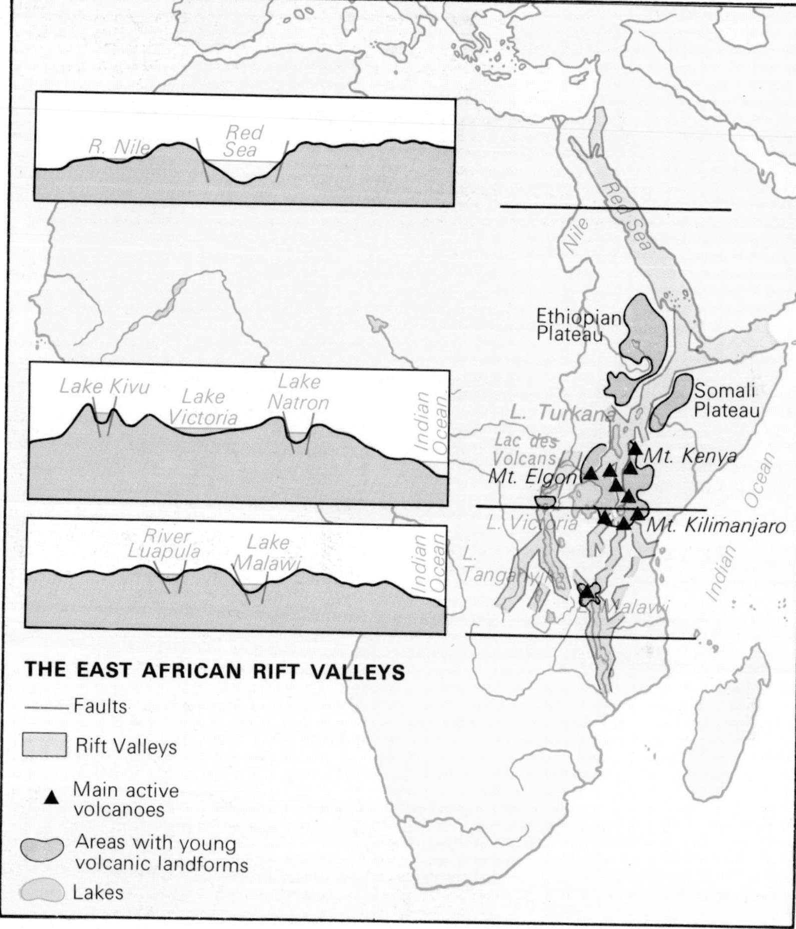

Below: The Great Pyramid and the Sphinx near Gizeh, in the Nile valley of Egypt. The Great Pyramid, about 450 ft (140 m) high, was built by the Pharaoh Khufu during the 2600s B.C. It was one of the wonders of the ancient world.

Priests of the Coptic Church in Addis Ababa, in Ethiopia. Christianity has strongly influenced life in Ethiopia since the A.D. 300s, when the kings of Axum, in the north-east of present-day Ethiopia, were converted by Christian missionaries. Today, although Christians are in the majority, there is also a large Muslim minority and there are smaller minorities of other religions.

Gabon

Country in western Africa, formerly a French possession, lying on the equator. In the W it has an Atlantic coastline. On the N it is bordered by Equatorial Guinea and Cameroon and on the E and S by Congo. The coastal areas are low-lying and there is a broad plain at the estuary of the River Ogooué, which crosses the country NE to W. Inland, and occupying most of the country, is a hilly plateau. Much of Gabon is covered by tropical forests. All regions have an equatorial climate and are hot and wet. Gabon has valuable mineral resources, particularly manganese, iron, uranium and oil. Its economic growth in the last decade has been very fast as a result of increased oil production, in which it ranks fourth in Africa. Its forests provide timber for export, the most important being mahogany, and okoumé, which is used for making plywood. Cash crops include coffee, cocoa, palm kernels and ground-nuts, but most people live by subsistence farming. There are few all-year roads. The chief towns are Libreville (the cap.), Port Gentil and Lambaréné. Area 103,100 sq miles (267,000 sq km).

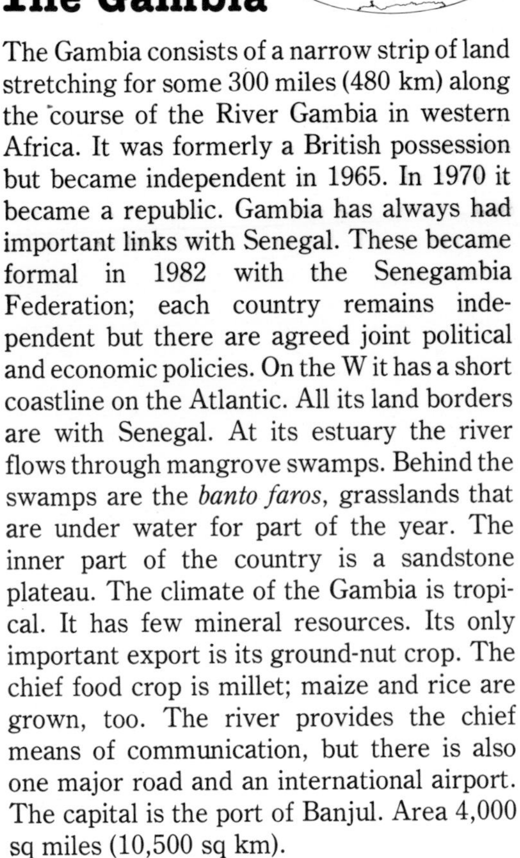

The Gambia

The Gambia consists of a narrow strip of land stretching for some 300 miles (480 km) along the course of the River Gambia in western Africa. It was formerly a British possession but became independent in 1965. In 1970 it became a republic. Gambia has always had important links with Senegal. These became formal in 1982 with the Senegambia Federation; each country remains independent but there are agreed joint political and economic policies. On the W it has a short coastline on the Atlantic. All its land borders are with Senegal. At its estuary the river flows through mangrove swamps. Behind the swamps are the *banto faros*, grasslands that are under water for part of the year. The inner part of the country is a sandstone plateau. The climate of the Gambia is tropical. It has few mineral resources. Its only important export is its ground-nut crop. The chief food crop is millet; maize and rice are grown, too. The river provides the chief means of communication, but there is also one major road and an international airport. The capital is the port of Banjul. Area 4,000 sq miles (10,500 sq km).

Ghana

Ghana, on the Gulf of Guinea in western Africa, was formerly the British colony of the Gold Coast. Most of Ghana is low-lying. In the S there is a broad coastal plain, and almost the whole of the eastern part of the country lies in the basin of the Volta River system. The White Volta flows southwards through the centre of the country. The Black Volta forms part of the western boundary before turning eastwards to join the White Volta. By the construction of a dam across the river at Akosombo, some 70 miles (180 km) from its estuary, a vast lake has been created, several hundred miles in length and covering an area of 3,270 sq miles (8,470 sq km). This great sheet of water, the largest artificial lake in the world, is used for irrigation and the production of hydro-electricity. The climate is hot in northern Ghana and warm in the centre. The N and E often suffer from long dry spells. The coastal plain is hot and wet.

Ghana has some mineral resources. Gold, which gave the country its former name, is still mined in sizeable quantities. Other minerals include diamonds, manganese, bauxite and graphite. Some of the bauxite is used to feed the important aluminium-smelting industry that has been developed using power from the Volta Dam installations. The bulk of the people live by agriculture. The chief food crops are rice, maize, cassava and yams. Cash crops include cocoa—the country's main export—coffee, copra and palm kernels. The chief towns are Accra (the cap.), Kumasi, Tamale and Sekondi-Takoradi. Area 92,100 sq miles (238,538 sq km).

Guinea

Country in western Africa, former French colony, with a coastline on the Atlantic. In the N it borders on Guinea-Bissau, Senegal and Mali, in the E on Ivory Coast, and in the S on Liberia and Sierra Leone. Its coastal plain, about 50 miles (80 km) wide, is fringed by mangrove swamps. Inland is the hilly plateau of Fouta Djallon, which is more than 3,000 ft (900 m) above sea-level in places. The forested Guinea Highlands in the SE rise to 5,748 ft (1,752 m) in the Nimba Mountains. The savanna lands of the E are drained by the River Niger and its tributaries. The Niger has its source in Guinea, as has the River Gambia. In most parts of the country the climate is hot and wet. Guinea is rich in mineral resources, the chief being bauxite and iron. But most of the people live by farming. Their crops include rice, bananas, maize, coffee, ground-nuts, palm kernels and citrus fruits. The chief towns are Conakry (the cap.), Kankan and Kindia. Area 96,900 sq miles (250,970 sq km).

Guinea-Bissau

Country in the NW bulge of Africa, lying between Senegal and Guinea. It has an Atlantic coastline and there are many off-shore islands. Formerly it was called Portuguese Guinea. The country is generally low-lying, but it rises in the SE to the hills of the Fouta Djallon. There are several rivers. Guinea-Bissau is hot and wet. The chief crop is rice; other crops include ground-nuts and cassava. The capital is Bissau. Area 14,000 sq miles (36,000 sq km).

Cassava being pounded in a wooden mortar by a two-man—or, in this case, a two-woman—pestle. In most of Ghana, vegetable products are the chief items of diet.

Above: Dancers at a festival in a prosperous Kenyan village. Africa's superb heritage of dance and music is endangered by the growth of industry and by the increasing modernization of towns and villages. But many people are aware of the need to preserve cultural traditions that can never be regained once they are lost.

Right: Drummers wait to provide the music at a celebration.

Ivory Coast

One of the richest countries of western Africa, and formerly a French colony. Mali and Volta border it on the N; Ghana on the E; and Liberia and Guinea on the W. A low, forested plain extends inland from the coast for some 150 miles (240 km). It rises to a plateau, averaging some 1,000 ft (300 m) above sea-level, that occupies the northern two-thirds of the country. The plateau vegetation is savanna. In the W the Guinea Highlands extend into Ivory Coast. The principal rivers, flowing N to S, are the Cavalla, Sassandra, Bandama and Comoé. Natural resources include timber, manganese, iron and diamonds. The most important crop is cocoa, grown mainly in the SE. Coffee, too, is produced in large quantities. Other crops include bananas, rice, maize, yams and cassava. The chief towns are Abidjan (the cap.), Bouaké and Grand-Bassam. Area 125,000 sq miles (323,700 sq km).

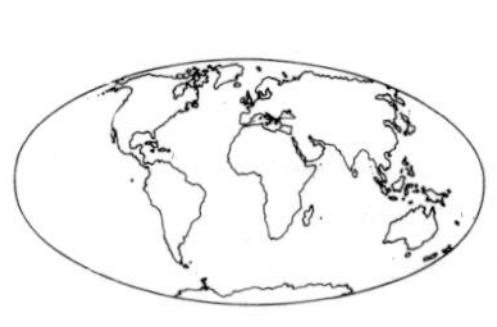

Kenya

Country of eastern Africa, on the equator. It was formerly a British colony. Much of the coastal plain is fertile, but it also has patches of tropical rain forest and mangrove swamp. Inland, the land rises to a vast, dry, grass-covered plateau, which occupies most of the northern part of the country and some of the S. The Kenya Highlands, in the SW, are the most heavily populated region. They have Kenya's two highest mountains: Mount Kenya (17,057 ft; 5,199 m), the second highest peak in Africa; and Mount Elgon (14,177 ft; 4,321 m). Both are extinct volanoes. W of the Highlands is the Nyanza Plateau, extending to the shores of Lake Victoria. This is one of the richest farming regions because of its volcanic soils. One branch of the Great African Rift Valley cuts through Kenya from N to S. Kenya's climate is largely dependent on altitude. The coastal plain is hot. The Highlands are relatively cool. Most of the country is dry, but the W has moderately heavy rainfall. The country is famous for its wild life and has a number of national parks. Kenya has very little mineral resources but development of hydro-electric power and industry continues. Most of the people live by agriculture. Half of the country is suitable only for livestock raising. Elsewhere people grow food crops. The most important cash crops are coffee, tea and sisal. The chief towns are Nairobi (the cap.), Mombasa, Nakuru and Kisumu. Area 224,960 sq miles (582,644 sq km).

A Kikuyu village in Kenya, showing the typical red laterite soil of some regions of equatorial Africa. The Kikuyu, one of Kenya's largest tribes, live in the highlands north-east of Nairobi. They are farmers, though many of them exist at a subsistence level.

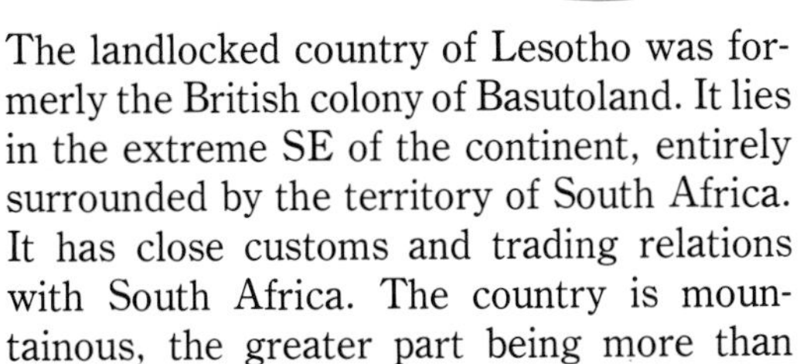

Lesotho

The landlocked country of Lesotho was formerly the British colony of Basutoland. It lies in the extreme SE of the continent, entirely surrounded by the territory of South Africa. It has close customs and trading relations with South Africa. The country is mountainous, the greater part being more than 6,000 ft (1,800 m) above sea-level. In the E, the Drakensberg Mountains rise to more than 10,000 ft (3,050 m). In general, the climate is moderate. But temperatures vary considerably according to altitude, and rainfall tends to be higher in mountainous areas.

The greatest concentration of population is in the S. Most people live by farming in organized agricultural communities. They grow wheat, maize, barley, peas and beans. Many are herdsmen and the rearing of sheep, goats and cattle is important to Lesotho's economy. Thousands of people find work in South Africa. The capital and largest town is Maseru. Area 11,716 sq miles (30,344 sq km).

Right: The port of Benghazi in the north-east of Libya, on the Gulf of Sidra. It stands on the site of the ancient city of Berenice, named in honour of the queen of the Pharaoh Ptolemy III.

Below: The ruins of Leptis Magna, on the Mediterranean coast of Libya, to the east of Tripoli. The city was founded by the Phoenicians in about 600 B.C. Later, under the Romans, it was an important port. Its ruined temples, houses, baths and forums make up the most impressive collection of Roman remains in Africa.

Liberia

Oldest independent country in W Africa. The republic was formed in 1847 from settlements for Negroes freed from slavery in the United States. Much of the country is tropical forest and the coastal plain is swampy.

There are valuable deposits of iron ore, and exports include rubber, coffee and cocoa. Many live by subsistence farming. Area 43,000 sq miles (111,000 sq km).

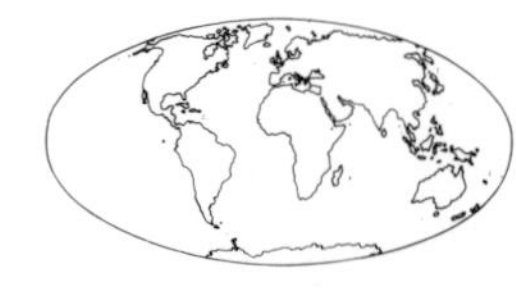

Libya

Country of northern Africa, with a 1,000-mile (1,600-km) coastline on the Mediterranean Sea. It is one of the continent's largest but most sparsely populated countries. It is bordered on the E by Egypt, on the SE by the Sudan, on the S by Chad and Niger, and on the W by Algeria and Tunisia.

At its centre, the coast has a broad indentation, the Gulf of Sidra. On both sides of the gulf there are upland areas. That in the E, the Jebel el Akhdar, rises to some 2,000 ft (600 m). Most of the fertile land in Libya lies along the coast. The rest of the country is in the Sahara: nine-tenths of it is desert or semi-desert. Just inland from the coastal strip in the W is a broad, arid plateau. Beyond lie great wastes of sand seas and barren rocks. In places there are gaunt mountains and escarpments. Several caravan routes and motor roads cross parts of the desert, mostly connecting oases with the coastal towns. Important oases include Ghat in the W, Jofra and Al Fuqaha in the centre, and Jalo and Kufra in the Libyan Desert in the E. Temperatures in the desert are often extremely

high. The coastal regions have a more moderate Mediterranean climate.

Libya's great source of wealth is petroleum; the country is a leading African oil producer. A major petro-chemical works and an iron and steel plant are to be built. Agriculture is confined mainly to the coastal areas and the oases. The chief crops are cereals, ground-nuts, olives, tobacco, citrus fruits and vegetables. In the desert, there are thousands of nomadic herdsmen. The chief towns are Tripoli (the cap.), Misurata, Benghazi, Homs-Cussabat and Zawia. Area 679,360 sq miles (1,759,534 sq km).

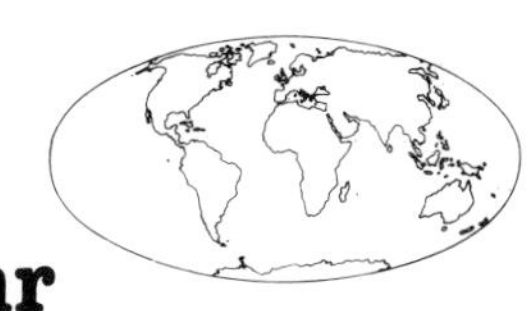

Madagascar

Island country in the Indian Ocean whose people are of mixed Asian and African descent. A former French colony, Madagascar became independent in 1960. French forces withdrew in 1973 and the French naval base was turned over to civilian use.

The main island and small off-shore islands of Madagascar lie about 250 miles (400 km) from the SE coast of Africa, across the Mozambique Channel. On its western side the island has a broad coastal plain. The plain rises to a vast, high, central plateau, which is greatly broken up. Lofty mountains rise in the N, centre and S. The highest point in the island is Mount Maromokotro (9,449 ft; 2,880 m) in the Tsaratanana Massif in the N. The escarpment of the plateau is much steeper on the eastern side, and the coastal plain is narrow. On this coast a shipping canal several hundred miles long has been constructed by connecting the many coastal lagoons. The island has a large number of rivers, the longest being the Betsiboka. Eastern regions are generally hot and wet. In the higher areas the climate is milder.

Madagascar has deposits of chrome, graphite, phosphates and mica, and has textile industries. But most people live by farming. Rice, millet, maize and cassava are grown for food. The chief cash crops are coffee, cloves and vanilla. Large numbers of cattle are reared.

The chief towns are Antananarivo (the cap.), Tamatave, Majunga and Fianarantsoa. Area 230,000 sq miles (595,700 sq km).

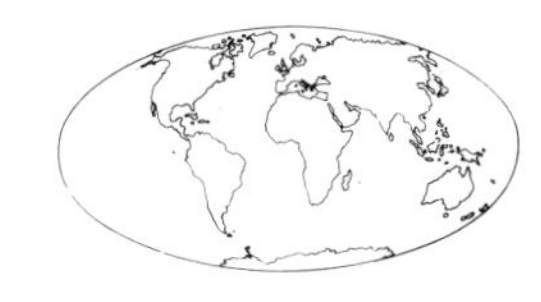

Malawi

Malawi, in southern Africa, was formerly the British protectorate of Nyasaland. It has no seacoast, but for hundreds of miles its eastern border runs through Lake Malawi (Lake Nyasa), the third largest lake in Africa. The country is long and narrow and the Great African Rift Valley runs through it for its whole length. In addition to Lake Malawi there are three smaller but still sizeable lakes —Malombe, Chiuta and Chilwa. Temperatures are high in most places and rainfall is high in the extreme N and the S. Most of the people live by farming. Important cash crops are tobacco, tea, cotton, sugar and groundnuts. Food crops include maize, millet and cassava. The chief towns are Lilongwe (the cap.), Blantyre, Zomba and Mzuzu. Area 45,500 sq miles (117,850 sq km).

A village on the open plateaux of Malawi. The wide 'mushroom' roofs of the circular mud huts protect the walls from destructive rains during the wet season.

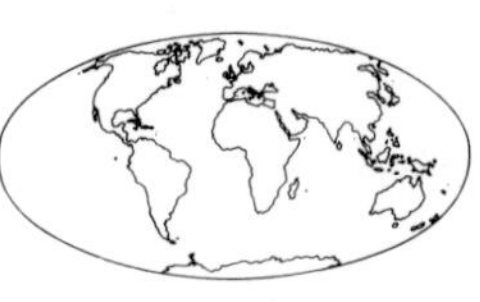

Mali

Large, inland country in western Africa, formerly the French colony of Soudan. In general the country is a flat, savanna-covered plateau, the northern third of which lies in the Sahara. But the S is mountainous and in the NE there are the highlands of the Adrar dea Iforas. In the N, in the desert, is the great oasis of Taoudenni, where salt is produced. The River Niger enters the country from Guinea, and flows north-eastwards and then south-eastwards in a great arc. In its central section it forms a number of swampy lakes. Another large river, the Senegal, waters the low-lying and relatively fertile areas of the SW. The highest temperatures are in the N, and the heaviest rainfall in the S. Mali has some mineral resources but they have not been exploited. Nearly all the people live by farming or fishing. Most of the Berbers, Fulani and Tuareg of the N are nomadic herdsmen. The Negroes of the more densely-populated S are mainly subsistence farmers, growing rice, millet or maize. But some cash crops, such as cotton and ground-nuts, are produced. Cattle are among Mali's exports. The chief towns include Bamako (the cap.), Gao, Kayes and Timbuktu. Area 465,000 sq miles (1,204,000 sq km).

Launderers at work beside the River Niger, near Bamako in Mali. Clothes are scrubbed and beaten in the river and then spread out on the bank to dry.

Mauritania

The Islamic Republic of Mauritania was formerly part of French West Africa. It is bounded on the NW by Western Sahara, on the NE by Algeria, on the E and SE by Mali, and on the SW by Senegal. In the W it has a coastline on the Atlantic. Much of the country consists of desert plateaux. There is only one permanent river, the Senegal, which forms the south-western border. The climate is hot and dry. The flood plain of the Senegal is fertile and farmers grow maize, millet, ground-nuts and other crops. In other parts of the country, the people are predominantly nomads. There are large deposits of iron and copper. Export of iron ore began in 1963 after construction of a railway. The copper mine re-opened in 1981. The chief towns include Nouakchott (the cap.) and Atar. Area 419,000 sq miles (1,085,000 sq km).

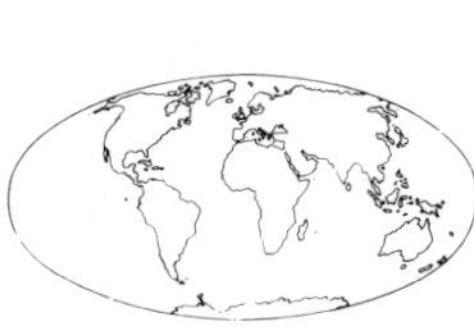

Mauritius

Republic consisting of a group of islands in the Indian Ocean some 500 miles (800 km) E of Madagascar. The largest islands are Mauritius and Rodrigues. Mauritius Island has a central plateau bordered by three mountain ranges. The chief industry is the production of cane sugar, but tea and tobacco are grown too. A bulk sugar terminal opened at Port Louis in 1980. The islands are densely populated. The chief towns include Port Louis (the cap.), Beau Bassin and Rose Hill. Area (total) 800 sq miles (2,000 sq km).

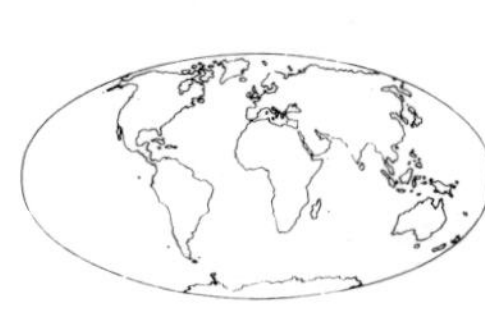

Morocco

Moghreb-el-Aksa, 'the Farthest West'. Kingdom in NW Africa. Morocco became independent in 1956. The Western (Spanish) Sahara came under joint Moroccan and Mauritanian control in 1976. Owing to strong opposition from Nationalist guerrillas, Mauritania gave up all claim, leaving the Western Sahara under Moroccan control. Its most northerly part, the peninsula of Tangier, is only some 9 miles (14 km) from Europe across the Strait of Gibraltar. Morocco has a coastline on the Mediterranean Sea in the N, and on the Atlantic in the W. The Mediterranean coast is rocky. The Atlantic coast is sandy, and inland from it there is a broad coastal strip. Parallel with the coastal strip extends the long, high range of the Atlas Mountains, rising to 13,661 ft (4,164 m) in Mt Toubkal. The mountains are dry and rocky and on the E side they slope down to the hot sands of the Sahara. Several rivers flow W from the Atlas range to the Atlantic. Those that flow down in an easterly direction dry up as they reach the desert sands. The country's longest river, the Moulouya, is on the desert side of the mountains, but it flows N to the Mediterranean. On the Atlantic coastal plain the climate is moderate. In the N the summers are hot, and in the desert areas they are extremely hot as well as long. The NW parts of the Atlas range have severe winters, with much snow. Most of the people live by farming, the productive areas being the coastal plains and the Atlantic slopes of the Atlas. The chief crops include wheat, barley, maize, beans, dates, nuts, olives, oranges, lemons and apricots. On the lower mountain slopes there are huge vineyards. Farmers and nomadic herdsmen keep sheep, goats and cattle. Forestry and fisheries are important. There are valuable mineral deposits, and Morocco is a leading producer of phosphates. The people of Morocco are mainly Arabs, Berbers and Moors. The chief cities are Casablanca, Marrakesh, Rabat (the cap.), Fès, Meknès and Tangier. Area 171,000 sq miles (442,888 sq km).

Mozambique

The People's Republic of Mozambique in SE Africa was, until 1975, an overseas province of Portugal. It has a 1,500-mile (2,400-km) coastline on the Indian Ocean. Mozambique provides access to the sea for the three inland countries of Zimbabwe, Zambia and Malawi to its W. On the N it is bordered by Tanzania, and on the SW by South Africa and Swaziland. Much of the country consists of a broad coastal plain, which is sandy and infertile in places. Many rivers flow eastwards across the country from the mountains and highlands of the W. They include the Zambezi and the Limpopo. The Cabora Bassa dam on the Zambezi produces hydro-electric power. The climate is tropical, with heavy seasonal rains. Subsistence farming is general, but there are good soils in the river valleys. Crops include cashew nuts, sisal, cotton, maize, sugar and tea. The chief towns are the capital Maputo (Lourenço Marques), Beira and Mozambique. Area 303,000 sq miles (705,000 sq km).

Above: The ancient Moroccan city of Marrakesh, at the foot of the Atlas Mountains.

Below: Nomads camp in the desert. Once, the nomads ruled the Saharan wastes. Today, they are herdsmen or merchants.

Niger

Large, landlocked country in north-western Africa, formerly part of French West Africa.

Right: Dromedaries—one-humped camels—being given a drink at a well in Nigeria. Camels can travel long distances without eating or drinking, and until recently were the chief means of transport in the desert. But today motor vehicles are taking their place.

Below: Traders in Niamey display their wares. Niamey, the chief commercial city as well as the capital of Niger, lies on the road between the historic settlements of Timbuktu, in Mali, and Parakou, in Dahomey. Parakou is connected by rail with Cotonou, which is Niger's nearest outlet to the sea.

Most of Niger is a plateau. The northern two-thirds is in the Sahara. The central Aïr region is in the desert but has valleys and plains where livestock can be pastured and millet and other crops grown. Southern Niger is covered with savanna. In the SW there is fertile, low-lying land in the valley of the River Niger. The SE is also low around Lake Chad. The whole of the country is hot and dry. In the S agriculture is aided by seasonal rains and seasonal flooding of the rivers, but regions in the N have only a few inches of rain a year. Niger has some mineral resources, including uranium, tin and iron. A traditional export is salt from an oasis in the Sahara. Many people, in both N and S, are nomadic herdsmen. Even in the more productive areas of the S, most people live by subsistence farming, growing millet, rice, or maize. Ground-nuts and livestock are the chief exports. There are no railways and very few roads. The capital is Niamey, Zinder the main trading town. Area 459,000 sq miles (1,189,000 sq km).

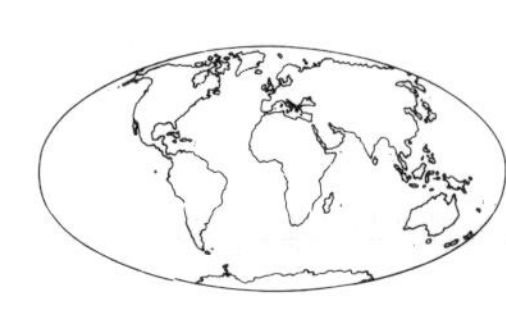

Nigeria

Nigeria, in western Africa, has twice as many people as any other country in the continent. It was formerly a British colony. Nigeria has a southern coastline on the Gulf of Guinea. It takes its name from the River Niger, which enters the country in the NW and flows generally southwards to its large marshy delta on the Gulf of Guinea. The coastal lowlands, up to 60 miles (96 km) wide in places, abound in marshes and mangrove swamps. Farther inland is a belt of dense, tropical rain forest. This gives way to the high table-land that forms the greater part of the country. The Niger and its tributaries break the table land up into distinct parts. To the W and S of the Niger are the Yoruba Highlands, the former kingdom of the Yoruba people. Most of the region lies at between 1,000 and 1,600 ft (300 to 480 m) above sea-level. To the N of the Niger and Benue are the Jos Plateau which rises to some 5,000 ft (1,500 m) and, E of it, the Bauchi Plateau which reaches some 4,000 ft (1,200 m). Both plateaux slope away northwards to about 2,000 ft (600 m). The extreme N of the country is semi-desert. In the NE, Nigeria includes part of the marshy Lake Chad. Some of the rivers of the N flow to the lake; others flow to the Niger. The climate is tropical, but more extreme in the N than in the S. The coastal region has very heavy rainfall from May to October. The country is relatively rich in minerals, the most important being petroleum. Tin and coal are also mined in large amounts, and there is an export trade in columbite, iron, gold, zinc and wolfram. Industries include the production of steel, cement and textiles; tin smelting; petrol refining; and food processing. But most people live by agriculture. The main cash crops are palm kernels, cotton, cocoa and ground-nuts. Other crops include maize, millet, rice, sorghum, cassava, rubber, tobacco, coffee and gum arabic. The chief towns are Lagos (the cap.), Ibadan, Ogbomosho, Kano, Oshogbo, Abeokuta, Port Harcourt, Zaria and Ilesha. Abuja is to become the new Federal capital. Area 356,669 sq miles (923,768 sq km).

Rwanda

Small independent country of E-central Africa, formerly part of the U.N. international trust territory of Ruanda-Urundi. It is poor but densely populated. Much of its western border runs through Lake Kivu. To the E of the lake is a ridge of volcanic mountains, the edge of the Great African Rift Valley, separating the basins of the Rivers Zaire and Nile. One peak, Mount Karisimbi in the Virunga Range, is 14,786 ft (4,507 m) above sea-level. A broken plateau extends eastwards, averaging 3,000 to 6,500 ft (900 to 2,000 m) in height. Most of the country is hot and wet. Many of the people live by subsistence farming, but some cash crops are grown, including coffee, tea and cotton. Some tin and tungsten are mined. There are no railways and roads are poor. The capital is Kigali. Area 10,169 sq miles (26,338 sq km).

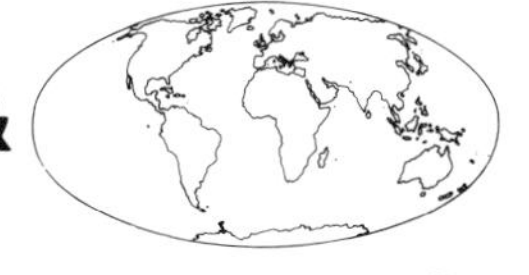

São Tomé & Principe

Republic consisting of the islands of São Tomé and Príncipe (and two small islets) in the Gulf of Guinea. The islands were first colonized by the Portuguese in the 1400s, and finally became independent in 1975. The cap. is São Tomé. Area 372 sq miles (963 sq km).

Left: A herd of elephants pursue their orderly way across the savanna. Despite their size, elephants make little noise as they move from place to place. The young elephants are playful, but they are strictly disciplined and they soon learn the danger of straying away from the herd.

Below: Mountains of ground-nuts awaiting transport from Kano, one of the chief commercial centres of the north of Nigeria. Ground-nuts and cotton are the most valuable cash crops of the northern region.

Return of the tuna fishermen to a coastal village near Cape Guardafui, in north-eastern Somalia. Fish is an important article of diet for those who live within reach of the sea in this arid corner of Africa.

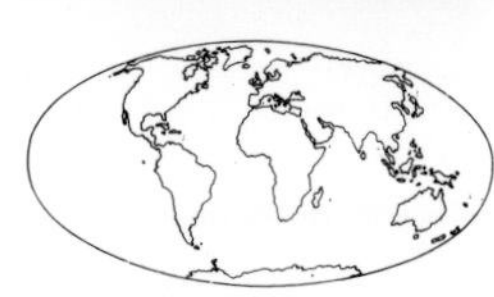

Senegal

Country on the western coast of Africa, formerly a French colony. In 1982 the Senegambia Federation was set up. (See Gambia p. 162.) The promontory of Cape Vert in Senegal is the continent's most westerly point. The country is bordered on the N by Mauritania, on the E by Mali and on the S by Guinea and Guinea-Bissau. The Gambia, an independent country, forms an enclave within the territory of Senegal. Most of Senegal consists of grasslands up to 500 ft (150 m) or so above sea-level. But in the SE, where the Fouta Djallon extends into the country from Guinea, there are places at heights of more than 1,600 ft (490 m). The many rivers include the Senegal and the Gambia. Coastal regions have a moderate climate, but inland temperatures are high. There is much rain in the S. The country is still almost entirely agricultural. Millet and sorghum are grown for food. The chief cash crop is ground-nuts. The chief towns are Dakar (the cap.), Kaolack and St. Louis. Area 77,800 sq miles (201,500 sq km).

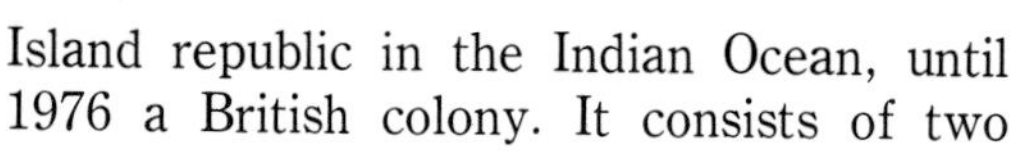

Seychelles

Island republic in the Indian Ocean, until 1976 a British colony. It consists of two groups of islands: the 45 mountainous Mahé islands and the 50 or so flat Coralline islands. Exports include copra and cinnamon bark. The cap., Victoria, is on Mahé, the largest island. Area 270 sq miles (700 sq km).

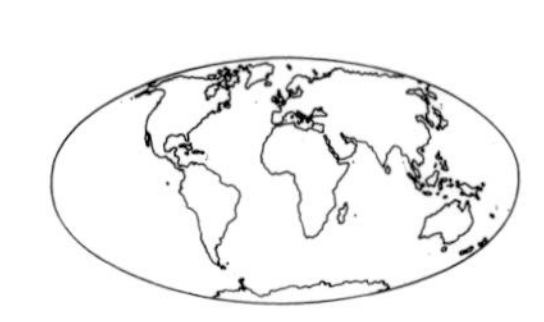

Sierra Leone

Country of western Africa, a former British possession. The broad coastal plain is swampy except for the Sierra Leone peninsula, where Freetown is situated. The typical vegetation on the plateau is savanna. In the low-lying areas there are still remains of the tropical rain forests that once covered them. Most of the country is hot and wet. The road network has been developed since the railway was phased out in 1974. Mineral resources include iron and diamonds. Most people live by farming; cash crops include cocoa and coffee. The chief towns are Freetown (the cap.), Bo and Makeni. Area 27,925 sq miles (72,325 sq km).

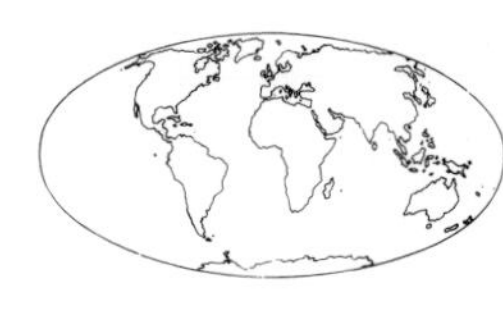

Somalia

The Somali Democratic Republic lies along the coast of the eastern 'horn' of Africa. It is made up of former British and Italian colonies. The broad, dry coastal plain in the E slopes upwards to the Ogaden plateau of Ethiopia. In the N a narrow coastal plain rises to an escarpment and the hilly northern fringe of the plateau. There are only two permanent rivers, the Juba and the Shibeli. The climate is hot and dry. Many of the people are nomadic herdsmen. Crops include cotton, maize, sorghum and ground-nuts. The chief towns are Mogadishu (the cap.), Merka and Hargeisa. Area 246,200 sq miles (637,700 sq km).

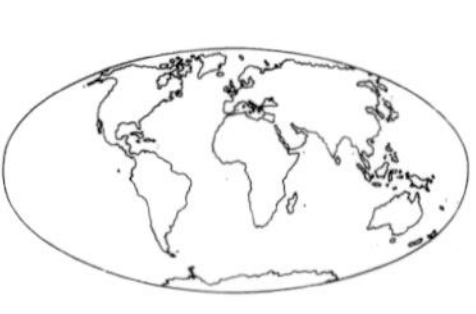

South Africa

South Africa is the southernmost country of the African continent, and the most prosperous. The country is ruled by its White inhabitants, who number about one-fifth of the population. Non-whites are excluded from many areas under a policy of apartheid which is widely condemned by other countries. South-West Africa (Namibia), a territory to the NW, is administered by South Africa. It became self-governing in 1978 but the United Nations is not yet satisfied with its constitution. South Africa has a coastline on the Atlantic in the W, and on the Indian Ocean in the E. It is bordered on the N by Angola, Botswana, Zimbabwe, Mozambique and Swaziland. Most of South Africa consists of broad plateaux, called *karroos*. The eastern and higher plateau, called the *High Veld*, is the richest and most fertile. It has great mineral resources, thriving industry and prosperous farms. The coastal areas of the W are partly in the Namib Desert. Most of the country has a temperate climate, though shortage of rain is a problem for farmers in many places. The country's rich natural resources include diamonds, gold, coal, iron, phosphates, uranium and copper. Important manufactures are steel, engineering products, textiles and processed foods. Agricultural products include wool, maize, wheat, tobacco, sugar-cane, potatoes, vegetables, fruit and grapes (for wine). The chief towns are Cape Town (legislative cap.), Pretoria (administrative cap.), Johannesburg, Durban, Port Elizabeth, Germiston and Bloemfontein. Area 472,360 sq miles (1,223,410 sq km).

The city of Cape Town, lying below the great rugged wall of Table Mountain. Cape Town, the oldest and second largest city in South Africa, is the country's chief passenger port.

Sudan

The Sudan, the largest country in Africa, was governed jointly by Egypt and the United Kingdom until 1956. The land consists largely of mountain-fringed plateaux. The southern, equatorial region contains a vast swamp, called the Sudd, 50,000 sq miles (130,000 sq km) in area. Northwards, a belt of savanna gives way to a wide belt of grassland and scrub. In the centre of this region rise the bare hills of Kordofan. The northern half of the country is desert, the most easterly part of the Sahara: W of the River Nile it is called the Libyan Desert, E of the Nile, the Nubian Desert. The Nile and its tributaries extend for thousands of miles through most of the country. The White Nile enters the country in the extreme S, the Blue Nile in the SE. The branches join at Khartoum, and flow northwards to Egypt. Successful irrigation works have been effected along the course of the river, the earliest being at Gezira, S of Khartoum. All parts of the country are hot, but there are considerable variations in the amount of rainfall, which is greatest in the S. The Sudan has few mineral resources. The economy is almost entirely agricultural. In the desert regions, most people are nomadic herdsmen. In more fertile areas, the crops include millet, sorghum, maize and cassava. Cotton and ground-nuts are grown as cash crops. Other exports are hides and gum arabic. The chief towns are Khartoum (the cap.), Omdurman, El Obeid, Port Sudan and Wad Medani. Area 967,500 sq miles (2,505,800 sq km).

Above: The strange landscape of the Nubian Desert, between the River Nile and the Red Sea.
Right: Zebras in one of Tanzania's national parks. Eastern Africa is exceptionally rich in the magnitude and variety of its animal life and many governments have laws to protect rare animals from hunters and further devastation of their natural habitats. But despite all efforts at conservation, animals in Africa as elsewhere seem to be fighting a losing battle against the destructive effects of technology.

Swaziland

Small landlocked kingdom in southern Africa, enclosed by South Africa and Mozambique. The climate is sub-tropical, with heavy rainfall in the higher areas. Agricultural products include maize, millet, sugar, rice and fruits. Cattle and sheep are raised. There are valuable mineral deposits; asbestos and iron are among those exploited. The people speak Bantu languages. Area 6,704 sq miles (17,363 sq km).

Tanzania

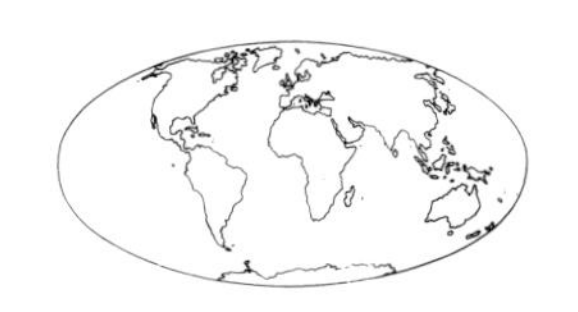

Large country of eastern Africa, formed of the former British trust territory of Tanganyika and a number of offshore islands, of which the chief are Zanzibar and Pemba. It is bordered on the N by Uganda and Kenya, on the S by Mozambique, Malawi and Zambia and on the W by Zaire, Burundi and Rwanda. The mainland part of Tanzania lies in two arms of the Great African Rift Valley. Much of its western border runs through Lake Tanganyika and Lake Malawi. Another great lake, Victoria, lies on the northern boundary. The 500-mile (800-km) long coastal plain is narrow except in the centre, near the mouth of the Rufiji, the largest river. The plain, swampy in many places, rises to the broken plateau that occupies most of the country. The plateau is covered with woods, scrub and grasslands. Kilimanjaro, Africa's highest mountain, rises to 19,340 ft (5,895 m) in the NE. Near to it is Mount Meru (14,979 ft; 4,566 m). The Livingstone Mountains, at the northern edge of Lake Malawi, rise to some 9,000 ft (2,750 m). The scenery of Tanzania is among the most varied and beautiful in Africa, and the country is famous for its wildlife. It contains Serengeti National Park. Eastern areas have a tropical climate. Inland areas are generally dry, but the mountains tend to have lower temperatures and more rain. Tanzania has good mineral resources and there are some light manufacturing industries. But the country is still predominantly agricultural. Maize, millet and cassava are grown for food, and many people in areas free from the tsetse fly keep cattle. The main cash crops are coffee, cotton, tea, sisal, cloves and cashew nuts. The chief towns are Dar es Salaam (the cap.), Zanzibar, Tanga, Arusha and Dodoma (which will become the cap. in the mid-1980s). Area 362,820 sq miles (939,700 sq km).

A typical stretch of savanna on the plateau that occupies much of continental Tanzania. The rough grass grows in clumps and patches and here and there spiky bushes and trees break the monotony. In most regions of savanna, farming consists chiefly of grazing.

Togo

Small republic in W Africa, long and narrow in shape and with a short coastline on the Gulf of Guinea. Along the coast and farther inland in the SE there are large mangrove swamps. The Togo-Atakora range of mountains runs diagonally across the country from SW to NE. In places, its peaks rise to more than 3,000 ft (900 m) above sea level. Many of the lower mountain slopes are covered with tropical rain forest. In the NW is an area of tropical savanna. Togo is extremely hot and humid, with heavy rainfall. It is a poor country. Its chief export crops are cocoa-beans, coffee and palm nuts. To a large extent, the people live by subsistence farming, growing such crops as maize, yams, cassava and millet. Phosphates are mined and the country also has deposits of iron, chromium and bauxite. The peoples of the S are mainly Bantu and many in the N are of Hamitic descent. They use a large number of languages. The chief towns are Lomé (the cap.), Sokodé and Palimé. Area 21,853 sq miles (56,599 sq km).

Right: A street in Tunis, an important trading city as well as Tunisia's capital. Tunis stands near the site of ancient Carthage.

Below: A tempting array of produce for sale in a Tunisian market. Shoppers consider each item carefully.

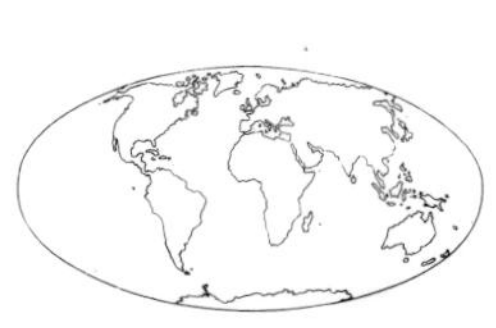

Tunisia

Tunisia has a Mediterranean coastline on both the N and the E. One of its promontories, Cape Blanc, is the northernmost point of Africa. Tunisia has land borders with Libya in the SE and Algeria in the W. The northern part of the country is in the Atlas Mountains. Just inland from the N coast, the Tell extends into Tunisia from Algeria. S of the Tell are hilly Atlas tablelands, rising in the Jebel Chambi to 5,138 ft (1,566 m). The tablelands fall away in the E to the coastal plain called the Sahel. The country's only important river, the Medjerda, flows across the Atlas region to the Gulf of Tunis. The S part of Tunisia lies in the wastes of the Sahara. There are some oases where cultivation is possible. In the W of this desert region are the great salt flats of Chott Djerid. In coastal and hilly areas, the climate is warm and pleasant, with moderate rainfall. Desert areas are hot and arid.

Tunisia has valuable mineral resources, including phosphates, iron, petroleum and natural gas. Its industries include food processing, sugar refining and the manufacture of chemicals and paper. Most people live by farming, the productive areas being in the N and NE. The main crops include barley, wheat, grapes (for wine), olives, citrus fruits, vegetables and dates. The chief towns are Tunis (the cap.), Sfax, Bizerta, Sousse and Kairouan. Area 45,000 sq miles (116,500 sq km).

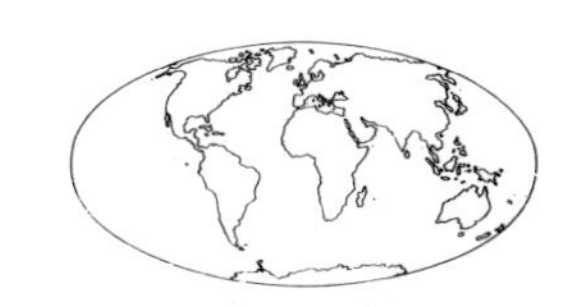

Uganda

Country of E-central Africa, lying on the equator. Until 1962 it was a British protectorate. Internal disorder and military dictatorship was a feature of the 1970s. A large part of the country is a high plateau, averaging about 4,000 ft (1,220 m) above sea level. Its characteristic vegetation is lightly-wooded savanna. Branches of the Great Rift Valley enclose the country. There are high mountains on both the W and the E borders. In the W the Ruwenzori Mountains rise to 16,794 ft (5,119 m). The headwaters of the White Nile drain the country. One of them, the Victoria Nile, flows NW from Lake Victoria and passes through the vast area of swamp called Lake Kyoga and from there to Lake Albert. The river then flows N from the lake as the Albert Nile and joins the White Nile in the Sudan. Because of Uganda's altitude, its climate is not tropical. But it is hot in some areas, and has heavy rainfall in the mountains and in the SSE around Lake Victoria. Most of the people are farmers. They raise cattle, sheep, goats and pigs, and grow plantains, yams, cassava and maize. The chief export crops are coffee and cotton, but tea, tobacco, ground-nuts and sugar are also grown. Uganda has many minerals, but only copper is mined in any quantity. The Owen Falls dam provides hydro-electricity for industry in its part of the country. In the S and centre, most of the people belong to Bantu groups, the largest tribe being the Baganda. In the N there are Nilotic tribes. The chief towns are Kampala (the cap.), Jinja, Mbale, Entebbe and Kabale. Area 93,981 sq miles (234,410 sq km).

A petrol station in Kampala, Uganda's principal commercial centre. Kampala is a well-planned, modern city with fine buildings and good roads. As industrialization advances in Africa, increasing numbers of people move from country areas into the towns.

Above: The irresistible, roaring waters of the Victoria Falls on Zambia's southern border.

Upper Volta

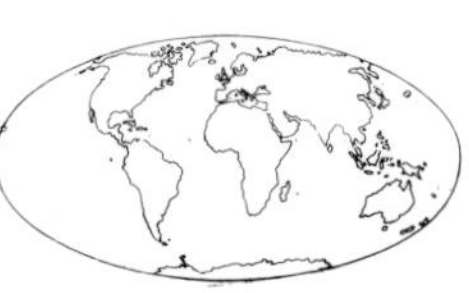

Republic of W Africa. Upper Volta consists almost entirely of a wide plateau, which slopes gently from N to S, averaging about 800 ft (244 m) above sea level. Most of it is covered with dry savanna. The country lies across the main headstreams of the River Volta.

The climate varies considerably. Half of the year is dry, the rest is hot and wet. In the northern savannas the rainfall averages less than 10 in (250 mm) a year, but the S averages more than 40 in (1 m).

Below: A boulevard in Kinshasa, formerly Léopoldville. The city, one of the most modern in Africa, lies on the Zaire (Congo) River, opposite Brazzaville.

Upper Volta is an almost completely agricultural country and most of its people live by subsistence farming. They grow sorghum, millet and guinea corn. Groundnuts, cotton and sesame are grown for sale, but the chief source of wealth is livestock. Sheep, goats and cattle are raised and meat is exported, chiefly to Ghana. There are some mineral deposits. The largest tribe is the Mossi, whose ancient settlement, Ouagadougou, is the country's cap. Other tribes include the Mande, the Gourma and the Bobo, who live in the district around Bobo-Dioulasso, the principal market town. Area 105,869 sq miles (274,199 sq km).

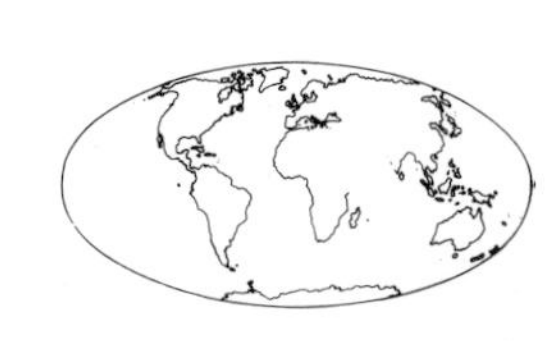

Zaïre

One of the largest countries in Africa, Zaïre lies in the S-centre of the continent, on the equator. It has only one outlet to the sea: a narrow corridor in the W gives it a 25-mile (40-km) coastline on the Atlantic. The country takes its name from the River Zaïre (Congo). It was formerly the *Belgian Congo* and then the *Democratic Republic of the Congo*. The greater part of Zaïre is a vast plateau, up to 3,000 ft (900 m) above sea level. Its high SE part is the Katanga plateau. On the E border is part of the Great Rift Valley, containing Lakes Albert, Edward, Kivu, Tanganyika and Mweru. Here, too, are great ranges of mountains. The Ruwenzori Mountains, between Lakes Albert and Edward, rise to 16,794 ft (5,119 m). The River Zaïre flows through the country for some 2,800 miles (4,500 km). It rises, as the River Lualaba, in the SE and flows northwards, then westwards and then southwestwards in a great half-circle to the Atlantic. Towards the end of its course, it forms the boundary with the Republic of Congo (Brazzaville) on the W. Its many tributaries include the Lufira, Lomami, Lulonga, Oubangui and Kasai Rivers.

Dense tropical forests cover much of the low-lying Zaïre basin in the W and NW. This region has a hot and humid climate. The higher regions are more temperate.

Zaïre is extremely rich in minerals, the most important being copper from the Katanga plateau. It also has deposits of cobalt, industrial diamonds, zinc, iron, manganese, tin and gold. Its plantations produce cotton, rubber, coffee and palm oil. The last of these is the main cash crop. Despite the country's valuable resources, many of its people live by subsistence farming, growing maize, cassava and millet. Most of the people speak Bantu languages. There are also Hamitic tribes, and small bands of pygmies live in the forests of the Zaïre basin. The chief towns are the cap. Kinshasa (formerly Léopoldville), Lubumbashi (formerly Elisabethville) and Kisangani (formerly Stanleyville). Area 905,565 sq miles (2,345,402 sq km).

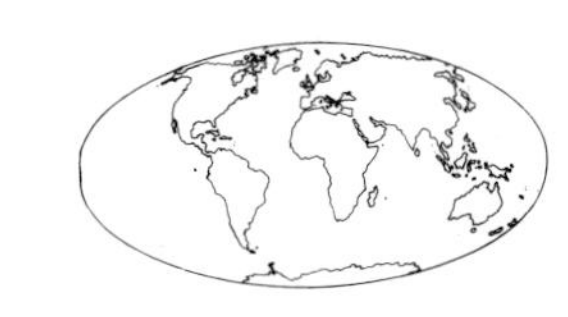

Zambia

Republic of S-central Africa, on the central African plateau. Before independence it was called Northern Rhodesia. Most of it lies between 3,000 and 5,000 ft (900 and 1,500 m) above sea-level. The largest rivers are the Zambezi, the Kafue and the Luangwa. The various tributaries of the great Zambezi fan out to almost every part of the country. There are many lakes. In the N, Lake Bangweulu lies completely within the country's boundaries. Another large lake, Kariba, lies on the S border. It is artificial, having been formed by the building of the Kariba dam. There are also many huge swamps. Much of the country is covered with lightly-wooded savanna. Although Zambia is in the tropics, its climate is relatively moderate because of its altitude. But there is a hot season from September to November. Zambia is one of the world's leading copper-producing countries. The so-called 'copper belt' is in the N and much of its electric power comes from the Kariba dam. Cobalt, vanadium, coal, lead, zinc and manganese are also mined. In general the country is thinly populated, its areas of heaviest population being in the mining districts. Tobacco, cotton and fruits are grown in some places, but most people who farm produce only subsistence crops, including millet, maize and yams. Almost the entire population belongs to Bantu groups. The chief towns include Lusaka (the cap.), Ndola, Kitwe, Mufulira, Luanshya and Chingola. Area 288,130 sq miles (746,253 sq km).

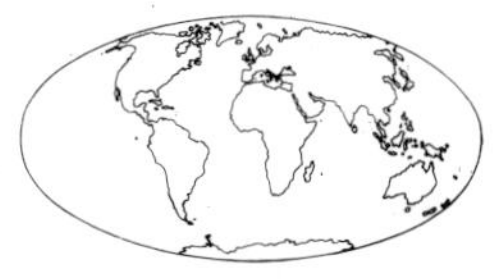

Zimbabwe

Inland republic of southern Africa. Formerly the British colony of Southern Rhodesia. The government was run by white Rhodesians—one-twentieth of the population. In 1965, this minority government declared Rhodesia independent to preserve their dominant position. Britain condemned this act and, with most other nations, imposed economic sanctions against Rhodesia. In the early 1970s, a war began between black guerrilla armies and government forces. This bitter war weakened Rhodesia's economy. A settlement was achieved in 1979, which restored British rule and ended the guerrilla war. British-supervised elections were held and a mainly black government came to power. The country became independent in April 1980 and was renamed Zimbabwe. The country is bordered on the NE and E by Mozambique, on the S by South Africa, on the W by Botswana and on the NW by Zambia. Zimbabwe consists mostly of a vast plateau, lying between the Zambezi River in the N and the Limpopo River basin in the SE. The centre of the plateau, which is known as the *High Veld*, is 400 miles (640 km) long and 70 miles (110 km) wide. It rises to over 4,000 ft (1,200 m) above sea-level. Most of the remaining country is over 2,000 ft (600 m) above sea-level. The lower regions are called the *Low Veld*. On the NW border is the 17-mile (280-km) long Lake Kariba formed by the building of the Kariba Dam, whose hydro-electric generators produce electricity for both Zambia and Zimbabwe. Because of its altitude, most of Zimbabwe has a moderate climate. Farming is a major economic activity. The chief crops are tobacco, maize, cotton, sugar-cane, wheat and fruits. Cattle products are also important, especially on white-owned farms. The three million black-owned cattle are mostly of poor quality. The country contains various minerals, including coal. Asbestos is the most valuable mineral. Chrome, gold and nickel are also mined. The economy's leading sector is manufacturing, which has expanded greatly since World War II. It is especially important in the towns, including Harare (the cap.), Bulawayo, Mutare and Gweru. Area 150,820 sq miles (390,622 sq km).

North and Central America

North America has only about one-tenth of the people of the world, but it is the richest of the continents both in natural resources and in industry. Factories in North America produce around one-third of the world's manufactures. Yet because of the extreme cold or desert conditions which exist in some regions, large parts of the continent are still undeveloped.

North America is the third largest continent. It extends from the Arctic Ocean in the N to the Isthmus of Panama in the S. It is bounded on the E by the Atlantic Ocean, and on the W by the Pacific. Most of its area is occupied by three countries—Canada, the United States and Mexico. Greenland—which is the world's largest island—the islands of the West Indies and Central America make up the rest of it.

In the NW, at the Bering Strait, North America approaches to within 56 miles (90 km) of Siberia, in Asia. In the E, Greenland is only about 1,000 miles (1,600 km) from Norway and Scotland, in Europe. The total area of the continent is about 9,600,000 sq miles (24,900,000 sq km).

From N to S, North America extends for about 4,500 miles (7,200 km), and the greatest distance W to E is about 4,000 miles (6,400 km).

COUNTRIES OF NORTH AND CENTRAL AMERICA POPULATIONS AND MONETARY UNITS

Country	*Population*	*Money*
Bahamas	237,000	Dollar
Barbados	249,000	Dollar
Canada	24,231,000	Dollar
Costa Rica	2,271,000	Colón
Cuba	9,706,000	Peso
Dominica	81,000	Dollar
Dominican Republic	5,648,000	Peso
El Salvador	4,938,000	Colón
Grenada	112,000	Dollar
Guatemala	6,044,000	Quetzal
Haiti	5,104,000	Gourde
Honduras	3,822,000	Lempira
Jamaica	2,220,000	Dollar
Mexico	71,193,000	Peso
Nicaragua	2,824,000	Córdoba
Panama	1,940,000	Balboa
Trinidad and Tobago	1,060,000	Dollar
United States	229,805,000	Dollar

Physical Features and Climate

North America has five major physical regions: the Eastern Highlands; the Western Highlands; the Canadian Shield; the Central Lowlands; and offshore, the islands of the West Indies.

The Eastern Highlands are formed by the Appalachian Mountains, which run parallel

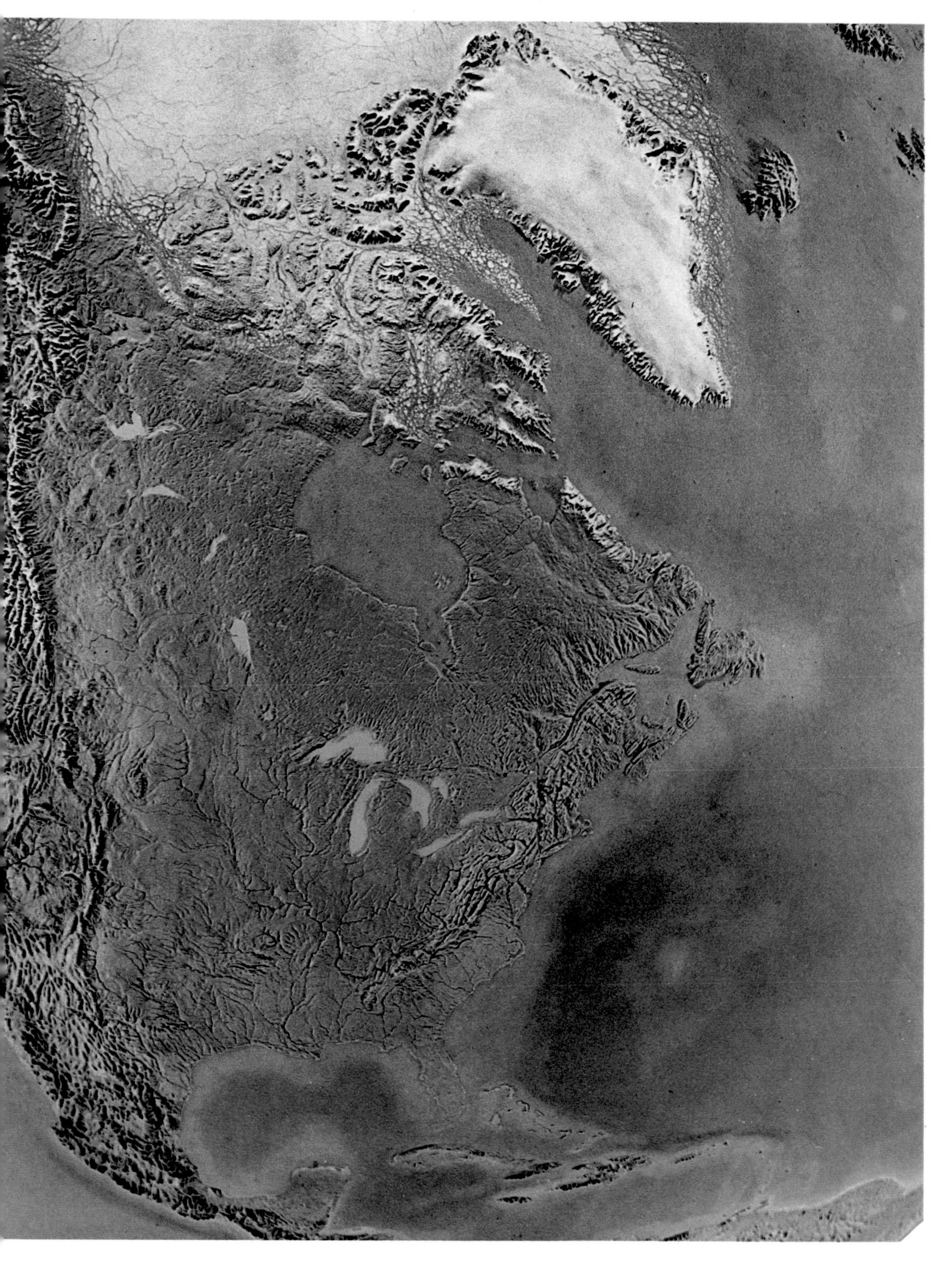

Above: Banff National Park in Alberta, Canada, lies on the eastern slopes of the Rocky Mountains. It is a favourite resort of climbers, fishermen, skiers and tourists. Glaciers feed its still, forest-fringed lakes; and hot mineral springs form pools at the foot of rugged, icy mountains.

Far right: Despite its general richness, North America has vast barren regions.

with the Atlantic coast, and the mountains of Newfoundland and Labrador, farther N. They are broken by the valley of the St. Lawrence River. The Appalachians formed a considerable barrier when the early colonists wanted to move westwards from the Atlantic coastal plain.

The Western Highlands are a complicated system of mountain chains, high plateaux and river basins running from Alaska in the N to the Isthmus of Panama in the S. The Rocky Mountains, in Canada and the United States, are the easternmost part of it.

The Canadian Shield is a low plateau of rock thinly covered with soil. It occupies most of central Canada, forming a great crescent-shaped area around Hudson Bay.

The Central Lowlands run S from the Canadian Shield to the Gulf of Mexico. They are drained by the Mississippi River and its many tributaries.

The Islands of the West Indies are the peaks of a mountain chain now submerged. They tail away from the bulk of the continent towards South America.

Most of North America has warm summers

IMPORTANT DATES IN THE HISTORY OF NORTH AMERICA

18000s B.C. Mongoloid peoples from Asia–the first American Indians—entered Alaska from Siberia.
600–900 A.D. The Maya civilization reached its height.
900s Eric the Red established a colony in Greenland.
1300s–1500s The Aztecs ruled the Valley of Mexico.
1492 Christopher Columbus reached the Americas.
1497 John Cabot landed on Canada's eastern coast.
1518–1521 Hernando Cortes conquered Mexico for Spain.
1607 Settlement established at Jamestown, Virginia.
1608 Samuel de Champlain founded Quebec city.
1620 Pilgrim Fathers landed at Plymouth, Massachusetts.
1642 French colonists founded Montreal.
1689–1763 The British captured the French colonies.
1773 The Boston Tea Party.
1774 The First Continental Congress in Philadelphia resolved to resist the 'Intolerable Acts'.
1775 First shots in the War of Independence, at Lexington, Massachusetts.
1776 The American colonies voted for independence.
1781 Lord Cornwallis surrendered to George Washington at Yorkstown—in effect, the end of the war.
1789 George Washington chosen as first President of the United States of America.
1803 The Louisiana Purchase doubled the size of the United States.
1812 War between the United States and Britain.
1821–1822 Agustín de Iturbide became Emperor of Mexico after defeating the Spanish garrison.
1803 The Monroe Doctrine proclaimed.
1824 Mexico declared a republic.
1830 The first American railway—the Baltimore and Ohio—began operation.
1845 Texas became the 28th state of the U.S.A.
1846–1848 The Mexican War: the U.S.A. defeated Mexico.
1849 The California gold rush.
1859 Petroleum pumped commercially in Pennsylvania.
1861 The French invaded Mexico.
1861–1865 Civil war in the U.S.A. between the northern Union states and the southern Confederate states. The Union, under President Abraham Lincoln, triumphed.
1864 The French made the Austrian prince, Maximilian, Emperor of Mexico
1867 Mexican troops executed Emperor Maximilian.
1867 British North America Act: established the Dominion of Canada.
1901 Theodore Roosevelt became American President; he offered the working man 'a square deal'.
1910–1920 Revolution in Mexico.
1914–1918 Canada and, later, the United States involved in World War I. President Wilson a chief architect of the peace. But the United States rejected the League of Nations.
1929 Stock market collapsed. Start of the Great Depression.
1931 The Statute of Westminster. The Canadian dominion became independent.
1933 Franklin D. Roosevelt became President of the United States: the 'New Deal'.
1939–1945 Canada and, later, the United States and Mexico involved in World War II.
1954–1959 Building of the St. Lawrence Seaway.
1962 The Trans-Canada Highway opened.
1969 America landed first man on the Moon.
1977 Panama Canal Agreement; Panama to take over by A.D. 2000.
1981 Space Shuttle launched.
1983 'Peace-keeping' invasion of Grenada.

and cold winters. But the west coast and the south are mild in winter while the north has weeks of persistent darkness and bitter cold.

Resources and Industry

North America's chief resources are its minerals, its fertile plains, its forests and its lakes and rivers.

The continent is very rich in minerals, particularly coal, iron and petroleum. Other minerals of which there are large deposits include asbestos, bauxite, copper, gold, lead, natural gas, nickel, silver and zinc. Some of these deposits are the largest, or among the largest, in the world. In many cases, the great industrial cities of Canada and the United States owe their development to the fact that they are within easy reach of valuable minerals. The continent's vast natural resources together with the industrial inventiveness of its people and their urge for ever-increasing efficiency have made it pre-eminent in industrial power.

The many lakes and rivers provide water

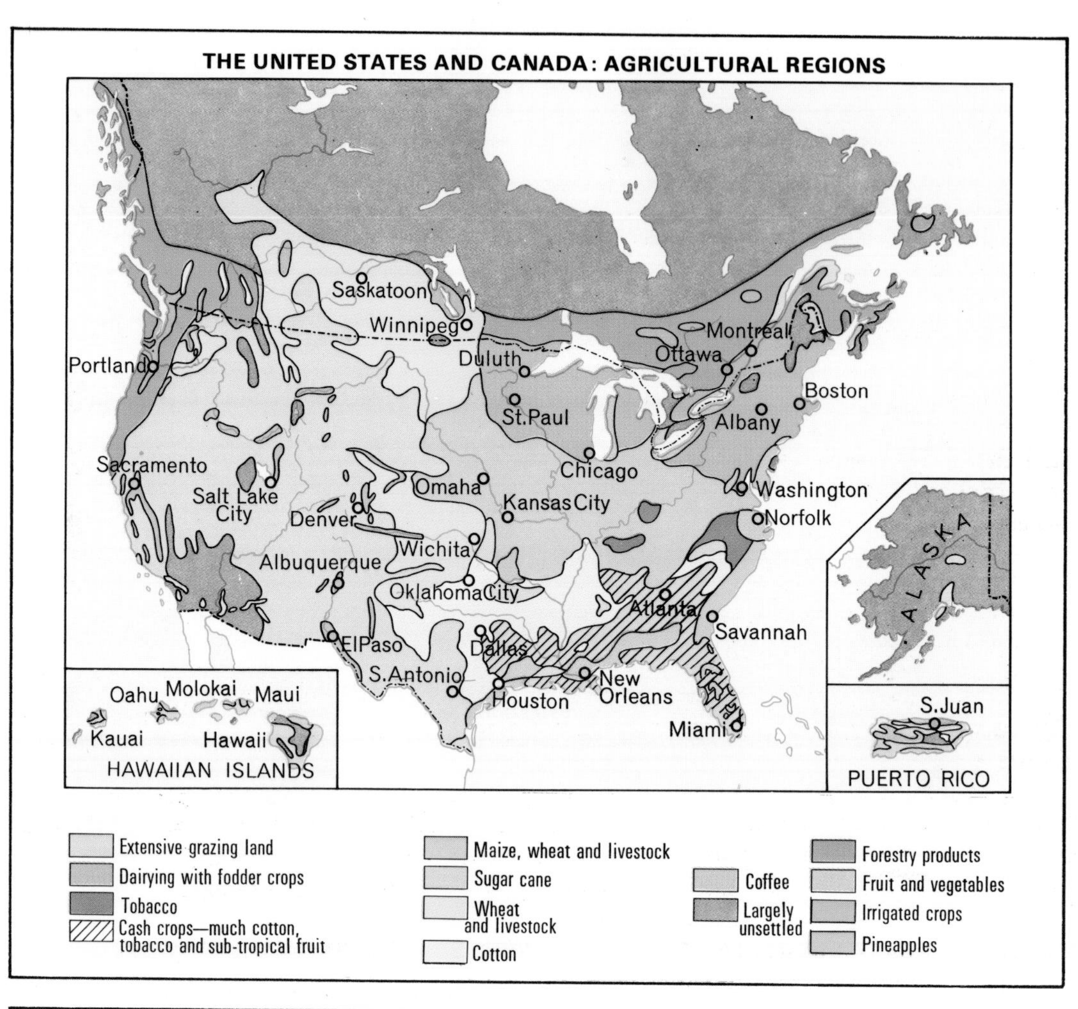
THE UNITED STATES AND CANADA: AGRICULTURAL REGIONS
Saskatoon
Winnipeg
Montreal
Ottawa
Portland
Duluth
Boston
St.Paul
Albany
Sacramento
Chicago
Washington
Salt Lake City
Omaha
Kansas City
Norfolk
Denver
Wichita
Albuquerque
Oklahoma City
Atlanta
Savannah
El Paso
Dallas
S.Antonio
New Orleans
Houston
Miami
ALASKA
Oahu
Molokai
Maui
Kauai
Hawaii
HAWAIIAN ISLANDS
S.Juan
PUERTO RICO
Extensive grazing land
Dairying with fodder crops
Tobacco
Cash crops—much cotton, tobacco and sub-tropical fruit
Maize, wheat and livestock
Sugar cane
Wheat and livestock
Cotton
Coffee
Largely unsettled
Forestry products
Fruit and vegetables
Irrigated crops
Pineapples

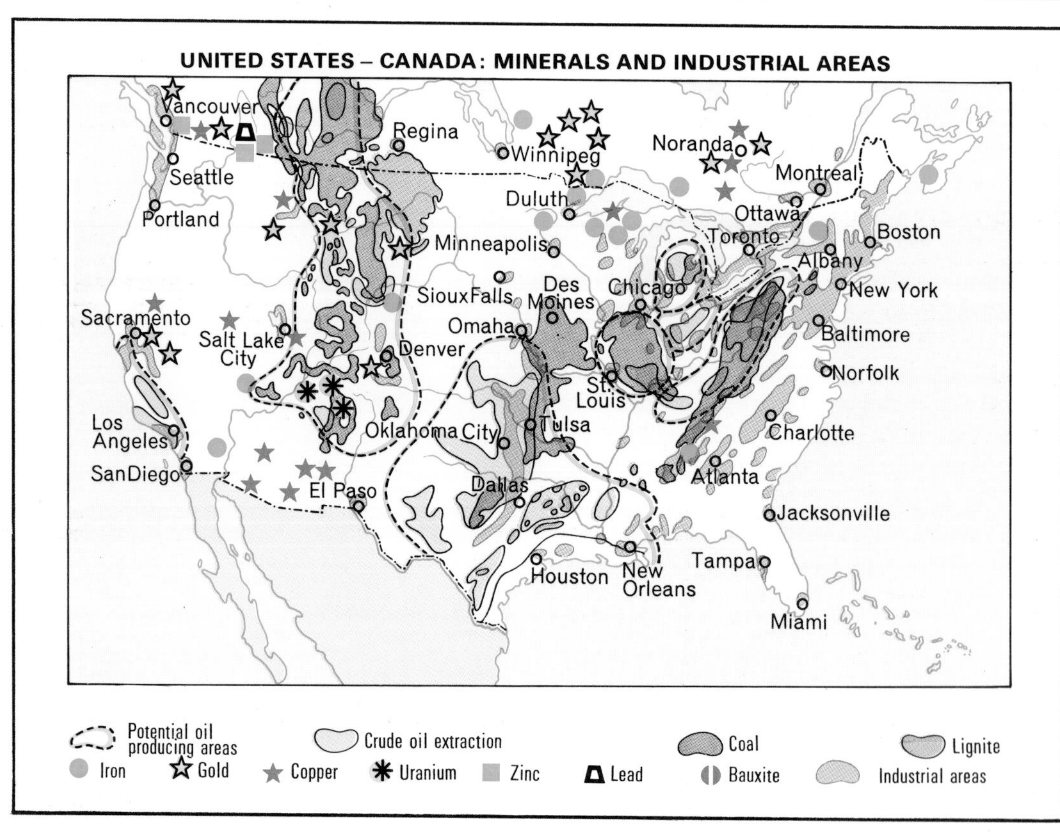
UNITED STATES – CANADA: MINERALS AND INDUSTRIAL AREAS
Vancouver
Regina
Winnipeg
Noranda
Seattle
Montreal
Duluth
Portland
Ottawa
Minneapolis
Toronto
Boston
Albany
Sioux Falls
Des Moines
Chicago
New York
Sacramento
Omaha
Baltimore
Salt Lake City
Denver
St. Louis
Norfolk
Los Angeles
Oklahoma City
Tulsa
Charlotte
San Diego
Atlanta
El Paso
Dallas
Jacksonville
Tampa
Houston
New Orleans
Miami
Potential oil producing areas
Crude oil extraction
Coal
Lignite
Iron
Gold
Copper
Uranium
Zinc
Lead
Bauxite
Industrial areas

for cities and for irrigation. Many of the bigger rivers, such as the Colorado, have been dammed for hydro-electric power and irrigation.

Communications

Southern Canada and the eastern and central parts of the United States have some of the world's best road and rail systems. The Pan American Highway runs from Alaska in the far NW to Panama in the S. Other highways cross the continent from the Atlantic Ocean to the Pacific. By means of the Great Lakes and the St Lawrence Seaway, cities in the industrial centre of the continent have direct access to the Atlantic. The Mississippi River gives central areas access to the Gulf of Mexico. Airlines connect all the major cities of the continent. The Panama Canal provides a short route between the Atlantic and the Pacific.

Agriculture, Fisheries and Land Use

The United States and Canada have some of the world's largest and most productive farms. On the fertile land wheat is grown in the N, maize (corn) in the midlands, and cotton in the S. The two countries produce much more food than they need, and are able to export vast quantities, particularly of wheat and maize. In the W, the great open ranges, with their scrubby vegetation, provide good grazing for cattle and sheep.

Mexico and Central America are less fortunate in their soil than the two northern countries. The agricultural output is generally only enough for the farmers' own needs—that is, it is subsistence farming.

Forests cover almost one-third of the continent. Their trees provide valuable timber, and also wood for paper-making.

People and Ways of Life

Four main groups of peoples live in North America. They are the American Indians; the Eskimoes; the settlers of European origin; and the black Americans.

American Indians were the first people to settle in the continent. Scientists believe that their ancestors may have crossed from Siberia about 20,000 years ago.

An oil refinery on Delaware Bay, on the Atlantic seaboard of the United States. The industrial might of North America is based on its rich natural resources. They include petroleum, coal, copper, nickel, iron, asbestos, bauxite and gold.

THE ST. LAWRENCE SEAWAY

The St. Lawrence Seaway has made it possible for ocean-going ships to sail for 2,300 miles (3,700 km) from the Atlantic to Lake Superior, in the heart of the North American continent. The seaway itself is 182 miles (293 km) long, stretching from Montreal to the entrance to Lake Ontario.

Canada and the United States co-operated in the seaway's construction. Work began in 1954 and the opening ceremony took place in 1959.

Power installations built at the same time supply hydro-electricity to Ontario in Canada and New York in the United States.

THE ST. LAWRENCE SEAWAY

L. Superior
CANADA
Sault St. Marie
Montreal
Ottawa
Duluth
Lake Huron
Superior
Toronto
Lake Ontario
Lake Michigan
Welland C.
Detroit
Buffalo
Chicago
Lake Erie
UNITED STATES

m.183 feet 598
Sault St. Marie
m.177 ft. 578
m.175 ft. 572
Welland Canal
Lake Superior
Lake Erie
m.75 feet 245
Sea level
Lake Michigan an Huron
Lake Ontario
Lake St. Louis

Iroquois
C A N A D A
Kingston
Prescott
Cornwall
Lake St. Francis
St. Lawrence River
Ogdensburg
Eisenhower
Snell
L. St. Louis
Montreal
Lake Ontario
Beauharnois
Côte St. Catherine
St. Lambert
U N I T E D S T A T E S

m.77 800 st.mls.48
m. 53 700 st.mls. 33
25 900 st.m.16
m. 40 700 st.mls. 25
27 800 st.m. 17
m.37 000 st.mls. 23
80 262
60 196
40 131
20 65 feet
0 metres
St. Lawrence River
Iroquois
Eisenhower
Canal
L. St. Francis
Canal
Côte St. Catherine
Snell
Beauharnois
Sea level
St. Lambert

Above: Breaking-in a young horse. Without horses, the great cattle ranches could not operate.

Right: An ornate Baroque church typifies the colonial architecture of Latin America.

Before the arrival of the white men, there were several hundred tribes of American Indians. Generally, each tribe had its own language. But the languages, and the tribes, can be grouped into about 50 families. In each family, the different languages are really dialects (regional variations) of the same basic tongue.

The Indians of the plains, the wide areas of the Central Lowlands, were *nomads* (wanderers) who lived mainly by hunting and fishing. Chiefly, they hunted the bison that roamed the plains in their millions. Before the arrival of the white men, the Indians had no horses, so they hunted on foot. For weapons they used bows and arrows, spears and tomahawks (axes). At this stage there were comparatively few Plains Indians, but once the Indians acquired horses many more tribes moved into the plains to hunt bison.

Forest Indians such as the Mohicans, Hurons and Iroquois also lived mainly by hunting and fishing. Some, however, were farmers.

The Indians of the SW lived in dry areas where game was scarce. They built villages, grew crops, and irrigated the dry land. In Mexico the Aztecs created a great civilization and built with stone.

The Eskimoes live in the cold lands of the N. They are also of Mongoloid origin. Over hundreds of years they have developed plump bodies whose extra fat helps to keep out the cold. The Eskimoes are found in Alaska, northern Canada and Greenland.

NORTH AMERICA AND CENTRAL AMERICA
COLONIES AND OTHER TERRITORIES

Name	*Location, etc*	*Area* *sq miles*	*sq km*	*Population*	*Capital*
Anguilla (U.K.)	Island in the Lesser Antilles.	35	90	6,000	
Antigua (U.K.)	Island in the Lesser Antilles.	108	280	77,000	St. John's
Belize (U.K.)	Caribbean coast of Central America. Formerly called British Honduras.	8,867	22,965	145,000	Belmopan
Bermuda (U.K.)	Group of 300 small Atlantic islands.	21	54	68,000	Hamilton
Cayman Islands (U.K.)	3 islands; S of Cuba.	100	259	18,000	George Town
Greenland (Den.)	World's largest island. *See Denmark.*	840,000	2,176,000	51,000	Godthåb
Guadeloupe (Fr.)	Island & group in the Lesser Antilles.	687	1,779	318,000	Basse-Terre
Martinique (Fr.)	Island in the Lesser Antilles.	435	1,127	308,000	Fort-de-France
Netherlands Antilles (Neth.)	Curaçao & other islands; West Indies.	391	1,013	261,000	Willemstad
Puerto Rico (U.S.)	Island in the West Indies.	3,435	8,897	3,199,000	San Juan
St. Kitts-Nevis (U.K.)	Islands in the Lesser Antilles.	101	261	67,000	Basseterre
St. Lucia	Island in the Lesser Antilles.	238	616	122,000	Castries
St. Pierre & Miquelon (Fr.)	Island group; S of Newfoundland.	93	241	6,000	St. Pierre
St. Vincent	Island group in the Lesser Antilles.	150	388	124,000	Kingstown
Turks & Caicos Islands (U.K.)	Island group in SE Bahamas.	193	500	7,000	Grand Turk
Virgin Islands (U.K.)	Part of West Indies island group.	59	152	13,000	Road Town
Virgin Islands (U.S.)	Part of West Indies island group.	133	344	114,000	Charlotte Amalie

The Europeans first arrived in North America in the late 1400s and early 1500s. The earliest settlers were Spaniards, who formed colonies in Central America, the West Indies and Mexico. Settlers from the British Isles occupied the eastern parts of what is now the United States, while French settlers moved into what is now southern Canada. During the 1880s and early 1900s, millions of people from Europe flocked to the United States.

The *black Americans,* who originate mainly from W Africa, were taken to America from the 1500s onwards. There they were made to work on plantations. Slavery was abolished in North America in 1865.

As a result of these different strains of settlement, the peoples of North America are today very varied in type. The Spaniards who colonized Mexico, Central America, and the West Indies often intermarried with the Indians who lived there. In Mexico, most of the people today are *mestizos,* people of mixed Spanish and Indian blood.

Most of the countries of Central America also have large numbers of mestizos in their populations. Some of the island countries of the West Indies, such as Jamaica and Haiti, have populations that are mainly black.

In the United States, very few settlers intermarried with the American Indians. Today, most of the Indians live in *reservations,* areas set aside for their exclusive use. A few have settled in other parts of the country and integrated with the rest of the population. The blacks, too, have largely remained a separate community. There are many *mulattos* (half black, half white), but they are generally grouped with the Negroes. The white people, who form the largest and most powerful section of the population, have tended to treat the blacks, and also the American Indians, as second-class citizens. This has led to racial conflict.

In Canada there is cultural conflict of a different kind. There, the tension is between the French-speaking Canadians of Quebec province and the English-speaking Canadians of the rest of the country.

In the United States and Canada, 70 out of every hundred people live in cities or towns. Only about 10 workers out of every hundred are engaged in farming. But in Mexico and Central America about 60 out of every hundred people are farmers or farm workers.

Top: The American Indians, immigrants from Asia, spread over North and South America and developed many different cultures. Today, Canada and the United States have about 700,000 Indians.

Centre: Eskimoes in the Arctic north. The Eskimoes of North America are believed to have crossed from Siberia, in Asia, thousands of years ago. For food they rely heavily on walrus, seals and fish. Their fragile boats, *kayaks* and *umiaks*, are made of walrus hide stretched over a wooden frame.

Below: An Unmistakable Skyline. Skyscrapers on Manhattan Island, New York City. In the foreground is the United Nations Secretariat Building. In the background: left, the Empire State Building; right, the Chrysler Building.

THE COUNTRIES OF NORTH AMERICA

Bahamas

The Bahamas are an archipelago in the north-eastern part of the West Indies, extending NW–SE for about 750 miles (1,200 km). The Bahamas include more than 2,700 islands, many of them little more than rocks in the sea. About 30 islands are inhabited. The chief islands include Grand Bahama, Abaco, Acklins, Eleuthera, New Providence, Andros, San Salvador, Exuma, Long Island, Mayaguana and Inagua. Most of the islands are low and flat, with fine beaches. The climate is warm and healthy. Tourism provides most of the country's income. Otherwise the main occupations are farming and fishing. The Bahamas are also a centre of international banking. Industries include cement, food processing and oil refining. Four-fifths of the people are of African descent and the rest of British or North American descent. English is spoken. The cap. is Nassau on New Providence Island. Area 5,400 sq miles (13,990 sq km).

Right: Niagara Falls on the Niagara River, which connects Lake Erie and Lake Ontario in Canada. The Falls are on the border between Canada and the United States. Horseshoe Falls, on the Canadian side, are 186 ft (56.7 m) wide. Below: Algonquin Provincial Park in Ontario, near the Ottawa River. It covers more than 2,500 sq miles (6,500 sq km), and has about 1,200 lakes.

Barbados

A small, flattish coral island, Barbados is the most easterly island of the West Indies. It lies about 250 miles (400 km) NE of mainland South America. From the W coast, terraces gently rise to a central ridge where Mount Hillaby reaches 1,104 ft (336 m). The climate is warm and healthy. Natural resources include fertile soil, limestone, coral and small amounts of natural gas. Tourism is the most important source of income. The chief crop is sugar cane and stock raising is important. The principal exports are sugar, molasses, rum and fish. Almost all the people are of African descent. The others are descended from European, mainly British, settlers. English is the official language. The cap. and largest city is Bridgetown. Area 166 sq miles (430 sq km).

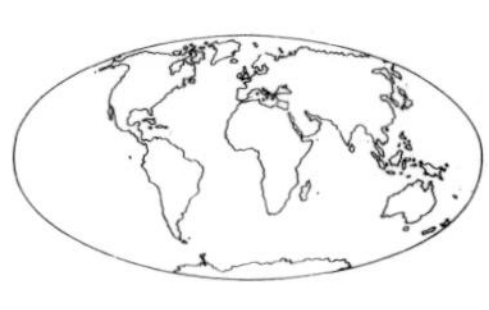

Canada

In area, the world's second largest country, covering 3,851,809 sq miles (9,976,139 sq km). Canada is in the northern part of the continent of North America, bounded on the S by the United States and on the N by the Arctic Ocean. Its capital is Ottawa. The country is a monarchy, acknowledging Queen Elizabeth II of the United Kingdom as its head of state. She is represented by a governor-general. Canada has 10 provinces, each with its own legislature and government, and two territories administered by councils. The 1982 Canada Act transfers to Canada full constitutional powers which were previously held by London. A federal parliament and government deals with national and international affairs. Canada is a member of the Commonwealth of Nations.

Land and Climate

The northern part of Canada, which includes many islands, lies within the Arctic Circle. This region, about a third of the country, is tundra—cold, treeless plains covered with moss, lichen and other dwarf plants. The ground is permanently frozen but for the top 3 ft (1 m). Half of the country is forested. Altogether, some three-quarters of Canada is too cold to be more than sparsely populated. Most people live in the southern 200-mile (320-km) strip along the border with the United States. Even there, cold and snow halt farming for several months in winter.

The southern part of the country has four main regions, marked by natural barriers. In the W is the high Pacific coast region of British Columbia, bounded on the E by the lofty, rugged Rocky Mountains. Most of Yukon Territory in the NW is also mountainous. The next region includes the prairie provinces, with wide, grassy plains sloping gently from W to E. In the plains there are two large lakes, Winnipeg and Winnipegosis. E of the prairies is the Canadian Shield, a horseshoe of hard rock comprising almost half the land area of the country.

The third and most important populated region is on the N bank of the St. Lawrence River and the borders of the five Great Lakes (Superior, Michigan, Huron, Erie and Ontario) which form part of the frontier with the United States. The fourth region is around the Gulf of St. Lawrence and along the Atlantic coast.

Agriculture and Industry

Canada has rich deposits of coal, petroleum and natural gas and also copper, gold, iron, lead, nickel, platinum, silver, uranium and zinc. From the forests come lumber and fur pelts. Cattle graze on the prairies and Canada is a leading producer of barley, oats and wheat. The manufacturing industries produce steel, paper, chemicals, textiles and many other types of products.

Cities and Communications

Besides Ottawa, Canada's leading cities are Montreal, Toronto, Vancouver, Winnipeg, Hamilton and Edmonton. The rail network has more than 43,000 miles (69,000 km) of track, linking the main centres of population, and the Trans-Canada Highway crosses the country from the Atlantic to the Pacific. The St. Lawrence Seaway links the Great Lakes and the Atlantic.

The People

The standard of living in Canada is one of the world's highest. About half of the people are descended from settlers from the British Isles. But more than a quarter are of French ancestry and speak French as their everyday language. Most of them live in Quebec province. One-fifth of all Canadians are from other European countries—Germany, Poland, the U.S.S.R., the Netherlands and the countries of Scandinavia. The rest are American Indians or Eskimos, descendants of the country's original inhabitants. English and French are the official languages.

Above: Toronto, the capital of Ontario. The second largest city in Canada, it has a fine harbour on the north-western shore of Lake Ontario. It is an important industrial and cultural centre.

Below: The Changing of the Guard in front of the Parliament Buildings in Ottawa. The sandstone buildings stand on three sides of a 35-acre (14-hectare) quadrangle. The cornerstone was laid by the Prince of Wales, later King Edward VII, in 1860.

Costa Rica

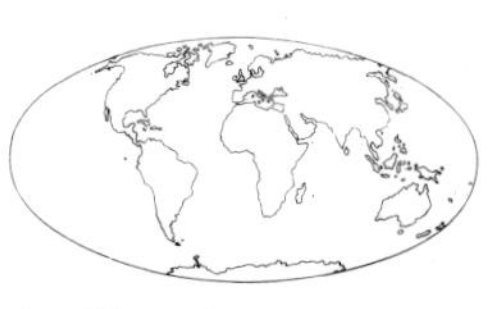

A wide coastal plain in the E and a narrower one in the W rise to volcanic mountain ranges crossing the country from N to SE. The climate is tropical in the coastlands, but cooler in the mountains. About two thirds of Costa Rica is forested. The main agricultural products are coffee, bananas, cocoa, sugar and rice. There are some manufacturing industries and some mineral deposits. Most of the people are of European descent, or of mixed Indian-European blood. They speak Spanish and are mainly Roman Catholics. The cap. is San José. Area 19,650 sq miles (50,890 sq km).

Right: Picking coffee berries on a large plantation in Costa Rica. The berries go to a pulping house where the coffee beans are extracted. Each berry contains two beans.

Below: The Thompson River valley in British Columbia. It rises in the Rocky Mountains, and flows 270 miles (434 km) south-west to the Fraser River.

Cuba

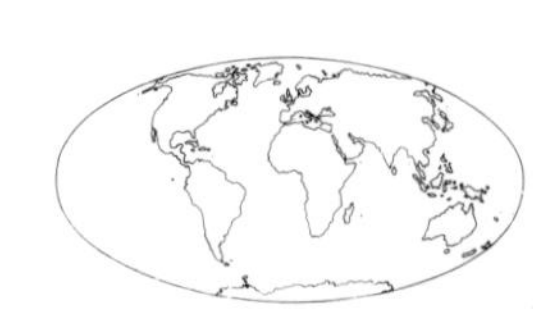

The largest and most westerly island in the West Indies, Cuba lies between the Caribbean Sea and the Atlantic Ocean. It is about 100 miles (160 km) S of Florida in the United States, at the mouth of the Gulf of Mexico. Gently rolling land, with many wide, fertile valleys and plains, makes up about 60 per cent. of Cuba's surface. The rest is hilly. The highest peak is Turquino (6,560 ft; 2,000 m) in the Sierra Maestra range in the SE. About 760 miles (1,220 km) long, Cuba varies in width from 25 to 125 miles (40 to 200 km), and has a much-indented coastline of about 2,000 miles (3,200 km). The climate is tropical in the lowlands, but cool and moderate in the hills. Cuba's rich mineral deposits include nickel, iron, manganese, cobalt, copper, chromite, silica, gold and silver.

About a third of the people live by farming or working in plantations. The main products are sugar, tobacco and coffee.

Industries include sugar refining and the manufacture of cigarettes and the world-famous Havana cigars, rum and textiles. Cuba also has a flourishing film industry.

More than two-thirds of the people are of European (mainly Spanish) descent. Of the rest about half are of African ancestry and half of mixed European-African blood. There is also a small Chinese minority. Spanish is the official language. Cuba has a communist government.

The chief cities are Havana (the cap.), Santiago de Cuba, Camagüey, Guantánamo and Santa Clara. Area 44,218 sq miles (114,524 sq km).

Dominica

Mountainous island in the Lesser Antilles. It became an associated state of the U.K. in 1967, and gained independence in 1978. It is very fertile and exports fruit, copra and cocoa. The cap. is Roseau. Area 290 sq miles (750 sq km).

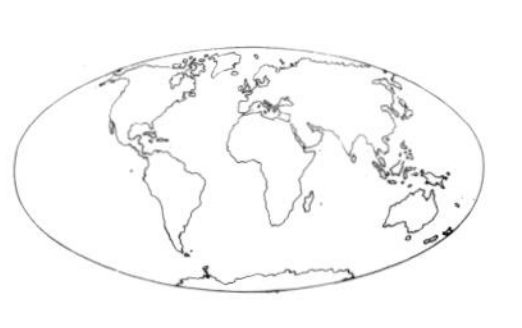

Dominican Republic

Situated in the West Indies, occupying the eastern two-thirds of the island of Hispaniola, which it shares with Haiti. Much of the land is mountainous, with four E to W ranges. The climate, tropical in low-lying areas, is cool in the mountains. Mineral deposits include bauxite, nickel and iron. Most of the people live by farming. Their main cash crops are sugar, coffee, cocoa, tobacco and bananas. They speak Spanish. The chief cities are Santo Domingo (the cap.) and Santiago. Area 18,800 sq miles (48,700 sq km).

El Salvador

El Salvador is in Central America. E to W mountain ranges divide the country into three main natural regions: a narrow Pacific coastal plain; a central region of plateaux and valleys; and mountains in the N. Mineral resources include quartz, china clay and gypsum. More than half the labour force works in agriculture. The main products are coffee, cotton, sugar and rice. Livestock are important. Most of the people are mixed Spanish-Indian and the remainder are either Spanish or Indian. The chief cities are San Salvador (the cap.), Santa Ana, San Miguel and La Union. Area 8,260 sq miles (21,400 sq km).

Grenada

The small island of Grenada in the West Indies is the most southerly of the Windward Islands, and the islands of the South Grenadines. In 1983 the U.S. made a controversial armed intervention at the request of neighbouring islands after the breakdown of democratic government in Grenada. Grenada is volcanic and mountainous, with fertile soil. It is famous for its spices. The people are of African, Carib Indian and European ancestry. They speak English and a French-African dialect. The cap. is St. George's. Area 133 sq miles (344 sq km).

Verandahs in a Cuban village street offer protection against sun and rain. The island lies in the tropics and the lowlands are extremely hot. The rainy season lasts from May to October.

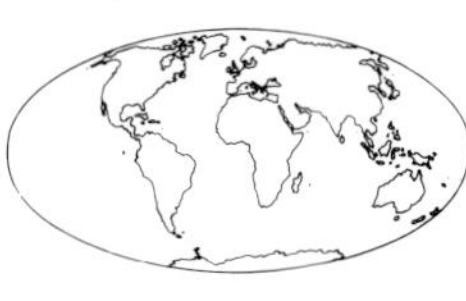

Guatemala

The republic of Guatemala in Central America is bordered on the N by Mexico, on the E by Belize and on the S by Honduras and El Salvador. In the E it has a coastline on the Caribbean Sea and in the SW on the Pacific. A highland region takes up about a fifth of the country in the SW. There is a narrow Pacific coastal plain, and a wider plain on the Caribbean side with fertile river valleys. The northern part of the country is a forested lowland, some of it thick jungle. The climate ranges from sub-tropical in the lower parts to temperate in mountain regions. The longest rivers are the Motagua, Usumacinta and Polochic. Guatemala's mineral deposits include zinc, silver, lead, mica, chromite, nickel and limestone. Among its crops are coffee, cotton, sugar cane, bananas, maize, beans, rice and wheat.

Industries include food processing and the manufacture of textiles, paper, chemicals and machinery.

About half of the people are Indians, descended from the Mayas, whose civilization is one of the oldest in the Western Hemisphere. The rest are mainly of mixed Indian and European descent. The official language is Spanish, but various Indian dialects are widely spoken. Most of the people are Roman Catholics. The chief cities are Guatemala City (the cap.), Quezaltenango, Mazatenango, Barrios and Antigua. The constitution was suspended after a military coup in 1982 and there is persistent trouble with guerrilla groups. Area 42,040 sq miles (108,900 sq km).

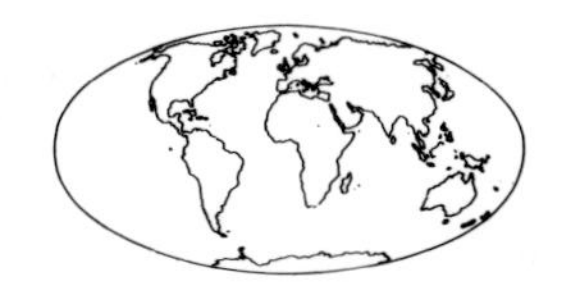

Haiti

Haiti in the West Indies occupies the western third of the island of Hispaniola, which it shares with the Dominican Republic. On the N is the Atlantic Ocean and on the S the Caribbean Sea. Much of the land is mountainous and difficult to farm. In lower areas the climate is tropical.

Bauxite and copper are mined, but most people live by subsistence farming. The main cash crops are coffee, sugar, sisal and cotton.

Almost all the people are of African descent. They speak French and Creole. The cap. and main city is Port-au-Prince. Area 10,700 sq miles (27,700 sq km).

CENTRAL AMERICA: ECONOMY

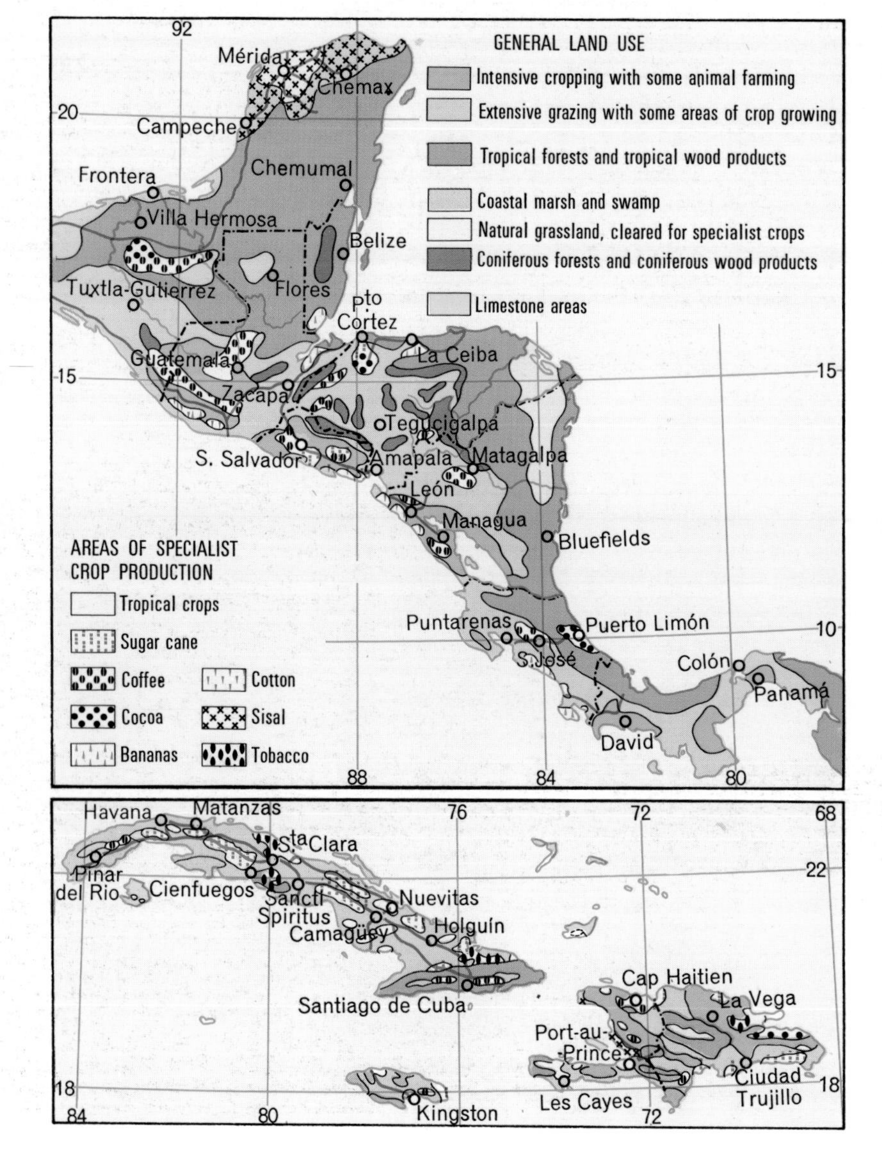

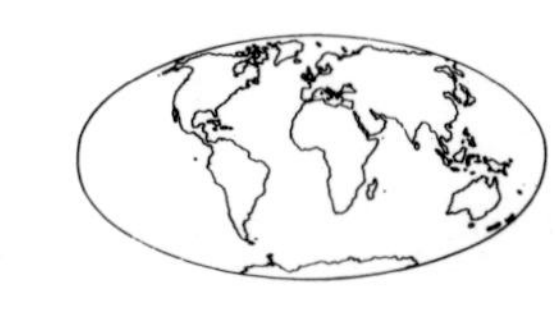

Honduras

Honduras is one of the largest and poorest of the countries of Central America. It is bordered on the S by Nicaragua and on the W by El Salvador and Guatemala. On the N it has a long coastline on the Caribbean Sea and on the SW a short Pacific coastline. Most of the country is mountainous except for a narrow Pacific coastal plain and a wider Caribbean one. There are many fast rivers. The climate is humid and subtropical in the lower parts. The chief products and crops are bananas, coffee, maize, meat, cotton and timber. The people speak Spanish and are mostly of mixed Indian-European blood. Railways serve the ports and banana plantations, 155 miles (250 km) of major highways have recently been opened and a new deep-water port has been opened at San Lorenzo. The cap. is Tegucigalpa. Area 43,300 sq miles (112,000 sq km).

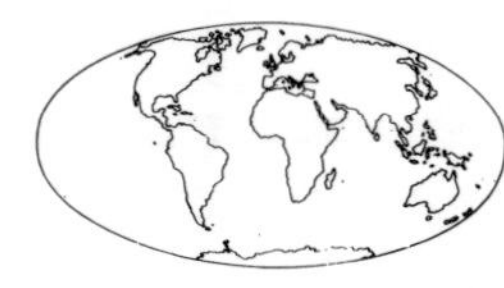

Jamaica

Island country in the Caribbean Sea, the third largest island in the West Indies. Most of the island is mountainous, with the main range running from E to W. The highest point is the Blue Mountain Peak (7,388 ft; 2,252m) in the E. There is a narrow coastal plain in the S and lowlands stretch across the W end of the island. The island is famous for its golden beaches. There are also many short streams. The climate is humid and subtropical. Jamaica's mineral resources include bauxite, of which it is the fourth largest producer, and gypsum, silica, marble and clays. The main agricultural products include sugar, bananas, cocoa, coffee, citrus fruits and copra. Forestry and fishing are also important. Food processing and the production of rum and beer are important industries. Most of the people are of African or mixed descent (Afro-Asian, Afro-European, or all three). English is the official language, but an Afro-English patois is also widely spoken. The chief cities are Kingston (the cap.), Montego Bay and Spanish Town. Area 4,440 sq miles (11,400 sq km).

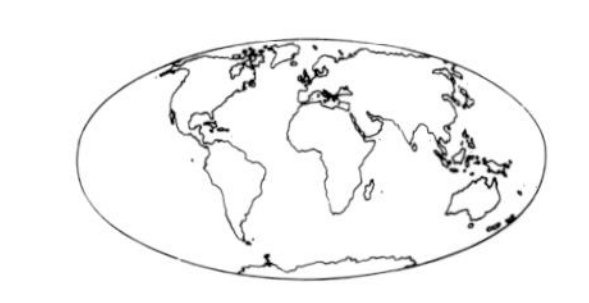

Mexico

The United Mexican States, in the southern part of the North American continent, has a long coastline on the Pacific Ocean in the W, and a coastline on the Gulf of Mexico in the E. Its capital is Mexico City. Mexico's area is 760,000 sq miles (1,968,000 sq km).

Land and Climate

Mexico has more than 6,000 miles (9,600 km) of coastline. The E coast has many sandbanks and lagoons and few natural harbours. The W coast has better harbours but is cut off from the interior by difficult terrain. Much of Mexico is mountainous, more than half the country being above 3,000 ft (900 m) in altitude. The main features are a large, steep-sided central plateau, rising to some 8,000 ft (2,400 m) in the S and flanked by the Western, Eastern and Southern Sierra Madre mountain ranges. The plateau has many volcanoes, including the active Popocatepetl (17,887 ft; 5,452 m), Iztaccihuatl (16,883 ft; 5,146 m) and Citlaltepetl (18,700 ft; 5,700 m), Mexico's highest peak. Other features include the rugged peninsula of Lower California, lying to the W of the Gulf of California; the Sonora Desert between the Western Sierra Madre and the coast; and the E coastal plain extending into the low Yucatán peninsula. Mexico's main rivers include the Rio Bravo (Rio Grande), the Lerma, the Santiago and the Usumacinta. The climate varies with altitude. At low altitudes it is subtropical, ranging from warm to cool temperate and Alpine with increasing altitude.

Agriculture and Industry

Half of the people live by farming. The main cash crops include cotton, coffee and sugar. Maize is the staple crop. Livestock raising, forestry and fishing are also important. Mineral resources include silver, of which Mexico is the world's leading producer, and sulphur, lead, zinc, oil, coal, tin, mercury, copper and iron. The chief industries are the making of textiles and steel.

Cities and Communications

Half of the population lives in towns and cities. Apart from Mexico City, the main centres are Guadalajara, Monterrey, León, Cuidad Juárez and Mexicali. Road, rail and air services are well developed.

The People

About two-thirds of Mexicans are *mestizos* of mixed Indian and European descent. Spanish is the official language, but many Indians speak only Indian dialects.

Below: Sightseers and merrymakers take part in a river fiesta. The decoration of the flat-bottomed boats recalls the art of the Aztecs. Bottom: Southern Mexico and Central America were the home of the remarkable Maya civilization that flourished from about the A.D. 300s to the 1400s.

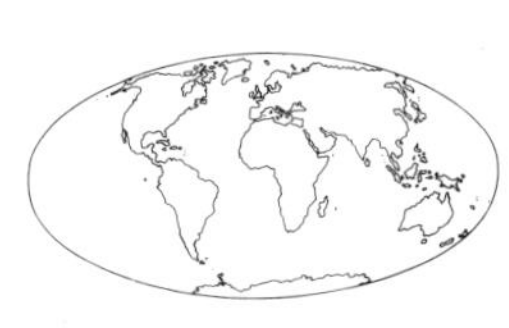

Nicaragua

The largest country of Central America. Left-wing guerrillas overthrew the government in 1979. The Pacific coastal plain rises gently towards a volcanic mountain range. Beyond is the forested interior of plains and hills, cut by rivers. The swampy Caribbean coastal plain—the Mosquito Coast—reaches about 50 miles (80 km) inland. The climate is humid and sub-tropical. Most people live by working on farms. The main products are cotton, coffee and sugar. Stock-raising, forestry and fishing are also important. Two-thirds of the people are of mixed European and Indian ancestry. Spanish is the official language. The chief cities are Managua (the cap.), León and Granada. Area 57,150 sq miles (148,020 sq km).

Right: The Panama Canal cuts through Central America to connect the Atlantic and Pacific Oceans. It saves ships the long journey round the southern tip of South America. By treaty with Panama, the United States governs the Panama Canal Zone, the strip of territory through which the canal runs, except for the cities of Panama and Colón.

Below: The luxuriant vegetation of Trinidad. The island, which lies 7 miles (11 km) from the coast of Venezuela, is celebrated for its beauty, its sugar and its asphalt.

ST. LUCIA AND ST. VINCENT

In 1979, two former British territories in the Lesser Antilles, St. Lucia and St. Vincent, became independent nations. The economies of both countries depend mainly on farming and tourism. Both islands are of volcanic origin.

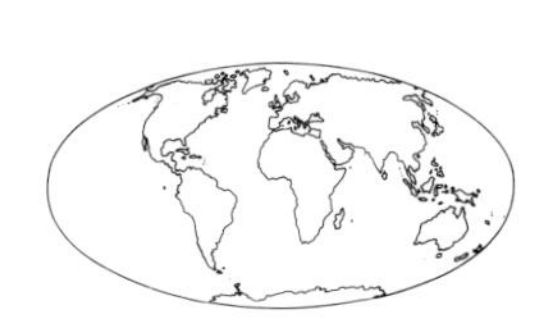

Panama

Panama, in Central America, occupies the southern part of the narrow Isthmus of Panama. Most of Panama is mountainous, with two main W to E ranges. The country is divided in two by the Panama Canal. The coastal regions are low, the E side being mostly covered by tropical forest. The climate is humid and subtropical. About half of the labour force works in agriculture. The chief crops are rice, sugar, bananas, cocoa, coffee and tobacco. Industries include food processing and the manufacture of textiles, cement and chemicals. Two important sources of income are Canal earnings and the registration of foreign ships under the Panamanian flag. Panama is to take over control of the Canal from the U.S. in the year 2000. Most people are of mixed Indian-European blood and speak Spanish. The chief cities are Panama City (the cap.) and Colón. Area 29,200 sq miles (75,600 sq km).

Trinidad and Tobago

The two most southerly islands of the Lesser Antilles, just off the coast of Venezuela in South America. Most of the people live on the larger island, Trinidad. It is crossed by three mountain ranges, the highest being in the N. There, El Cerro del Aripo rises to 3,085 ft (940 m). Tobago is also mountainous. The islands lie near the equator and have a tropical climate—hot and wet. Their prosperity is based on oil and asphalt resources, and sugar and cocoa. The people are mainly of Indian and African ancestry. The cap. is Port of Spain, on Trinidad. Area 1,980 sq miles (5,130 sq km).

United States of America

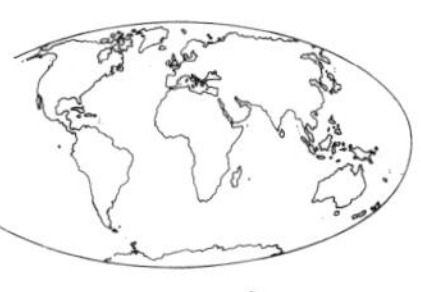

The world's fourth largest country in area and population, the United States is the richest and most powerful of all countries. Its highly developed industries are based on immense natural resources.

The United States consists of the central part of the North American continent, plus Alaska in the NW and the islands of Hawaii in the Pacific Ocean. Its total area is 3,615,211 sq miles (9,363,353 sq km).

The United States is a federation of 50 states, one of which is Alaska and one Hawaii. In addition, there is the federal District of Columbia (D.C.), in which Washington is situated. Each state has its own elected legislative assembly, governor and state government. Control over the whole country is exercised by the federal government in Washington, D.C.

Land and Climate

Along the Pacific coast there is a narrow lowland strip rising sharply to the Western Highlands, a region of mountain ranges and high plateaux. The Pacific Coast Ranges lie very close to the sea, and inland and parallel to them comes a mountain chain made up of the Cascade Mountains and the Sierra Nevada. Still farther inland are the lofty Rocky Mountains, which extend N–S through most of the North American continent. Between the Cascade-Sierra Nevada chain and the Rockies lies the Colorado Plateau, parts of which are more than 6,000 feet (1,830 m) above sea-level. There are also basins that once contained salt lakes.

The Great Plains—the prairie country—are the heart of the United States. These vast rolling expanses are watered by the Mississippi River and its tributaries, the Missouri and the Ohio. To the N of the Great Plains lie the five Great Lakes—Superior, Michigan, Huron, Erie and Ontario.

East of the Great Plains is another barrier, formed by the Appalachian Mountains, which for a long time limited the westward expansion of the early colonies. From the Appalachians, the Eastern Coastal Plain slopes away to the Atlantic.

Earthquake activity is associated with the

The Empire State Building in New York City was for long the world's tallest building and is still probably the most famous of all skyscrapers. It has 102 storeys.

Right: The Statue of Liberty on Liberty Island in New York harbour. The largest statue ever made, it is 151 feet (46 m) high. It is made of beaten copper over an iron framework. The statue was designed by the French sculptor Frédéric Auguste Bartholdi, and was a present from the people of France to the people of the United States. It was unveiled in 1886.

Below: The San Francisco–Oakland Bay Bridge spanning San Francisco Bay between the cities of San Francisco and Oakland. With its approaches, it is some 8 miles (13 km) long, and has a number of sections. Some sections are connected by a tunnel on Yerba Island in the bay. The bridge has two decks, the upper one being a roadway with six traffic lanes. It was opened in 1936, and is one of the longest bridges in the world.

San Andreas Fault in California. In 1980 there was a major volcanic eruption of Mount St. Helens.

The climate of the United States varies considerably, partly because of the wide differences in elevation of various regions of the country and partly because the United States stretches from cool northern latitudes almost to the Tropic of Cancer. Along the Atlantic coast the climate is hot and humid in summer. But it is cold in winter because of the influence of a cold Atlantic current. On the other hand, Florida, in the SE of the country, is famous for its warm winters. In the Great Plains most areas enjoy hot summers and cold winters. This area has between 20 and 40 inches (500 mm and 1 m) of rainfall each year, about half the rainfall of the E coast. The Western Highlands have less than 20 inches (500 mm) of rain and much of the region is desert, or at any rate too dry to be of any use as farmland. Along the NW coast winds blowing in from the Pacific bring the country's highest rainfall.

Alaska, the cold north-western part of North America, has an area almost one-fifth that of the rest of the United States. It is a land of icy mountains and forests. A third of it is N of the Arctic Circle. It has two great mountain ranges. The Alaska Range curves round from the Aleutian Islands in the W to the south-eastern coast. The Brooks Range, in the Arctic N, is part of the Rocky Mountain system. Between the two ranges is the wide Yukon River Valley, a swampy, hilly region with lichen, shrubs and a few forests. Reindeer and moose graze in this region.

Hawaii consists of a group of more than 20 volcanic islands, lying in the tropics. The largest of the group is Hawaii Island, but most of the people live on Oahu. Rainfall on the W side of these mountainous and beautiful islands is more than 100 inches (2.5 m) a year, while in the E it is only about one-third of that.

Agriculture and Industry

The United States is one of the world's chief agricultural countries, with millions of acres of rich land. On the Great Plains in the centre, a variety of crops is grown and huge herds of cattle roam. The principal crop in many areas is maize (corn). The United States grows about half of the world's maize. In the N, spring wheat is the chief crop, together with fodder and hay for dairy cattle. In the S, where there are more than 200 frost-free days every year, cotton is the principal crop. Rice is grown on the S coast, along the Gulf of Mexico.

In the eastern states, farmers raise cattle and other livestock and grow fruit, vegetables, tobacco and, in the S, cotton. The mountainous regions of the W have little soil suitable for crops, but cattle and sheep graze on the poor pastures. A large area of pasture is needed for a few animals. By irrigation schemes from the rivers, such as the Colorado, the amount of arable land has been greatly increased.

The mighty industries of the United States are based on its abundant mineral resources. The chief of these are coal, iron and petroleum and the country also has rich supplies of bauxite, copper, gold, lead, phosphorus, silver, uranium and zinc. But some of these minerals are not plentiful enough to feed the country's developed industries and the United States imports copper, iron ore, uranium and petroleum, as well as minerals which it does not possess in great quantities, such as tin.

About one-third of all the world's manufactured goods are produced in the United States. The bedrock of American industry is steel, of which American plants produce as much as the rest of the world put together. It was because of its enormous industrial output and strength that the United States was able to put men on the Moon and lead the world in the 'Space race'.

In addition to its actual manufactures, the country operates a continuing programme of research and development, which keeps it in the forefront of the world's scientific discoveries. The programme for the building and development of nuclear power stations received a setback in 1979 when a radiation

Left: A steamboat on the Mississippi River. Today, such craft carry mainly tourists, but once they travelled up and down the river transporting cotton, farm produce and other goods between the cities and small towns along the Mississippi's banks. The author Mark Twain was a pilot on the old Mississippi paddleboats.

Below: Chicago, from the Lake Michigan shore. The city, the second largest in the country, is the commercial and industrial centre of the Midwest. It is also a major port on the Great Lakes.

Right: Mount Rushmore National Memorial, a huge carving on a granite cliff in the Black Hills of South Dakota. The carving depicts the faces of four American presidents: George Washington, Thomas Jefferson, Abraham Lincoln and Theodore Roosevelt. It was designed by the American sculptor Gutzon Borghum and work on it began in 1927.

leak caused an emergency at the Three Mile Island nuclear plant.

Other resources include the great forests that still cover part of the country. Many forests were cleared in the course of settlement, but some are now being replanted.

Cities and Communications

The United States has more than 50 cities with populations exceeding 250,000. New York City is one of the largest cities in the world, and its metropolitan area may have the greatest population of any city. It is the leading manufacturing city, closely followed by Chicago, Illinois; Los Angeles, California; Philadelphia, Pennsylvania; and Detroit, Michigan. Detroit is the centre of the world's automobile industry and that industry is one of the country's largest.

Very often, American cities are not both administrative centres and manufacturing centres. The federal capital, Washington, D.C., has relatively little industry and many of the state capitals are also basically non-industrial.

Highly-developed networks of roads, railways, waterways and airlines provide speedy and efficient communication between one part of the country and another. There are more than 3 million miles (5 million km) of roads, including a 41,000-mile (66,000-km) system of inter-state highways. Railways total more than 200,000 miles (320,000 km) but they are declining in importance. Internal airlines link all the major cities of the United States and through international airports provide services to all parts of the world.

America's waterways are based on two great river systems. The St. Lawrence Seaway, a combination of river and canals, links the Great Lakes with the Atlantic Ocean. Other canals link the Great Lakes with the Mississippi River, which with its tributaries provides water transport over a great deal of the central plains, linking the Great Lakes also to the Gulf of Mexico. These waterways provide a cheap and efficient way of moving bulk cargoes.

Below: The Navajo Indian Tribal Park of Monument Valley in south-eastern Utah.

THE STATES OF THE UNITED STATES

The states of the United States (excluding Alaska and Hawaii) are grouped in seven geographical regions. This table shows the states in each region, with the abbreviations of their names that are commonly used and their capitals.

NEW ENGLAND	
Connecticut (Conn)	Hartford
Maine (Me)	Augusta
Massachusetts (Mass)	Boston
New Hampshire (N.H.)	Concord
Rhode Island (R.I.)	Providence
Vermont (Vt)	Montpelier
MIDDLE ATLANTIC STATES	
New Jersey (N.J.)	Trenton
New York (N.Y.)	Albany
Pennsylvania (Penn)	Harrisburg
MIDWESTERN STATES	
Illinois (Ill)	Springfield
Indiana (Ind)	Indianapolis
Iowa (Ia)	Des Moines
Kansas (Kan)	Topeka
Michigan (Mich)	Lansing
Minnesota (Minn)	St Paul
Missouri (Mo)	Jefferson City
Nebraska (Nebr)	Lincoln
North Dakota (N.D.)	Bismarck
Ohio (O.)	Columbus
South Dakota (S.D.)	Pierre
Wisconsin (Wis)	Madison
SOUTHERN STATES	
Alabama (Ala)	Montgomery
Arkansas (Ark)	Little Rock
Delaware (Del)	Dover
Florida (Fla)	Tallahassee
Georgia (Ga)	Atlanta
Kentucky (Ky)	Frankfort
Louisiana (La)	Baton Rouge
Maryland (Md)	Annapolis
Mississippi (Miss)	Jackson
North Carolina (N.C.)	Raleigh
South Carolina (S.C.)	Columbia
Tennessee (Tenn)	Nashville
Virginia (Va)	Richmond
West Virginia (W. Va)	Charleston
SOUTH-WESTERN STATES	
Oklahoma (Okla)	Oklahoma City
Texas (Tex)	Austin
ROCKY MOUNTAIN STATES	
Arizona (Ariz)	Phoenix
Colorado (Colo)	Denver
Idaho (Ida)	Boise
Montana (Mont)	Helena
Nevada (Nev)	Carson City
New Mexico (N. Mex)	Santa Fe
Utah (Ut)	Salt Lake City
Wyoming (Wyo)	Cheyenne
PACIFIC COAST STATES	
California (Calif)	Sacramento
Oregon (Ore)	Salem
Washington (Wash)	Olympia
OTHER STATES	
Alaska (no abbreviation)	Juneau
Hawaii (no abbreviation)	Honolulu

The People

Most people in the United States are immigrants from Europe, or the descendants of immigrants. About one in every 10 is of black ancestry, descendants of slaves brought over from Africa. There are small numbers of American Indians and Eskimos.

The first settlers were mostly British and Irish, with some French and Spaniards. Later settlers included a great many Germans, Italians, Poles, Scandinavians and Russians. All these peoples brought with them their own cultures and traditions, which helped to make the American culture of today.

Above: The United States Capitol in Washington, D.C. The U.S. Senate meets in its northern wing, and the House of Representatives in its southern wing. George Washington laid the cornerstone of the gleaming white building in 1793.

Below: Death Valley in east-central California. The parched desert valley lies 282 feet (86 m) below sea-level and is about 130 miles (210 km) long. Parts of it are spectacularly beautiful.

South America

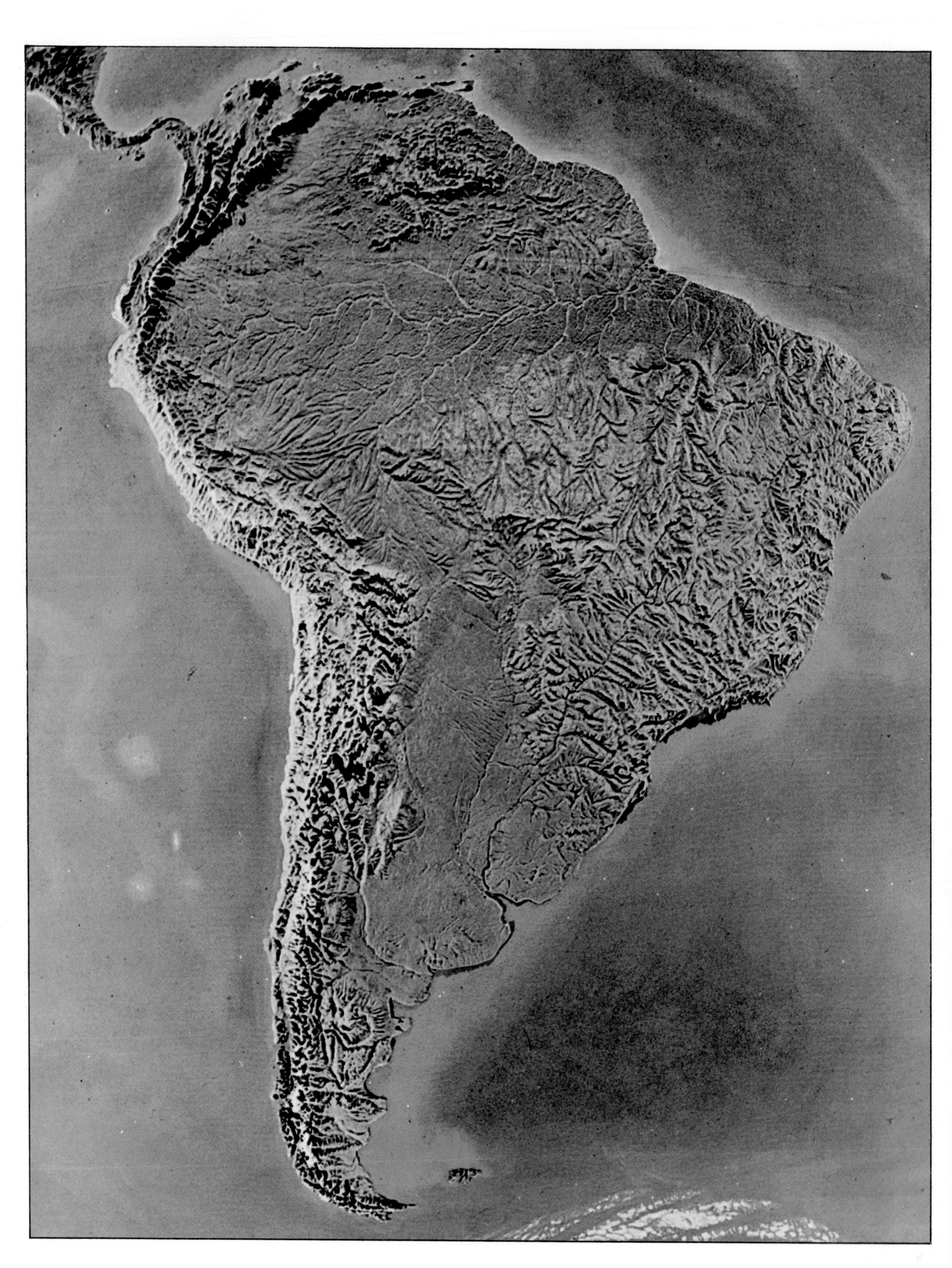

South America is the fourth largest of the continents. Vast areas of it are still largely unexplored. Potentially, it is very rich, but most of its immense mineral resources have yet to be exploited. In some of its countries, politics tend to be violent and revolutions are not uncommon.

South America is linked to North America by the narrow Isthmus of Panama. It is roughly triangular in shape, tapering to a point in the S. It reaches farther S than any other continent. Cape Horn, its southernmost tip, is notorious among sailors because of the violence of its storms. At this point, the continent of Antarctica is only about 400 miles (640 km) away. South America is bordered on the W by the Pacific Ocean, and on the E by the Atlantic Ocean. To the N lies the Caribbean Sea.

The coastline of the continent is comparatively smooth except in the SW, where it is broken by the many inlets and the islands of the Chonos Archipelago. The large island of Tierra del Fuego lies at the extreme S. Other island groups are the Galápagos Islands off the north-western coast, and the Falklands off the south-eastern shore.

The total area of South America is about 6,772,000 sq miles (17,540,000 sq km). The greatest distance N to S is 4,750 miles (7,640 km) and W to E 3,200 miles (5,150 km).

Physical Features and Climate

The equator crosses South America just to the N of its widest part. As a result, more than three-quarters of the continent lies between the tropics of Cancer and Capricorn. There are eight well-defined physical regions, but two of them—the Andes Mountains and the basin of the River

COUNTRIES OF SOUTH AMERICA POPULATIONS AND MONETARY UNITS

Country	*Population*	*Money*
Argentina	28,085,000	Peso
Bolivia	5,755,000	Peso
Brazil	121,547,000	Cruzeiro
Chile	11,406,000	Peso
Colombia	28,776,000	Peso
Ecuador	8,644,000	Sucre
Guyana	903,000	Dollar
Paraguay	3,268,000	Guarani
Peru	17,031,000	Sol
Surinam	352,000	Guilder
Uruguay	2,927,000	Peso
Venezuela	14,313,000	Bolivar

Top: An Indian from a forest tribe makes fire by rubbing two sticks together. South America once had a highly-developed Indian civilization, but most tribal Indians today live in primitive simplicity.

Below: The national congress buildings in Brasilia, the capital of Brazil. Brasilia has some of the most imaginative of contemporary architecture.

Top: Angel Falls in Venezuela, with a drop of 3,212 ft (979 m), is the highest waterfall in the world. It is located on Mount Auyantepui in the Guiana Highlands. South America also has the world's most powerful waterfall, the Guaíra on the River Paraná.

Amazon—are the dominant features of the continent.

The Andes, often referred to as the Western Cordillera, are a range of high, jagged mountains close to the Pacific Ocean coast. They stretch along the coast for more than 4,000 miles (6,450 km), and at their widest point are 300 miles (480 km) from W to E. The highest point in the range is Mount Aconcagua in Argentina, which rises to 22,835 ft (6,960 m) above sea-level. More than 50 other mountains rise above the 20,000 ft (6,100 m) level. Many Andean mountains are active volcanoes, and the whole region is shaken by earthquakes from time to time. The western slopes rise precipitously from the narrow coastal plain to heights of more than 3 miles (4.8 km). On the eastern side, the mountains slope more gradually down to the plains. In the S, the Andes form a single ridge, but farther N they broaden out into two, sometimes three, parallel ridges, with high, level plateaux between them.

The Pacific Coastlands lie between the Andes and the sea. They are very narrow, ranging between 5 miles and 50 miles (8–80 km) wide. They are widest in the N, where they are swampy.

The Orinoco Lowlands lie in the N, around

Below: The Amazonian forest seen from the air. South America's tropical rain forest covers an area of one million sq miles (2.5 million sq km)—more than 10 times the area of the United Kingdom. Movement through the forest is extremely difficult, and a newly-cut path may disappear in a few weeks. The Trans-Amazonia Highway is the first major road.

the basin of the Orinoco River. Much of this area is swamp. It is known as the *Llanos*, a word meaning 'plains'.

The Guiana Highlands occupy most of Guyana, Surinam and Guyane (French Guiana) in the northern part of the continent. Their highest peak, Mount Roraima, reaches 8,620 ft (2,627 m) above sea-level. The Highlands slope gently to the N and drop sharply to the Amazon basin in the S.

The Amazon River Basin covers more than 2 million sq miles (5 million sq km). Most of the region is very low lying, and even at the foothills of the Andes some land is no more than 50 ft (15 m) above sea-level. For almost the whole extent of the Amazon and its tributaries, the region is covered by dense, tropical rain forest, much of it only partly explored.

The Brazilian Highlands, a triangular region of high ground, lie to the E and S of the Amazon basin. Most of the region is less than 5,000 ft (1,500 m) above sea-level. But the tallest peak, Pico da Bandeira, reaches 9,480 ft (2,890 m).

The Paraná-Paraguay Plains form the basin of the Paraná River and its tributaries, of which the Paraguay is the most important. The plains are low-lying, sloping gently to the S. Part of them is covered by the *Gran Chaco* ('great hunting ground'), a region of forest and swamp which is still mostly a wilderness.

S of the Gran Chaco lies the eastward-sloping Pampas of Argentina, 250,000 sq miles (650,000 sq km) of rich, fertile land.

The Patagonian Plateau lies in the southernmost part of the continent. It is a flat, irregular rocky tableland, mostly bare and windswept, nowhere more than 3,000 ft (900 m) above sea-level.

Tierra del Fuego is separated from the mainland by the stormy Magellan Strait. The southernmost point of the continent, Cape Horn, is on Horn Island, one of the many smaller islands S and W of Tierra del Fuego. The Falkland Islands, 300 miles (480 km) to the NE of Tierra del Fuego, have deep fiords and wild, grassy moorland.

The Galápagos Islands, 500 miles (800 km) W of Ecuador, consist of 12 large islands and several hundred smaller ones. They are noted for their plant and animal life, which includes many species unique to the islands.

The climate of South America is mostly warm, except in the high Andes plateaux and the temperate south. The Amazon basin has

Top: Mount Condoriri (21,998 ft; 6,705 m), north of La Paz in Bolivia. It is in the Cordillera Real, one of the great Andean ranges.

Centre: The ruined Inca city of Machu Picchu, about 50 miles (80 km) from Cuzco, in Peru. It stands between two sharp peaks, about 6,750 ft (2,000 m) up in the Andes Mountains. The ruins of other cities have been discovered near-by.

Below: Cattle-herding in southern Chile. In the background is the symmetrical cone of Mount Osorno. Most South American farms are small, but the continent also has some of the world's largest plantations and ranches.

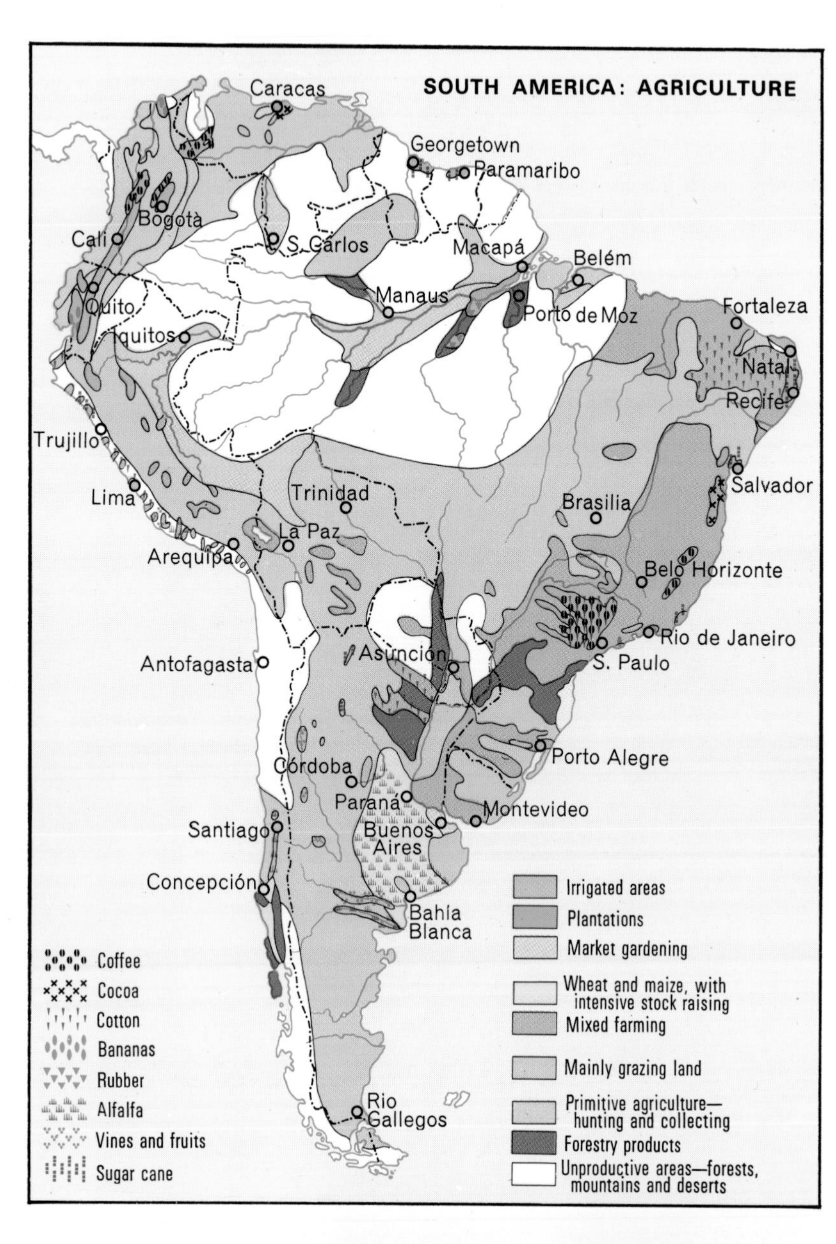

Right: A cattle market in Argentina. Huge amounts of beef are frozen or canned for export to other countries.

heavy rainfall, but most of the continent has moderate or light rainfall.

Resources and Industry

South America has many valuable deposits of minerals, particularly of iron, manganese, petroleum, copper, tin and bauxite. There are some active mines producing silver and gold, the lure that drew the early Spanish conquerors to the continent. Colombia is the world's chief source of fine emeralds, and there are diamonds in Brazil. Northern Chile has the world's only natural deposits of sodium nitrate, an important ingredient of fertilizers. The continent has little coal, which is still one of the mainstays of industrial economies.

One under-developed resource is water power, which can be exploited for generating electricity. But in its rivers South America possesses about a quarter of the world's total hydro-electric potential.

South America has tapped some of its other resources, such as timber. It produces many materials for supplying industry, including cotton, wool and hides.

Industry in South America has developed relatively slowly. Argentina, Brazil and Chile are the most highly-developed countries industrially. Some other countries are very backward. Until World War I the continent exported most of its mining production, and large amounts of minerals, particularly of petroleum, copper and iron, are still exported.

Manufacturing industries consist mainly of the processing of raw food materials, such as meat, wheat and sugar, and textile production. Heavy industries, such as iron and steel production, are still slow to develop, partly because of the coal shortage. Light manufacturing industries producing goods for home consumption, such as clothing, soap, ceramics, furniture and cigarettes, are flourishing in some countries. And many craftsmen produce fine handmade goods such as leatherwork, silverware, baskets and wood carvings. The continent has an abundance of labour, but lacks the capital needed to set up new industries on a large scale.

Communications

Communications are difficult in many parts of South America. Vast areas, such as the forested parts of the Andes, have few roads and no railways. The chief railway networks are in Argentina and SE Brazil.

Seven different railway gauges are in use—Colombia, for example, with only about 2,000 miles (3,200 km) of track, has three different gauges; while Venezuela, with less than 1,000 miles (1,600 km), has five gauges.

Most of the continent has no motor roads, and even in areas where there are roads many stretches are unpaved or in poor condition. A major road-building programme is going on, but the majority of cities even in Argentina are linked by tracks that are dusty in summer and quagmires in winter.

By contrast, air services are comparatively well developed. Planes can fly over the vast regions of forest and mountain, and air services have done a great deal to open up the continent. Though there are rivers in most areas, many of them have rapids or become shallow in dry weather and consequently are of little use for transport. Some areas with usable rivers have few people or goods to carry.

Only a minority of the people of South America can afford radio, television, or telephones, even when they live within range of such services. About half the people cannot read. News spreads slowly in many areas.

Agriculture, Fisheries and Land Use

More than half the people of South America live by farming. For most of them, subsistence farming is the rule. They produce just enough food for their own needs. Some primitive Indian tribes in the forests still live by hunting and seed-gathering, as people did in the Stone Age. And huge areas of the continent are useless for agriculture. Some are covered by forests, like most of the Amazon basin; others are mountainous and difficult if not impossible to farm.

Only three countries have well-developed agriculture—Argentina, Uruguay, and Brazil. In those countries and in parts of Chile there are large farms and ranches—some of the biggest in the world. On the Pampas cattle and sheep are raised, while in other parts cash crops such as bananas, coffee, cocoa-beans, cotton, linseed, sugar, sisal, tobacco, wheat and sunflower seeds are grown. Some countries devote so much attention to growing cash crops that they have to import food, even though there is enough land to grow the food as well. Poor farming methods are largely to blame for this

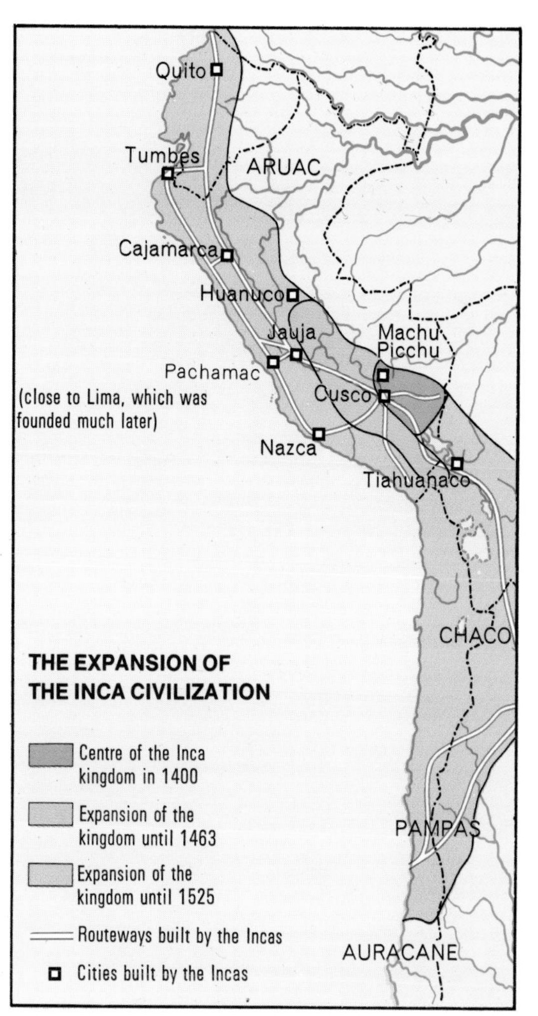

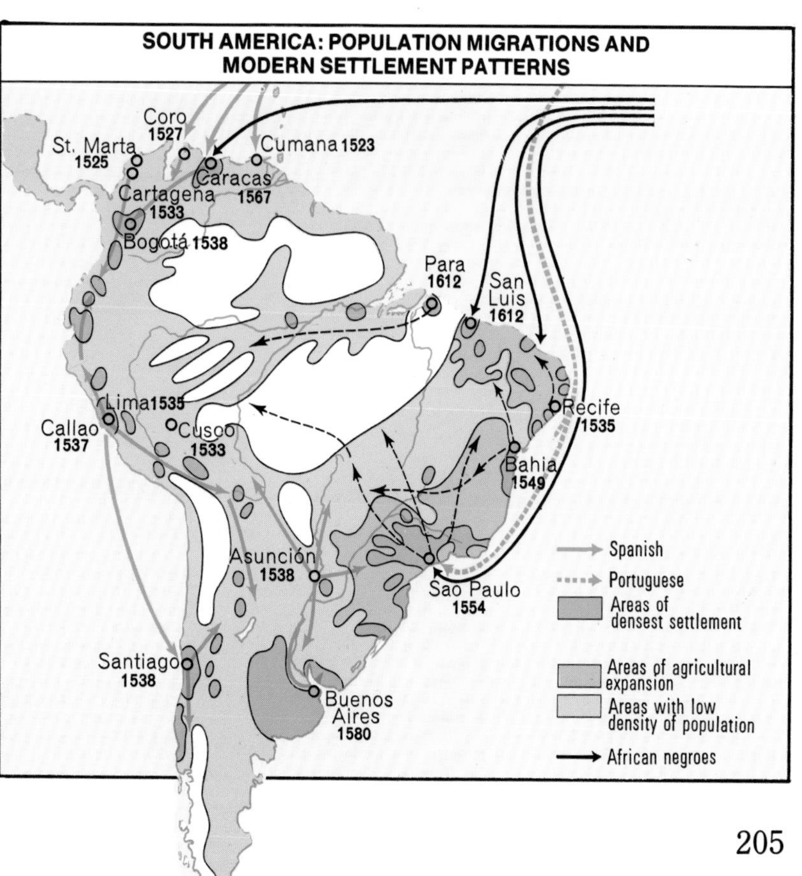

Top and centre: Birthplace of the Incas. Lake Titicaca, 12,500 ft (3,800 m) above sea-level in the Andes, on the boundary between Peru and Bolivia. According to legend, the Incas originated on an island in the lake. The Indian boats are made of reeds.

Below and opposite page: The city of Rio de Janeiro, formerly Brazil's capital. Like other major South American cities, it has wide boulevards and huge modern buildings as well as narrow old streets and stucco colonial houses.

situation.

Another problem is unequal distribution of land. In many countries most of the good land is held by a few rich people, while the rest of the people are poor, landless workers.

People and Ways of Life

South America contains the world's most mixed population. Before Spanish explorers reached the continent in the late 1400s, about 15 million American Indians lived there. These *Amerindians*, as they are often called, were descended from Asian peoples who crossed over what is now the Bering Strait when land still joined Asia and North America. They probably reached South America between 10,000 and 15,000 years ago.

Today the people of South America are of three main stocks: European, Amerindian and Negro. The Negroes are descended from the millions of slaves imported from Africa. These three stocks have inter-married to produce *mestizos*, people of mixed white and Amerindian blood; *mulattos*, of mixed white and Negro blood; and *zambos*, people of mixed Amerindian and Negro blood. In addition, the mestizos, mulattos and zambos have also intermarried, producing a racial hotch-potch that will, in time, lead to a completely new racial type.

There is little racialism as such in South America. Any differences are social, and are largely a matter of wealth. As it happens, social divisions are often on racial lines, but there are many exceptions to the rule. The wealthy landowners are mostly white; the professional and business people are white

THE INCAS

The Inca civilization and empire developed in the Andes Mountains and reached its peak in the late 1400s. Its capital was Cuzco, in present-day Peru. The Incas were skilled builders, weavers and jewellers. Their language was Quechua. The Spaniards conquered them in the 1500s.

and mestizo (mixed white and Amerindian), and the bulk of the workers are mestizos, Amerindians, and Negroes, or mixtures of these groups.

The proportions of the different racial groups vary from country to country. For example, in Argentina about 90 per cent. of the people are white, descendants of Spanish and other European settlers. The rest are mostly mestizos, and there are a few pure-blooded Indians. In Brazil about 20 per cent. of the people are mestizos and 5 per cent. are Indians or Negroes.

Most of the countries of South America have much smaller populations than they are capable of supporting, given proper use of the available land. But the population of some parts of the continent are growing faster than almost anywhere else in the world. This has several causes, the main one being the continued decline in the death rate due to better hygiene and medical services. Some experts estimate that by the year 2000 the total population of the continent may approach 300 million.

Most of the people of South America live near the coast, except in the NW. There, the tropical heat and humidity of the coastlands has driven people into the highlands, where the climate is cooler and more healthy. Peru and Bolivia are populated chiefly in the high plateau lands of the Andes. Difficulties of travel have hindered much settlement inland, and where it has occurred it has been mainly along navigable rivers, which provide the only effective means of transport.

Spanish is the principal language spoken in South America, because most of the countries were formerly Spanish colonies. Brazil was a Portuguese colony, and there Portuguese is the main language. In Guyana, a former British colony, the main language is English. Many Indian languages are spoken in the continent, but in only one country, Paraguay, is an Indian language sufficiently common as to be regarded as a second tongue. Most of the people of South America are members of the Roman Catholic Church.

About two-thirds of the people live in rural areas, in simple houses made of timber, stone, or brick with thatched roofs. They work on the land. In cities, many people live in cramped, overcrowded conditions, with poor sanitation and water supplies. The better-off people live very comfortably with all modern conveniences.

THE COUNTRIES OF SOUTH AMERICA

Argentina

Argentina occupies most of the southern part of South America. It is the second largest and second most populous country in the continent. Only Brazil is larger. Its northern regions swelter in the tropics, while in the S, the coastal areas lie just a few hundred miles from Antarctica. Although it ranks with Brazil as one of the two leading manufacturing nations of South America, Argentina's reputation and wealth are based mainly on livestock and grain.

In addition to its mainland territory, which includes disputed border lands with Chile in the W, Argentina also shares the large island of Tierra del Fuego with Chile in the S. It also claims possession of the offshore British Falkland Islands, which Argentinos call *las Islas Malvinas*. In support of this claim Argentina invaded and occupied the Falklands in April 1982. Britain responded within a few days by sending a task force. All attempts by the United Nations, the United States and others to achieve a peaceful settlement having failed, British forces attacked the Falklands. Despite considerable loss of lives and equipment the Falkland Islands were recaptured by mid-June.

The capital of Argentina is Buenos Aires, the largest city in the Southern Hemisphere. Area 1,073,000 sq miles (2,779,000 sq km).

The Iguaçu Falls, on the boundary between Argentina and Brazil, consists of hundreds of separate falls. They extend for more than 2 miles (3 km) across the Iguaçu River. The surrounding countryside is richly wooded and is celebrated for its exotic plant and animal life.

Land and Climate

Argentina is bordered on the N by Bolivia and Paraguay, on the NE by Brazil and Uruguay and on the W by Chile. In the E and S it has a coastline on the South Atlantic.

In the N of the country, heavily wooded plains form the *Gran Chaco*. To the S and SW of Buenos Aires lie the vast stretches of the Pampas, an almost treeless expanse of fertile grasslands. To the W of the Pampas is the Andine region in the shadow of the Andes Mountains. There the people grow citrus fruits and vines, or work in the mines. In the S lies Patagonia, a bleak, windy plateau where the main sources of income are sheep and petroleum.

The towering Andes Mountains, separating Argentina from Chile in the W, include Mount Aconcagua, 22,835 ft (6,960 m) high, the highest peak in the Western Hemisphere.

The two main rivers are the Paraná and Uruguay. They flow into the Rio de la Plata, a wide freshwater estuary on whose banks stand two capitals—Buenos Aires and Montevideo (Uruguay). In the N, the climate is hot and humid in summer and mild in winter. In central areas, it is mild all the year round. Patagonia has hot summers and cold winters. Rainfall is heaviest in the W of the country.

Agriculture and Industry

Argentina's agricultural wealth is based mainly on the products of the Pampas. It exports more beef than any other country and is among the world's leading exporters of sheep and wool. Other important agricultural products include wheat, cotton, barley, rye, oats, alfalfa, linseed, sugar, cotton and fruit. The principal minerals include petroleum, coal, lead, zinc, iron, copper, silver and gold. Meat processing and canning is the main industry, although flour-milling, chemicals, textiles and sugar-refining are growing in importance.

Cities and Communications

The chief cities are Buenos Aires, Córdoba, La Plata, Rosario, Tucumán and Santa Fé. About a third of the population lives in Buenos Aires and its suburbs. There is an excellent, if ageing, rail network, and waterways and airlines provide access to inland areas.

The People

There are few Indians, the original inhabitants of Argentina, left. Most of the people are descendants of Spanish and Italian settlers. The official language is Spanish. About 80 per cent. of Argentinos are Roman Catholics. The most highly-developed art forms are painting and sculpture, and music and dance, particularly popular dances.

Bolivia

Bolivia is a landlocked country that straddles the Andes Mountains in the heart of South America. It is bordered by Brazil on the N and E, Paraguay and Argentina on the S and Chile and Peru on the W. Between parallel mountain ranges in the W is the Altiplano, a high, bleak plateau where most of the people live. In the N and E there are forested lowlands, which turn to scrub and desert farther S. Lake Titicaca, the highest navigable stretch of water in the world, lies on the Peruvian border 12,507 ft (3,812 m) above sea-level. The climate is hot in the E and cooler in the W, with heavy rains from December to February. Tin, silver, lead, oil, natural gas, antimony, bismuth and coffee are the main products. Gas provides about a quarter of Bolivia's exports and there are plans to build a pipeline to São Paulo, Brazil. Large numbers of cattle are exported. Bolivia is rich in natural resources, but the difficult nature of the land has hindered their exploitation.

Almost all the people are Indians or of mixed Indian and European blood. Most of the Indians are very poor, in spite of the fact that many work in the mines, which are the country's main source of wealth. The official language is Spanish, but Indian tongues are widely spoken. La Paz, the highest capital city in the world, is the seat of government, but Sucre is the official capital. Area 415,000 sq miles (1,075,000 sq km).

Left: Sheep penned on a ranch in the Pampas. Though Argentina is best known for its cattle, it is also one of the world's principal sheep-rearing countries.

Below: Llamas are used as beasts of burden on the high, sparsely-vegetated plateau of Bolivia. Their wool—and the wool of their near relatives, alpacas and vicunas—is highly prized. They belong to the same family as the camel.

TIERRA DEL FUEGO
The archipelago of Tierra del Fuego consists of one large island (also named Tierra del Fuego) and several small islands at the southern tip of South America. Part of it belongs to Chile and part to Argentina. Cape Horn, the most southerly point of the continent, is on Horn Island, which is Chilean. Tierra del Fuego is separated from the mainland by the Strait of Magellan. Drake Passage separates it from the Antarctic Peninsula. Total area 27,500 sq miles (71,000 sq km).

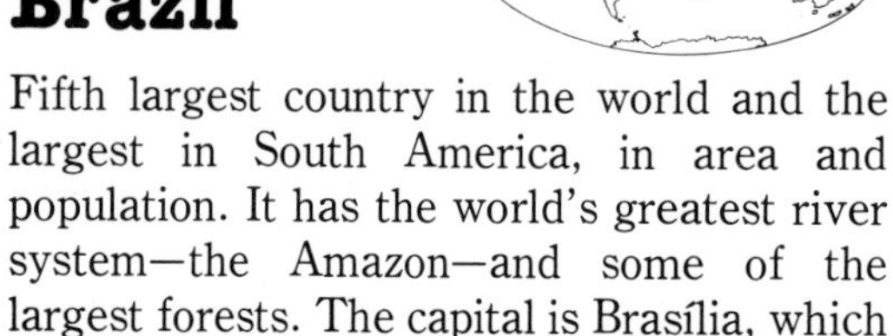

Brazil

Fifth largest country in the world and the largest in South America, in area and population. It has the world's greatest river system—the Amazon—and some of the largest forests. The capital is Brasília, which superseded Rio de Janeiro in 1960. Area 3,286,488 sq miles (8,511,964 sq km).

Right: Rio de Janeiro, on Guanabara Bay, Brazil, has one of the finest sites of any of the world's cities. Beside it rises the unmistakably-shaped Pão de Açúcar—Sugar Loaf Mountain. It is believed that the city's name—meaning 'January River'—derives from the fact that the Portuguese explorer and navigator Gonçalo Coelho first sailed into the bay on January 1, 1502 and decided that it was the mouth of a hitherto-unknown river.

Below: A house in the Amazon forest. The great, slow Amazon River—fed by more than 500 smaller streams—flows through many almost-inaccessible regions. Most of the Indians of Amazonia live simple tribal lives. Others are more sophisticated. But many 'centres of population' marked on maps of the Amazon are only the most rudimentary settlements.

Brazil occupies the NE and much of the central part of South America. The equator passes through the N of the country. The mouth of the Amazon is on the equator.

Land and Climate

The land can be divided into three main regions. In the N is the immense, heavily forested Amazon region. It includes highlands near the Venezuelan border. In the NE, some farming is possible on the coastal strip, but most of the area is scrub and farmers suffer severely from drought. The central and southern regions are a series of plateaux crossed by fertile river valleys. Almost half of Brazil's population lives in the S and 75 per cent. of the country's agriculture and 80 per cent. of its industry are located there.

The Amazon River discharges more water than any other river on earth and is navigable for its entire length of 1,962 miles (3,158 km) within Brazil.

The climate is generally hot. It is humid in the Amazon region, hot and dry in the NE, and cooler and dry in the central plateaux and the S.

Agriculture and Industry

Brazil is a great storehouse of mineral wealth, still largely untapped. But already it supplies much of the world's quartz crystal, beryl, sheet mica, manganese, columbium and tantalum. It also has valuable deposits of iron, gold, diamonds, nickel, chrome, tungsten and oil. The chief agricultural product is coffee, followed by cotton, cocoa, sugar and soya beans, together with fruit.

The Itaipú dam on the Paraná River on the Brazil–Paraguay borders will result in the world's largest hydro-electric power scheme on completion of the power station.

Cities and Communications

The chief cities are São Paulo (one of the fastest growing cities in the world), Rio de Janeiro, Belo Horizonte, Recife, Fortaleza, Porto Alegre and Brasília. Communications are poor but improving, being confined mainly to sea and air routes. But there are 26,000 miles (41,800 km) of navigable waterways and new highways are being carved.

The People

Most are descended from early Portuguese

settlers or later arrivals such as Spaniards, Germans, Italians, Japanese and Lebanese. The official language is Portuguese.

Chile

The Republic of Chile is like a long, narrow ribbon running down the W coast of South America. It stretches about 2,800 miles (4,500 km) from Peru in the N to within 400 miles (640 km) of Antarctica in the S. Hemmed in between the Andes Mountains and the Pacific Ocean, it is nowhere more than 250 miles (400 km) wide. On the E it borders Bolivia and Argentina. Because Chile is so long, it has many different geographic zones, each with its own climate. The far N, where most copper and nitrate resources are found, is hot desert. Farther S, this becomes semi-desert with a little vegetation and some crops under irrigation. Then comes a region of fertile valleys washed by rains in winter and warmed by hot sun in summer. This is Chile's heartland, where most of the people live. Still farther S is a region of huge lakes, forests, rivers and mountains. This part is very popular with tourists. In the far S, stretching to Cape Horn, is a desolate land of forests, fiords, islands and glaciers, which is frequently swept by storms. Chile's main products are copper, nitrates, iron, oil, fishmeal, wool and fruit. Copper provides over 40 per cent. of exports by value. About half Chile's oil is produced on Tierra del Fuego or in its off-shore waters, and there is considerable production of natural gas. The people are of European descent, mainly Spanish. There are also about 150,000 Araucanian Indians. Spanish is the official language. The capital is Santiago. Other large towns include Valparaíso, Antofagasta and Concepción. Area 292,000 sq miles (756,000 sq km).

Bleak, windswept terrain on Tierra del Fuego, the archipelago south of the Strait of Magellan.

Indian market in Ambato, a city of Ecuador high in the Andes. Ambato is a holiday resort and is known for its trade in fruit.

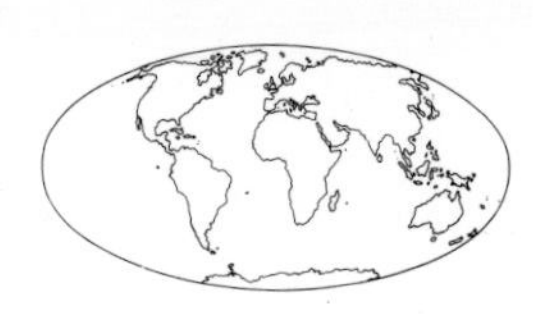

Colombia

Colombia is the fourth largest country in South America. Four great mountain ranges, all part of the Andes, run parallel N from the border with Ecuador almost to the Caribbean coast. To the W of them is the Pacific and to the E the country's capital, Bogotá. Between the ranges are high, narrow valleys where most of the people live. The mountains seriously hamper transport and communication. As a result, air transport is important. To the S and E of the country are grassy plains called *llanos*. They eventually merge into the tropical rain forests of the Amazon basin. The fertile river valleys are rich farming regions.

Climate varies with altitude. It is tropical in the steaming jungle lowlands and freezing above the snowline of the Andes. The chief products are oil, gold, emeralds, coal, platinum, coffee, meat and bananas. Colombia supplies almost all the world's emeralds. Many Colombians are of mixed blood. About a fifth of them have European ancestry, but there are also nearly 400 Indian tribes. The official language is Spanish. Bogotá is the largest city; Barranquilla and Cartagena are the next most important towns. Area 439,737 sq miles (1,138,914 sq km).

Ecuador

Ecuador is one of the smallest countries in South America. Its name means 'equator', and it is so named because it straddles the equator on the NW coast of the continent. Ecuador's mountainous 'backbone' is made up of two main ranges of the Andes. These run from NE to SW for about 425 miles (680 km). A great valley, about 30 miles (48 km) wide, lies between them. Most of the people live there. The mountains include more than 30 volcanoes, many of them active. Cotopaxi, 19,344 ft (5,896 m) high, is the highest active volcano in the world. Rivers flow eastwards down the mountain slopes to join the Amazon system or westwards to empty into the Pacific. To the E of the mountains lies the *Oriente*, a sparsely populated, jungle-clad region of lowlands. To the W is the Pacific lowland, about 60 miles (96 km) wide. Climate ranges from the steaming tropics of the lowlands to the freezing air of the high mountain peaks. The main products are bananas, cocoa, coffee, oil, straw ('Panama') hats, fish, gold and silver. Ecuador exports more bananas than any other country, though in 1982 fish was the leading export. Oil may soon prove a more important source of income. About 40 per cent. of the people are of pure Indian blood. Slightly more than this have mixed Indian and Spanish blood. Most of the remainder are of Spanish descent. Spanish is the official language. Quito is the country's capital and Guayaquil the main port. Area, 109,500 sq miles (283,600 sq km).

Guyana

Guyana, formerly British Guiana, is located on the NE coast of S America. It is bordered by Surinam, Brazil and Venezuela. On the N it has a coastline on the Atlantic. The land is hot and humid, with dense forests. The people are a mixture of East Indians, American Indians, Chinese, Europeans and Africans. The official language is English. Georgetown, the capital, is the only large city. Area 83,000 sq miles (215,000 sq km).

Left: The city of Quito, in the Andes, is Ecuador's capital and one of the world's highest cities. Though it lies almost on the equator, its climate is relatively mild because of its altitude.

Below: Traders reach a bargain on the bank of the Paraguay River, near Asunción. Paraguay, which under Jesuit rule in the 1600s and 1700s was one of the most advanced regions of South America, is today backward and impoverished.

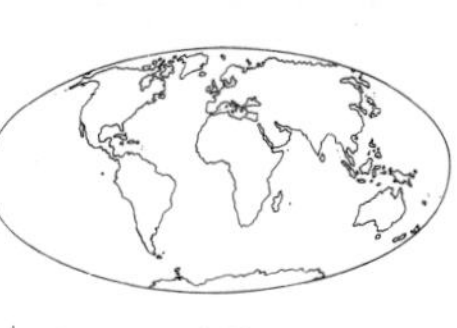

Paraguay

Paraguay is a landlocked republic in the centre of the South American continent. The River Paraguay, flowing N and S, cuts the land in two. To the W lies the *Chaco*, a region of jungle, scrub and marshes. Immediately to the E of the river is a fertile area of gently rolling grasslands that rises to a high, heavily forested plateau. Almost all of the people live in the region to the E of the river. The climate is subtropical, with mild, fairly dry winters. The economy depends almost entirely on crops and livestock. Major exports are timber, *quebracho* (a wood extract used for tanning), meat, hides and *yerba mate* (Paraguayan tea). Surplus hydro-electric power is exported to neighbouring countries. (See reference to Itaipú Dam on page 210.) Spanish is the official language. The only large city is Asunción, the capital. Area 157,000 sq miles (406,600 sq km).

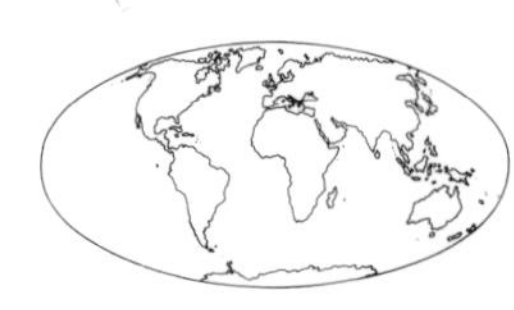

Peru

Peru, the third largest country in South America, lies on the W coast of the continent. It is a land of mountains, deserts and jungles. On the W, Peru has a 1,400-mile (2,250-km) coastline on the Pacific Ocean. A narrow desert coastal area, sandwiched between the sea and the mountains inland, contains most of the country's major towns. The mountainous 'backbone' of Peru is made up of several ranges of the Andes Mountains, running N and S, and divided by deep valleys. To the E of the mountain peaks, the lower slopes flatten out into plains and heavily forested, largely unexplored stretches of territory. More than 50 short, swift rivers flow across the coastal plains from the mountains into the ocean. But the two longest rivers, the Marañón and Ucayali, flow on the eastern side of the Andes and are among the headwaters of the Amazon. Lake Titicaca, on the border with Bolivia, lies 12,507 ft (3,812 m) above sea-level. It is the highest navigable body of water in the world. Peru's climate ranges widely, from the icy, snow-covered peaks of the Andes to the almost tropical climate of many of the valleys. The Humboldt Current, flowing northwards from the Antarctic, brings a cold touch to Peruvian shores. Amounts of rainfall vary from 50 inches (1,270 mm) a year to 10 inches (254 mm) in the dry southern part of the country. Peru is rich in antimony, copper, bismuth, gold, lead, silver, zinc and other minerals. As a result, mining and the processing of minerals are major industries. Minerals, chemicals and fertilizers (especially fishmeal)—together with some sugar, cotton, wool and coffee—form the bulk of the exports.

About half of the people are Indians, a third are *mestizos* (part Indian and part European), and a small group are *criollos*. Criollos are descendants of the original Spanish settlers and most of Peru's wealth is still in their hands. The official language is Spanish, but Quechua, the language of the conquered Incas, is also widely spoken. The main cities are Lima, the capital, Callao, Arequipa, Trujillo, Chiclayo and Cuzco (the old Inca capital). Area 496,000 sq miles (1,285,000 sq km).

Above: The ruins of the Inca stronghold of Sacsahuamán, at Cuzco in Peru. It was built in the 1400s and was captured in the 1530s by the Spaniards besieged in Cuzco by Manco Capac, the last Inca emperor.

Centre: The Incas built in granite and other stones, fitting the unmortared blocks together with extraordinary precision.

Below: In most rural communities, ponchos (capes) and blankets—woven locally to traditional designs—add colour to the scene.

SURINAM

Surinam, formerly Dutch Guiana, is on the north-eastern coast of South America. It became independent in 1976. Rice is the chief crop, but bauxite is the leading export.

A street in Caracas, Venezuela, named after 'the Liberator', Simón Bolivar, who was born in the city. Although very near the Caribbean coast, Caracas stands some 3,100 ft (945 m) above sea-level.

Uruguay

Uruguay is one of the smallest republics of South America. The land is made up of rolling, grassy plains, low hills and clumps of trees on the banks of many rivers and streams. Uruguay is unique among South American countries in that virtually the whole of its land is populated and there is no great contrast between one part of the country and another. The main rivers are the Rio de la Plata and the Uruguay, which form the boundary with Argentina. Another major river is the Negro, which flows through the centre of the country. The largest of many shallow lakes and lagoons is Lake Mirím, which is about 110 miles (177 km) long and 25 miles (40 km) wide. It lies partly in Uruguay and partly in Brazil. The climate is temperate, pleasant and healthy. Uruguay is poor in mineral resources. Its greatest natural resources are the fine climate and rich soil. Agriculture is the main industry and there are more cattle and sheep in Uruguay than people. About nine-tenths of the land is devoted to rearing livestock. The remainder is used for growing crops. The main exports are wool, beef and hides. There are also large plants for processing agricultural products. The people are almost all of European ancestry, mainly Spanish and Italian. Spanish is the official language. The only large city is Montevideo, the capital. Area 68,536 sq miles (177,507 sq km).

Venezuela

Venezuela is the richest country in South America. Its name means 'Little Venice'. It was given this name by Spanish explorers in 1499. They came across an Indian village built on stilts in the middle of Lake Maracaibo and were reminded of Venice. Venezuela lies on the N coast of South America, facing the Caribbean Sea. It is bordered on the E by Guyana, on the S by Brazil and on the W by Colombia. The land has four main regions. To the S of the River Orinoco lie the Guyana Highlands. This wild, sparsely populated region covers more than half the country. The *llanos* (flat grasslands) of the Orinoco occupy a huge central plain. The third region is made up of the Maracaibo lowlands. This hot, humid area fringes Lake Maracaibo—a vast, shallow, freshwater lake open to the sea. Its bed is the source of Venezuela's petroleum. The Andean or Venezuelan Highlands make up the fourth region. They are an offshoot of the Andes and run across the NW corner of the country. The Orinoco, some 1,600 miles (2,575 km) long, is the eighth longest river in the world. Other major rivers include the Caroni, Caura and Apure. Angel Falls, with a drop of 3,212 ft (979 m), is the highest waterfall in the world. Venezuela lies in the tropics and the temperature varies more between night and day and according to altitude than between seasons. The country's wealth is derived from its petroleum deposits, and it exports more oil than any other country. But exploitation of other mineral resources is increasing. Among these are bauxite, iron, asbestos, asphalt, nickel and lead. Farmers produce only enough food for themselves and their families, although cattle are reared on the llanos. Most of the people are of European descent. There are also some Negroes and American Indians. The official language is Spanish. The chief town is Caracas, the capital, and the chief port is La Guaira. Area 352,145 sq miles (912,051 sq km).

Australasia

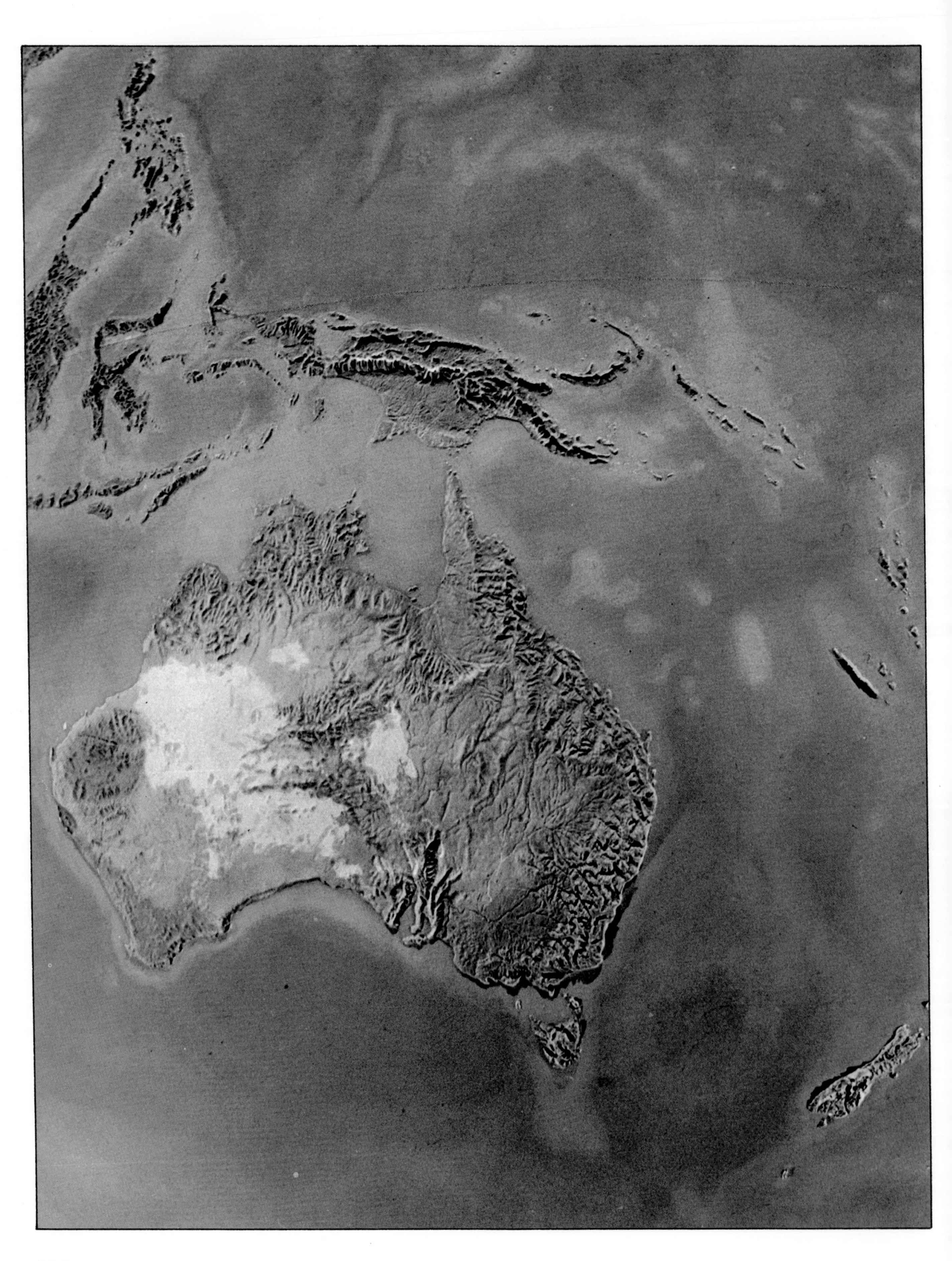

Australasia is the region that includes Australia, New Zealand and Papua New Guinea. But the term *Australasia* is not a precise one: it can be defined in several different ways. Sometimes it is used in a way that includes a number of the countries of Asia—such as Malaysia and Indonesia—as well as Australia, New Zealand and Papua New Guinea. Sometimes it includes some of the more important Pacific island groups, or even territories in Antarctica. To zoologists and botanists, Australasia may mean all the land and sea areas east and south of Wallace's Line—an imaginary line drawn N–S through the Celebes Sea and between Borneo and Bali, on one side and Sulawesi and Lombok, on the other. The line is named after a British naturalist, Alfred Wallace, and it is said that plant and animal life on one side of it differs significantly from life on the other. The great wealth of wildlife helped to lead Wallace, by a remarkable coincidence, to the same conclusions about evolution as Charles Darwin.

Left: Sydney, the oldest and biggest city in Australia. The beautiful harbour has an area of more than 20 sq miles (50 sq km).

Below: Sheep-shearing in New Zealand. Sheep-farming is the country's main industry.

Above: Ayers Rock, south-west of Alice Springs in central Australia, rises 1,100 ft (335 m) above the plain. It is more than a mile and a half (2.5 km) long.

Centre: The kangaroo, the largest of the marsupials. It bounds along on its powerful hind legs at speeds of up to 25 m.p.h. (40 k.p.h.).

Below: The famous 'sails' of Sydney Opera House, at Bennelong Point in Sydney Harbour. They cover two main auditoriums.

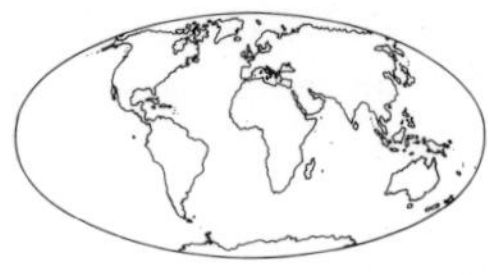

Australia

The only country that is also a continent, Australia has an area of 2,968,000 sq miles (7,687,000 sq km).

Australia lies to the SE of Asia. Its W coast is washed by the Indian Ocean and its E coast by the Pacific. The country is a monarchy, acknowledging Queen Elizabeth II of the United Kingdom as queen of Australia. She is represented by a governor-general, who acts as head of state. Australia is a federation of six states, each with its own government and legislature, plus two territories and some overseas territories. The federal government deals with national and international matters. The federal capital is Canberra. The country's official name is the Commonwealth of Australia.

Land and Climate

A mountain chain, the Great Dividing Range, lies close to Australia's E coast. This range traps the rain-bearing winds that blow in from the Pacific Ocean. As a result, the E coastal area is fertile and green. Part of the land to the W of the Great Dividing Range is also fertile, and is watered by three great rivers, the Murrumbidgee, the Darling and the Murray. The central part of Australia is a vast arid desert, where few plants can grow and few people live. It covers almost two-thirds of the continent. Alice Springs, a town in the centre of the country, is one of the few places with enough rainfall to support life.

A large part of Australia to the W of the Great Dividing Range is watered by artesian wells—wells that draw on underground sources of water.

Along the S coast is another dry desert area, the Nullarbor Plain. The name *Nullarbor* comes from Latin words meaning 'no tree'. The N coastal area is lined with tropical woodlands, parts of the NE coast having dense tropical rain forest. Australia's SW region is also fertile, particularly around Perth.

Australia has a great many lakes, but nearly all of them are dry for part of the year, as are the streams that feed them. The largest is Lake Eyre, in South Australia, 39 ft (12 m) below sea-level.

There are several islands off the coast of the Australian mainland. The largest is Tasmania, 124 miles (200 km) to the SE across the Bass Strait. Tasmania is one of the states of the Commonwealth of Australia. It has plenty of rain. In its forested mountains there are many rare animals.

Animals that people particularly associate with Australia are the marsupials. These animals bear their young only partly formed and rear them in pouches on their mothers' abdomens. Marsupials include the kangaroo, the koala, the wombat, the phalangers and also the Tasmanian wolf. Two other Australian animals, the duck-billed platypus and the spiny anteater, are mammals that lay eggs. Other animals found only in Australia include black swans, and two flightless birds, the emu and the cassowary.

Agriculture and Industry

Farming is the chief source of Australia's wealth. There are more sheep than in any other country, and also great numbers of cattle. Australia exports meat, wool and dairy products—mainly butter and cheese. It also produces a great amount of wheat, as well as sugar-cane and fruits. Tasmania has long been famous for its apples. The continent is extremely rich in minerals, and new deposits are being found every year. Australia produces and exports gold, lead, silver and zinc, and has rich reserves of iron, coal, petroleum, natural gas, copper, uranium and tin.

Manufacturing industry has grown rapidly since World War II, and employs more than a quarter of the country's workers. But

Top: The Aborigines are thought to have arrived in Australia about 16,000 years ago. They moved around from place to place, and lived by hunting, fishing and plant-gathering. Today, few follow their traditional way of life. Most live on special reserves.

Centre: The Macdonnell Ranges extend for about 100 miles (160 km) across central Australia. Some peaks rise more than 4,000 ft (1,200 m). The ranges are celebrated for their strangely-shaped chasms and bluffs and for their colour changes.

Below: Herding sheep. Australia, the world's leading sheep-raising country, produces more than one-quarter of the world's wool. Most of the country's sheep are Merinos.

Statue of James Cook (1728–1779) in Sydney. Captain Cook, the greatest explorer of the Pacific Ocean, made three long voyages of discovery. He charted the coasts of New Zealand and the eastern coast of Australia, and visited many hitherto 'unknown' islands. He was killed by Hawaiian islanders while trying to recover a stolen boat.

AUSTRALIA, NEW ZEALAND: AGRICULTURE

Darwin
Townsville
Selwyn
Brisbane
Cue
Cobar
Perth
Elliston
Canberra
Sydney
Adelaide
Melbourne
Auckland
Hobart
Wellington
Christchurch

Dairy cattle
Dairy cattle and sugar cane
Wheat growing and sheep raising
Sheep farming
Beef cattle
Forestry products
Unproductive areas: desert and mountains

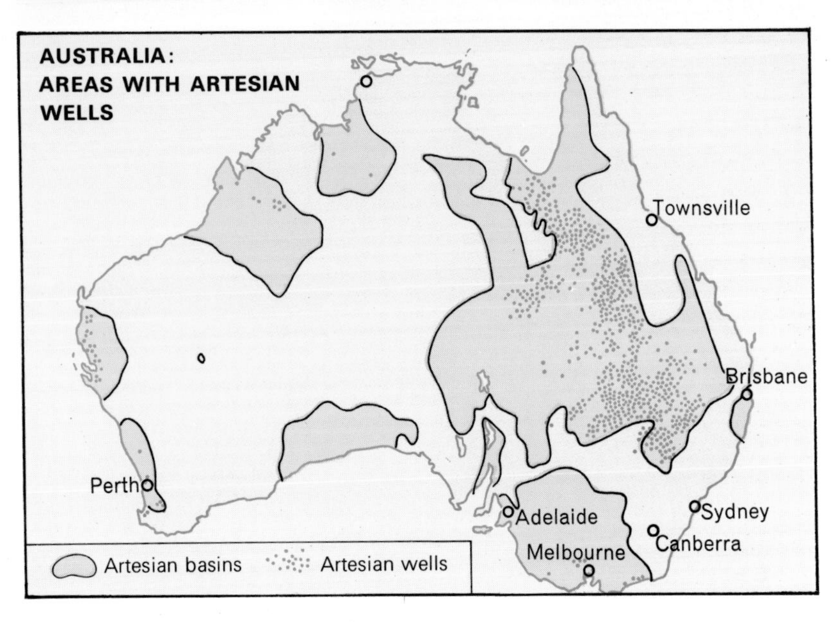

Australia still imports a high proportion of its manufactured goods.

Cities and Communications

Although Australia is renowned for its wide open spaces, most of its people live in towns and cities. Sydney, in New South Wales, with a population of some 3 million, is the largest city, followed by Melbourne, Brisbane, Adelaide and Perth.

There are more than 25,000 miles (40,000 km) of rail lines, but the tracks were laid in three different gauges, which makes rail communication difficult on some routes. Many broad and narrow gauge lines have been converted to standard gauge since 1957.

Australians make much use of airlines and private airplanes to link the big cities and the lonely parts of the outback. Medical treatment is provided by the Flying Doctor Service. Remote stations are linked by radio, and children living outside the cities receive much of their education by radio.

The People

The original inhabitants of Australia were the Aborigines, dark-skinned people related to some of the people of S India and Malaysia. About 40,000 live in Australia today. Most of the people are descended from settlers from the British Isles, but since World War II more than a million people from continental Europe have emigrated to the country.

The language of Australia is English, but some immigrants from countries outside the English-speaking world still use their own languages among themselves.

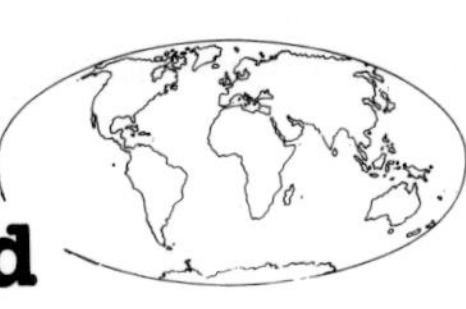

New Zealand

New Zealand, in the South Pacific, lies about 1,200 miles (1,900 km) SE of Australia across the Tasman Sea. It consists of two main islands, the North Island and the South Island, together with Stewart Island, the Chatham Islands, and outlying small islands. The country's capital is Wellington, on the North Island. Its area is 103,736 sq miles (268,675 sq km).

Land and Climate

The country's climate varies from subtropical in the far N to temperate in the S.

The two main islands extend more than 1,000 miles (1,600 km) from N to S. They are separated by the Cook Strait. Their coastlines are irregular and deeply indented in places. Both are mountainous and have fertile coastal plains. The high central plateau of the North Island is of volcanic origin. Its active peaks include Mount Ruapehu (9,175 ft; 2,797 m) and Mount Ngauruhoe (7,515 ft; 2,290 m). The glaciated mountains of the Southern Alps form the backbone of the South Island. Mount Cook (12,349 ft; 3,764 m), at the centre, is New Zealand's highest peak. To the E of the Southern Alps is the Canterbury Plain.

New Zealand has many lakes, the largest being Lake Taupo (238 sq miles; 616 sq km) on the North Island. From it flows the Waikato (270 miles; 435 km), the longest river. The Rotorua area, near the Bay of Plenty on the North Island, is famous for its hot springs and geysers.

Agriculture and Industry

New Zealand's agriculture is among the world's most efficient and productive. Farm products—mainly meat, dairy produce and wool—account for most of the country's exports. Industries include food processing and the manufacture of textiles, paper and steel. There are coal, iron, limestone and oil deposits.

Cities and Communications

New Zealand's chief cities include Auckland, Christchurch, Wellington, Dunedin and Napier-Hastings. Road and rail systems are highly developed. Internal air services link the country's main centres.

The People

Nine out of ten New Zealanders are descended from British settlers. About 8 per cent. are Maoris, the country's original Polynesian inhabitants. The country's language is English, but the Maoris also speak Maori. Two-thirds of the people live on the North Island.

Top: New Zealand. Queenstown and Lake Wakatipu, in the South Island. The lake is about 50 miles (80 km) long, and its water level changes considerably. Queenstown is the South Island's most popular tourist resort.

Below: Wellington, the country's capital and second-largest city, on the southern coast of the North Island. Much of it is built on hills overlooking Port Nicholson, usually called Wellington Harbour. The principal business area, however, stands on land reclaimed from the harbour.

Papua New Guinea

The country of Papua New Guinea, formerly a trust territory of Australia, consists of the E half of the island of New Guinea and a number of nearby islands. The W half of New Guinea is West Irian, part of Indonesia. Lying just N of Australia across the Torres Strait, Papua New Guinea forms part of Melanesia. It is a wild region of mountains, jungles and swamps. Mount Wilhelm, in the Central Range of New Guinea, rises to 15,400 ft (4,694 m). Most of the people live by subsistence farming, but there are rich mineral resources. The country's capital is Port Moresby, and its area 183,000 sq miles (474,000 sq km).

Pacific Islands

Palm trees and surf in the Hawaiian Islands.

Below: Ancient stone carvings on Easter Island. The island has more than 600 of such figures, some more than 40 ft (12 m) high. Their origin is unknown.

The Pacific Ocean, the largest of the oceans, also has the largest number of islands—more than 30,000 of them. Most are in the South Pacific, the part of the ocean that lies S of the equator. The South Pacific islands together with the islands of the central Pacific are often known as *Oceania*.

In this area there are three main island groups: Melanesia, Micronesia and Polynesia. These groups are based on the different physical characteristics of the original islanders.

Melanesia ('the black islands') is a large island group in the SW Pacific. It includes New Guinea, the Bismarck Archipelago, the Solomon Islands, New Caledonia and the Fiji Islands. The people of Melanesia are dark-skinned and have frizzy hair. Generally, the climate on this group of islands is hot and wet.

Micronesia ('the little islands') lies NE of Melanesia, and includes the Bonin, Mariana, Caroline, Marshall and Kiribati Islands. On the ocean floor between the Mariana and Caroline Islands is the Marianas Trench, which has the lowest known point on the earth's surface. The people of Micronesia are copper-coloured and have straight black hair.

Polynesia ('many islands') includes many widely-scattered islands and archipelagos. The chief of these are the Hawaiian, Line, Marquesas, Tuvalu, Phoenix, Tokelau, Cook, Society, Tubuai and Kermadec Islands. New Zealand, too, is usually included in Polynesia. The indigenous peoples of Polynesia are tall and black-haired and have light skins.

Geologically, the Pacific islands are of two principal kinds. Some are *volcanic islands*, formed by subterranean action, and are mountainous and have fertile soil. Others are *atolls*, formed by coral growth.

Left: A village in Fiji, an island group consisting of two large islands and some 300 small islands. Fiji lies just west of the International Date Line. About half of its people are immigrants from India.

Below: An ancient tomb on the island of Upolu in Western Samoa, reminiscent in style of monuments in South America. There is much evidence of links between Polynesia and Asia and South America.

This table lists the principal Pacific islands and island groups, excluding those—such as New Zealand, Indonesia and Japan—that have individual entries. The islands listed vary in status. Some, such as Nauru and Western Samoa, are independent states controlling their own affairs. Most others are dependencies of large countries, chiefly Australia, France, New Zealand, the United Kingdom and the United States. Since 1959, the Hawaiian Islands, in Polynesia, have been a state of the United States.

Name	*Area*		*Population*
	sq miles	*sq km*	
ISLANDS OF MELANESIA			
Bismarck Archipelago	22,920	59,363	169,000
D'Entrecasteaux Is.	1,200	3,108	40,000
Fiji	7,040	18,234	641,000
Louisiade Archipelago	600	1,550	10,000
New Caledonia	7,340	19,010	143,000
Schouten Is.	1,230	3,186	33,000
Solomon Is.	16,120	41,750	237,000
Trobriand Is.	170	440	10,600
Vanuatu Republic	5,700	14,763	122,000
ISLANDS OF MICRONESIA			
Bonin Is.	40	104	220
Caroline Is.	500	1,295	59,735
Kiribati Republic	166	430	59,000
Mariana Is.	452	1,170	78,275
Marshall Is.	70	181	18,205
Nauru	8	21	8,000
Wake I.	3	8	2,000
ISLANDS OF POLYNESIA			
Chatham Is.	372	963	300
Cook Is.	93	241	19,000
Easter I.	46	119	1,000
Gambier Is.	12	31	2,000
Kermadec Is.	13	34	10
Line Is.	240	622	1,000
Marquesas Is.	480	1,243	3,000
Midway Is.	2	5	2,000
Phoenix Is.	19	49	1,200
Pitcairn I.	2	5	58
Samoa, American	76	197	32,000
Samoa, Western	1,097	2,841	157,000
Society Is.	650	1,680	40,000
Tokelau Is.	4	10	2,000
Tonga (Friendly) Is.	270	700	99,000
Tubuai Is.	115	298	3,600
Tuvalu	14	36	7,000
Wallis Is.	40	104	1,000

Polar Regions

The Arctic. Few Eskimoes now spend winter in igloos. They are built of blocks of frozen snow, 'mortared' together with loose snow. In summer, Eskimoes live in tents, which they also call igloos.

Opposite page: Antarctica. Penguins, the inquisitive, endearing birds of the icy South. They use their wings to swim in the cold sea, and live by catching fish.

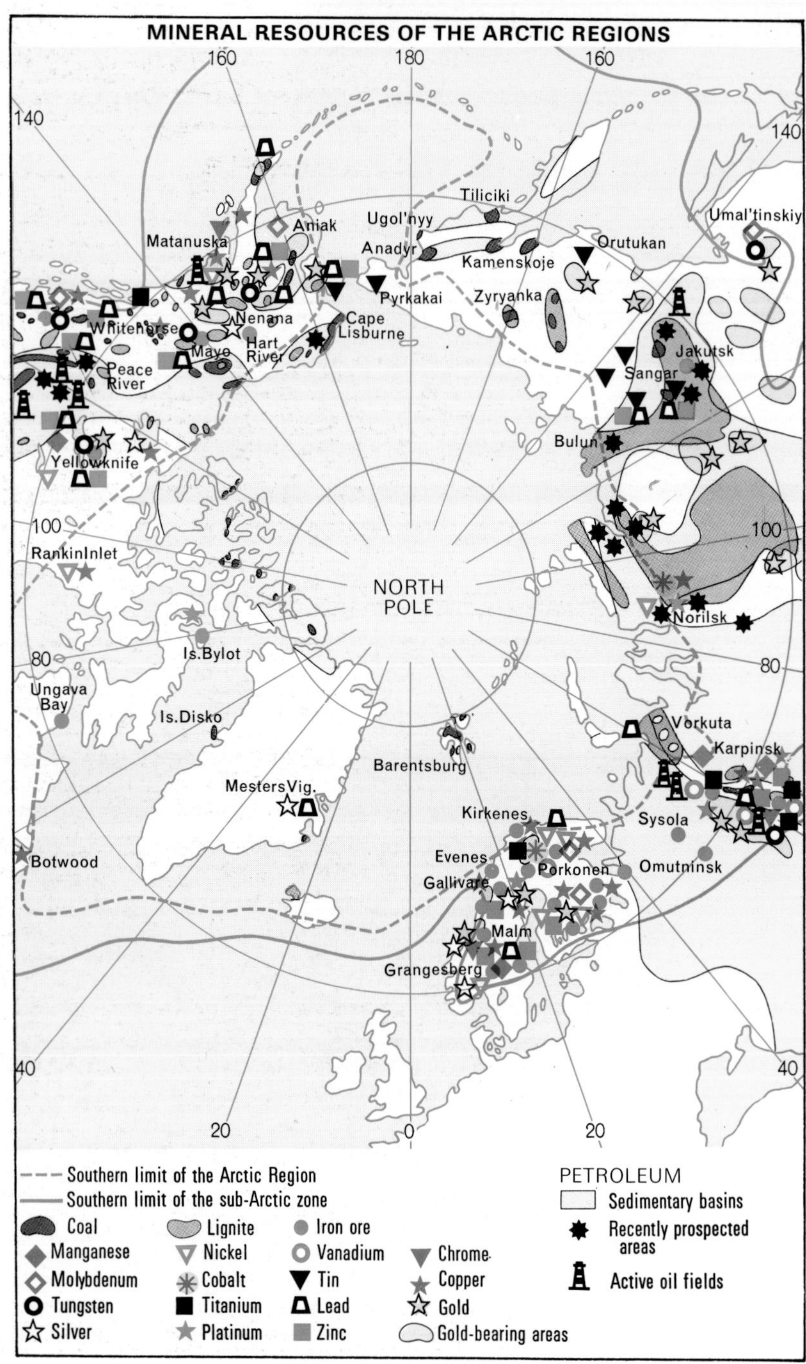

The Arctic is the cold region around the North Pole, extending from the Pole to the Arctic Circle, latitude 66.5°N. At its centre is the Arctic Ocean, which has an area of some 5.5 million sq miles (14 million sq km). Bordering the Ocean are the most northerly parts of Europe, Asia and North America. Within North America, the greater part of the huge, icy island of Greenland (a territory of Denmark) lies N of the Arctic Circle. Though Greenland is always icy, most of the Arctic lands have no snow or ice in summer.

Much of the Arctic Ocean is covered with a permanent sheet of ice, called the *ice pack*. Other parts are frozen only in winter. In spring, the ice starts to melt and break up to form *floes*. Icebergs are formed by glaciers entering the Ocean from the Arctic lands.

During the winter there is little sunshine in Arctic regions. But from March to September there is sunshine every day, and on one day the sun never sets.

Sometimes the term 'Arctic' is used to include regions farther from the Pole than the Arctic Circle—all regions N of the *tree line*, the line N of which forests do not grow. This definition limits the animal life of the Arctic, since many animals (including many birds) are found only where there are trees.

Just S of the Arctic is the *subarctic*, a region that is equally cold in winter, but is warmer in summer. Both the Arctic and the subarctic are within the area of *permafrost*, the regions in which the ground is frozen for some distance down even though the topsoil may thaw during the summer months.

Despite the severe conditions, some people live in the Arctic. They include the Lapps (who live in Lapland—northern Norway, Sweden and Finland), the Yakuts and other peoples in the U.S.S.R., and the Eskimoes in Greenland, Alaska and Canada.

Antarctica, the frozen continent around the South Pole, lies under a great blanket of snow and ice. The continent is larger than Europe or Australia, but most of it is without life of any kind, animal or plant. Its surface consists of mountains, enormous glaciers and seemingly limitless snow plains buffeted by blizzards. The surrounding seas are sections of the Atlantic, Pacific and Indian Oceans and are dangerous to shipping, when not actually impassable. In winter they are frozen over with pack ice, and in summer the seas become a heaving mass of icebergs and floes.

Many kinds of animals live in the coastal waters and on the shores. They include various species of fish, whales and seals, and penguins and other sea birds. A few plants and some insects and spiders also manage to exist in coastal areas.

It is believed that the continent is rich in mineral resources, though much of it has yet to be explored. Several countries, including Australia, Argentina and Norway, claim territory in Antarctica and many have research stations there. As in the Arctic, there is at least one day in the Antarctic year on which the sun does not shine, and one on which it does not set. The land area of the continent is about 5 million sq miles (13 million sq km).

Antarctica is roughly onion-shaped and has two huge bays, the Weddell Sea and the Ross Sea. They are so choked with ice on their southern sides that their precise extent is not always clear.

The edges of the continent are marked by bare mountains and cliffs, some of which form part of a mountain chain that divides Antarctica into two parts. The smaller part is called *West Antarctica.* From it a long, mountainous tongue of land, the Antarctic Peninsula, extends towards Cape Horn, the tip of South America. *East Antarctica,* almost twice the size of West Antarctica, faces Africa, Asia and Australia. A vast and relatively smooth snow plateau is known to conceal a rugged and mountainous terrain. The mountains and valleys are hidden under an ice cap which is several kilometres deep in places.

The South Pole is about 300 miles (480 km) from the great trans-Antarctic mountain chain. It was reached for the first time in 1911 by the Norwegian explorer Roald Amundsen.

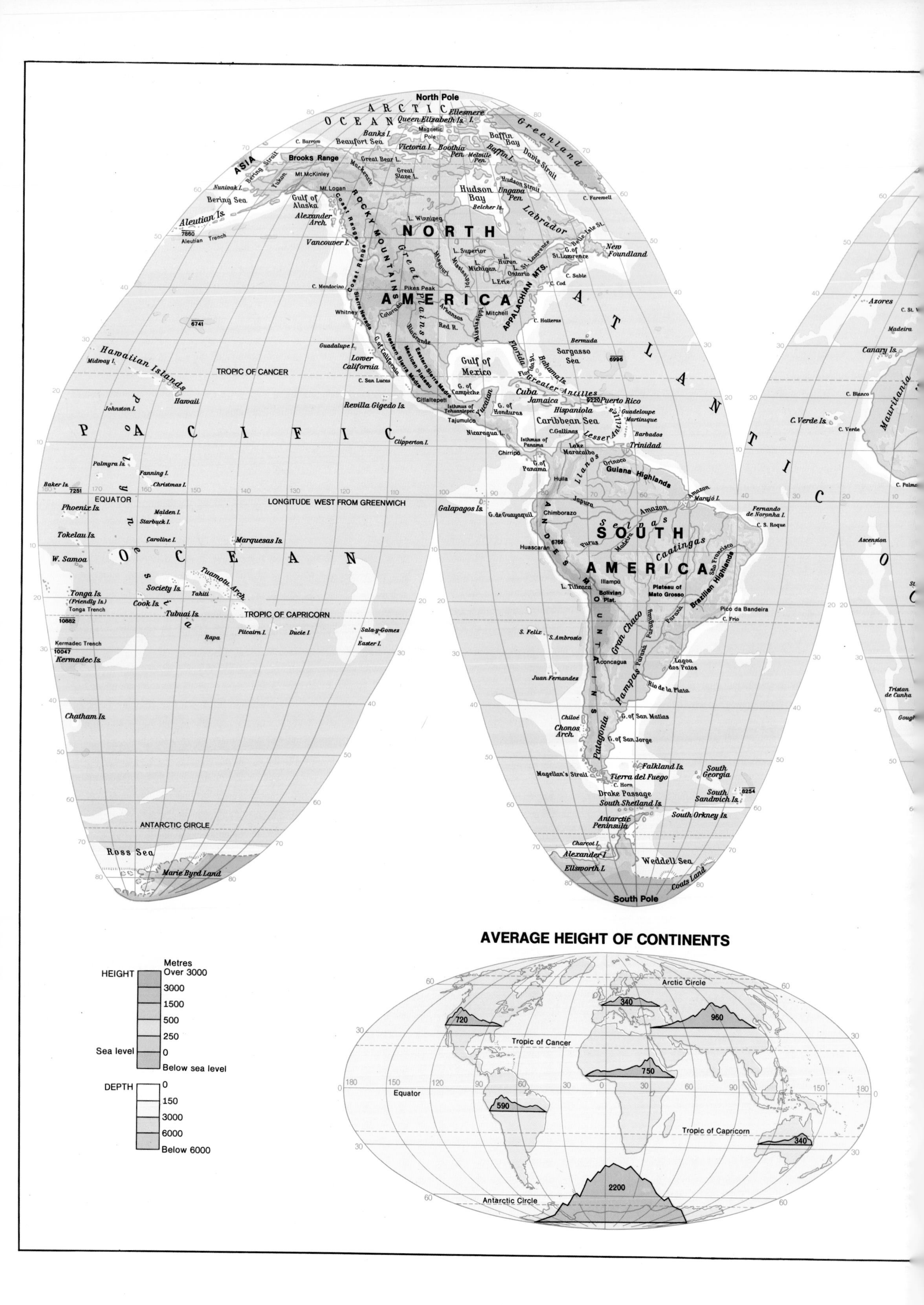
North Pole
ARCTIC OCEAN
Queen Elizabeth Is.
Ellesmere I.
Greenland
Banks I.
Magnetic Pole
Baffin Bay
Baffin I.
Davis Strait
C. Barrow
Beaufort Sea
Victoria I.
Boothia Pen.
Melville Pen.
ASIA
Bering Strait
Brooks Range
Great Bear L.
Mackenzie
Great Slave L.
Yukon
Mt.McKinley
Mt.Logan
Hudson Strait
Ungava Pen.
Hudson Bay
Belcher Is.
Labrador
C. Farewell
Nunivak I.
Bering Sea
Gulf of Alaska
Alexander Arch.
Aleutian Is.
7860
Aleutian Trench
L. Winnipeg
NORTH AMERICA
ROCKY MOUNTAINS
Coast Range
Great Plains
Vancouver I.
Belle Isle Str.
New Foundland
G. of St.Lawrence
St. Lawrence
L. Superior
L. Michigan
L. Huron
L. Ontario
L. Erie
Missouri
Mississippi
APPALACHIAN MTS.
C. Sable
C. Cod
C. Mendocino
Pikes Peak
Colorado
Arkansas
Sierra Nevada
Whitney
Mitchell
Red R.
C. Hatteras
6741
Rio Grande
Western Sierra Madre
Mexican Plateau
Eastern Sierra Madre
Bermuda
Sargasso Sea
6996
Guadalupe I.
Lower California
G. of California
Gulf of Mexico
Florida
Bahama Is.
Hawaiian Islands
Midway I.
TROPIC OF CANCER
C. San Lucas
G. of Campeche
Greater Antilles
Cuba
Jamaica
Hispaniola
Puerto Rico
6220
Guadeloupe
Martinique
Hawaii
Johnston I.
Revilla Gigedo Is.
Citlaltepetl
Isthmus of Tehuantepec
Tajumulco
Yucatan
G. of Honduras
Caribbean Sea
C.Gallinas
Lesser Antilles
Barbados
Trinidad
Nicaragua L.
PACIFIC OCEAN
Clipperton I.
Isthmus of Panama
Chirripo
Lake Maracaibo
Orinoco
Llanos
G. of Panama
Guiana Highlands
Amazon
Marajó I.
Palmyra Is.
Fanning I.
Christmas I.
Baker Is.
7251
EQUATOR
LONGITUDE WEST FROM GREENWICH
Huila
Japura
Galapagos Is.
G. de Guayaquil
Chimborazo
Amazon
Fernando de Noronha I.
C. S. Roque
Phoenix Is.
Malden I.
Starbuck I.
Selvas
Tokelau Is.
Caroline I.
Marquesas Is.
SOUTH AMERICA
Purus
Madeira
Caatingas
Huascaran
6768
ANDES MOUNTAINS
W. Samoa
Polynesia
Tuamotu Arch.
San Francisco
Illampu
Society Is.
Tahiti
L. Titicaca
Bolivian Plat.
Plateau of Mato Grosso
Brazilian Highlands
Tonga Is.
(Friendly Is.)
Tonga Trench
10882
Cook Is.
Tubuai Is.
TROPIC OF CAPRICORN
Pico da Bandeira
C. Frio
Paraná
Paraguay
Gran Chaco
Rapa
Pitcairn I.
Ducie I.
Sala-y-Gomez
Easter I.
S. Felix
S. Ambrosio
Kermadec Trench
10047
Kermadec Is.
Aconcagua
Lagoa dos Patos
Juan Fernandez
Pampas
Rio de la Plata
Chatham Is.
Chiloé
G. of San Matias
Chonos Arch.
Patagonia
G. of San Jorge
Falkland Is.
South Georgia
Magellan's Strait
Tierra del Fuego
C. Horn
Drake Passage
South Shetland Is.
South Sandwich Is.
8254
South Orkney Is.
Antarctic Peninsula
ANTARCTIC CIRCLE
Charcot I.
Alexander I.
Ellsworth L.
Weddell Sea
Coats Land
Ross Sea
Marie Byrd Land
South Pole
ATLANTIC OCEAN
Azores
Madeira
Canary Is.
C. Blanco
Mauritania
C. Verde Is.
C. Verde
C. Palmas
Ascension
Tristan de Cunha
AVERAGE HEIGHT OF CONTINENTS
Arctic Circle
720
340
960
Tropic of Cancer
750
Equator
590
Tropic of Capricorn
340
2200
Antarctic Circle
HEIGHT
Metres
Over 3000
3000
1500
500
250
Sea level
0
Below sea level
DEPTH
0
150
3000
6000
Below 6000

MAP 1

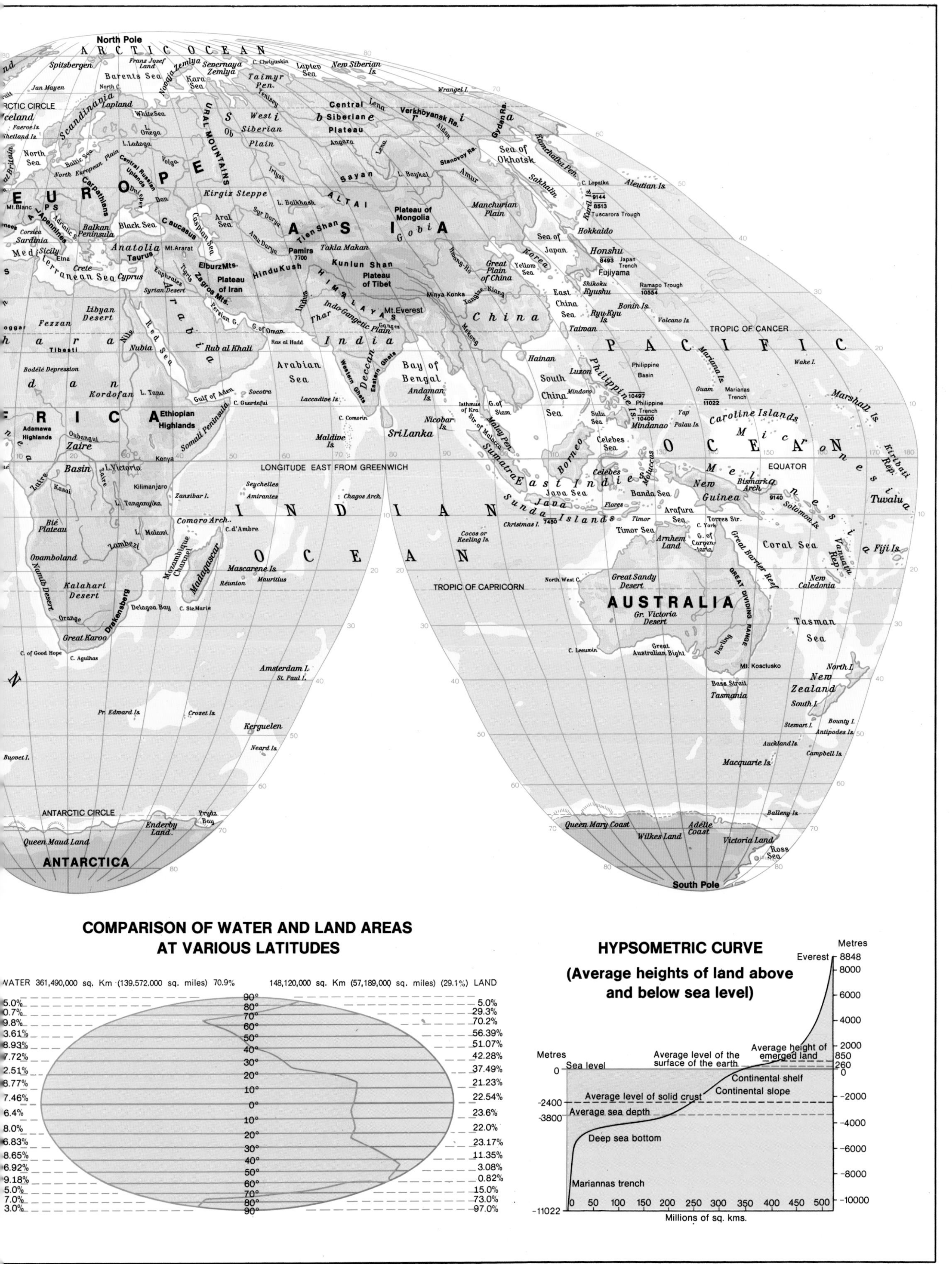

North Pole
ARCTIC OCEAN
Spitsbergen
Franz Josef Land
Severnaya Zemlya
Barents Sea
Kara Sea
Taimyr Pen.
Laptev Sea
New Siberian Is.
Wrangel I.
Jan Mayen
ARCTIC CIRCLE
Lapland
Scandinavia
White Sea
Central Siberian Plateau
Verkhoyansk Ra.
West Siberian Plain
URAL MOUNTAINS
North Sea
Baltic Sea
Sea of Okhotsk
Kamchatka Pen.
Sakhalin
Aleutian Is.
Kuril Is.
Tuscarora Trough
EUROPE
Carpathians
Kirgiz Steppe
L. Balkhash
ALTAI
Plateau of Mongolia
Manchurian Plain
Hokkaido
Mt.Blanc
ALPS
Apennines
Corsica
Sardinia
Balkan Peninsula
Black Sea
Caucasus
Caspian Sea
Aral Sea
ASIA
Gobi
Sea of Japan
Sicily
Etna
Anatolia
Mt.Ararat
Taurus
Pamirs 7700
Takla Makan
Korea
Honshu
Japan Trench
Fujiyama
Mediterranean Sea
Crete
Cyprus
Elburz Mts.
Plateau of Iran
Zagros Mts.
Hindu Kush
Kunlun Shan
Plateau of Tibet
HIMALAYAS
Mt.Everest
Great Plain of China
Yellow Sea
Shikoku
Kyushu
Ramapo Trough 10554
East China Sea
Bonin Is.
Ryu-Kyu Is.
Volcano Is.
Taiwan
Syrian Desert
Arabia
Persian G.
G. of Oman
Indo-Gangetic Plain
India
China
TROPIC OF CANCER
Libyan Desert
Fezzan
Sahara
Tibesti
Nubia
Red Sea
Rub al Khali
Ras al Hadd
Arabian Sea
Deccan
Western Ghats
Eastern Ghats
Bay of Bengal
Andaman Is.
Hainan
Luzon
Philippine Basin
Philippine Is.
PACIFIC
Mariana Is.
Wake I.
Bodélé Depression
Kordofan
L. Tana
Gulf of Aden
Socotra
C. Guardafui
Laccadive Is.
South China Sea
Mindoro
Philippine Trench 10400
Mindanao
Guam
Marianas Trench 11022
Yap
Palau Is.
Caroline Islands
Marshall Is.
AFRICA
Ethiopian Highlands
Adamawa Highlands
Ubangui
Zaire
Somali Peninsula
C. Comorin
Nicobar Is.
Sri Lanka
Maldive Is.
Isthmus of Kra
G. of Siam
Malay Pen.
Str. of Malacca
Sulu Sea
Celebes Sea
Borneo
Moluccas
Micronesia
OCEANIA
EQUATOR
Kiribati Rep.
Kenya
LONGITUDE EAST FROM GREENWICH
Zaire Basin
L. Victoria
Kasai
Kilimanjaro
Zanzibar I.
Seychelles
Amirantes
Chagos Arch.
Sumatra
East Indies
Celebes
Java Sea
Banda Sea
New Guinea
Bismarck Arch.
Melanesia
Tuvalu
L. Tanganyika
INDIAN OCEAN
Sunda Islands
Java
Flores
Timor
Timor Sea
Arafura Sea
Torres Str.
C. York
G. of Carpentaria
Solomon Is.
Bié Plateau
L. Malawi
Comoro Arch.
C. d'Ambre
Christmas I.
Cocos or Keeling Is.
Arnhem Land
Coral Sea
Vanuatu Rep.
Fiji Is.
Zambezi
Ovamboland
Mozambique Channel
Madagascar
Mascarene Is.
Réunion
Mauritius
North West C.
Great Sandy Desert
Great Barrier Reef
New Caledonia
Namib Desert
Kalahari Desert
Drakensberg
Delagoa Bay
C. Ste.Marie
TROPIC OF CAPRICORN
AUSTRALIA
Gr. Victoria Desert
GREAT DIVIDING RANGE
Orange
Great Karoo
Tasman Sea
C. of Good Hope
C. Agulhas
C. Leeuwin
Great Australian Bight
Darling
Mt. Kosciusko
Amsterdam I.
St. Paul I.
North I.
New Zealand
Bass Strait
Tasmania
South I.
Pr. Edward Is.
Crozet Is.
Stewart I.
Bounty I.
Antipodes Is.
Kerguelen
Auckland Is.
Campbell Is.
Heard Is.
Bouvet I.
Macquarie Is.
ANTARCTIC CIRCLE
Prydz Bay
Enderby Land
Balleny Is.
Queen Maud Land
Queen Mary Coast
Adélie Coast
Wilkes Land
Victoria Land
Ross Sea
ANTARCTICA
South Pole
COMPARISON OF WATER AND LAND AREAS AT VARIOUS LATITUDES
WATER 361,490,000 sq. Km (139.572.000 sq. miles) 70.9%
148,120,000 sq. Km (57,189,000 sq. miles) (29.1%) LAND
5.0% 90° 5.0%
0.7% 80° 29.3%
9.8% 70° 70.2%
3.61% 60° 56.39%
8.93% 50° 51.07%
7.72% 40° 42.28%
2.51% 30° 37.49%
8.77% 20° 21.23%
7.46% 10° 22.54%
6.4% 0° 23.6%
8.0% 10° 22.0%
6.83% 20° 23.17%
8.65% 30° 11.35%
6.92% 40° 3.08%
9.18% 50° 0.82%
5.0% 60° 15.0%
7.0% 70° 73.0%
3.0% 80° 97.0%
90°
HYPSOMETRIC CURVE
(Average heights of land above and below sea level)
Metres
Everest 8848
8000
6000
4000
2000
850
260
0
-2000
-4000
-6000
-8000
-10000
Metres
0 Sea level
Average level of the surface of the earth
Average height of emerged land
Continental shelf
Continental slope
-2400 Average level of solid crust
-3800 Average sea depth
Deep sea bottom
Mariannas trench
-11022
0 50 100 150 200 250 300 350 400 450 500
Millions of sq. kms.

Scale 1:14 000 000
0 140 280 420 560 700 Kms.
0 140 280 420 St.mls.
ICELAND
Reykjavik
Isafjördhur
Akureyri
Seydhisfjördhur
Vestmannaeyjar
Arctic Circle
Norwegian Sea
ATLANTIC OCEAN
Faeroe Is. (Den.)
Shetland Is.
Orkney Is.
UNITED KINGDOM
Scotland
England
Wales
IRELAND
DUBLIN
Cork
Limerick
Galway
Sligo
Belfast
N. Irel.
Londonderry
GLASGOW
Edinburgh
Dundee
Aberdeen
Inverness
Newcastle
Carlisle
Hull
LEEDS
LIVERPOOL
MANCHESTER
SHEFFIELD
BIRMINGHAM
Norwich
Ipswich
LONDON
Cardiff
Bristol
Plymouth
Portsmouth
Penzance
English Channel
NORTH SEA
NORWAY
SWEDEN
Bergen
OSLO
Stavanger
Kristiansand
Trondheim
Kristiansund
Ålesund
Bodö
Lofoten Is.
Vesterålen
Skagerrak
Kattegat
Göteborg
Jönköping
Kalmar
Öland
Karlskrona
Malmö
Uppsala
Örebro
DENMARK
COPENHAGEN
Aalborg
Aarhus
Odense
Zealand
Flensburg
Kiel
Lübeck
Rostock
HAMBURG
BREMEN
HANOVER
BERLIN
Szczecin
Bydgoszcz
Poznań
Wrocław
DRESDEN
LEIPZIG
Magdeburg
Halle
Erfurt
Münster
ESSEN
DORTMUND
DÜSSELDORF
COLOGNE
Bonn
FRANKFURT
Mannheim
Nuremberg
STUTTGART
Augsburg
MUNICH
Freiburg
AMSTERDAM
THE HAGUE
ROTTERDAM
NETHERLANDS
BRUSSELS
BELGIUM
Antwerp
Liège
LUXEMBOURG
PRAGUE
Pilsen
Brno
Ostrava
VIENNA
Salzburg
Innsbruck
Linz
Graz
AUSTRIA
BUDAPEST
FRANCE
PARIS
Le Havre
Cherbourg
Brest
Rennes
Le Mans
Nantes
Angers
Orléans
Tours
Rouen
Amiens
Lille
Calais
Boulogne
Reims
Nancy
Metz
Strasbourg
Mulhouse
Dijon
LYONS
Clermont-F.
St. Etienne
Limoges
Bordeaux
Toulouse
Montpellier
Nimes
Avignon
MARSEILLES
Toulon
Nice
MONACO
Bay of Biscay
SWITZERLAND
Basle
Bern
Geneva
Zürich
MILAN
TURIN
GENOA
Brescia
Padua
Venice
Trieste
Parma
Bologna
Florence
Leghorn
Perugia
Ancona
ROME
NAPLES
Salerno
Bari
Taranto
Foggia
ITALY
Corsica
Sardinia
Cagliari
Sicily
PALERMO
Messina
Catania
Syracuse
Tyrrhenian Sea
Adriatic Sea
MALTA
Valletta
YUGOSLAVIA
Ljubljana
ZAGREB
Rijeka
Split
Sarajevo
Dubrovnik
SPAIN
MADRID
BARCELONA
VALENCIA
Saragossa
Valladolid
Salamanca
Burgos
Bilbao
Santander
Oviedo
Gijón
Corunna
Vigo
Toledo
Cuenca
Teruel
Lérida
Tarragona
Albacete
Murcia
Cartagena
Córdoba
Seville
Granada
Málaga
Almería
Cádiz
Badajoz
Cáceres
ANDORRA
PORTUGAL
LISBON
Oporto
Coimbra
Braga
Setúbal
Faro
Balearic Is.
Majorca
Minorca
Ibiza
Str. of Gibraltar
Gibraltar (Br.)
MEDITERRANEAN
MOROCCO
Tangier
Tetuan
Rabat
DAR EL BEIDA (CASABLANCA)
Meknès
Fès
Oudjda
Marrakesh
Safi
Mogador
Agadir
ALGERIA
EL DJEZAIR (Algiers)
Ouahran
Tlemcen
Sidi Bel Abbès
Constantine
Annaba
Biskra
TUNISIA
TUNIS
Sousse
Sfax
Gabès
Djerba I.
West from Greenwich 0 East from Greenwich

North C.
Vadsö
Pechenga
Murmansk
Kola Pen.
Kolguyev I.
Kanin Pen.
Naryan-Mar
Vorkuta
Salekhard
Ob
Berezovo
Khanty-Mansiisk
Pechora
Kozhva
Mezen
Kandalaksha
White Sea
Lapland
Rovaniemi
Tornio
Oulu
Karelia
Kem
Arkhangelsk
Onega
N.Dvina
Ukhta
Syktyvkar
Kotlas
Vel Ustyug
Konosha
L.Onega
Petrozavodsk
Kuopio
FINLAND
Tampere
L. Ladoga
Vyborg
HELSINKI
G.of Finland
LENINGRAD
Vologda
Cherepovets
L.Rybinsk
Tallin
ESTONIA
Chudskoye L.
Novgorod
Tartu
Pskov
Bologoye
YAROSLAVL
Ivanovo
GORKI
Kalinin
MOSCOW
Arzamas
Volga
Oka
RIGA
LATVIA
Velikie Luki
Daugavpils
Vitebsk
Smolensk
Vyazma
Dnieper
Ryazan
Tula
Saransk
Penza
Kaunas
Vilnius
WHITE RUSSIA
MINSK
Baranovichi
Bialystok
Bryansk
Orel
Yelets
Tambov
VORONEZH
Gomel
Kursk
Brest
Pripet
Konotop
Belgorod
KHARKOV
Millerovo
Lublin
Rovno
KIEV
Zhitomir
Poltava
UKRAINE
LVOV
DNEPROPETROVSK
DONETSK
Shakhty
ZAPOROZHYE
Taganrog
ROSTOV
Chernovtsy
Dniester
Prut
Krivoy Rog
Botoşani
MOLDAVIA
Kishinev
Nikolayev
Kherson
Zhdanov
Yeysk
Manych
Sea of Azov
Iaşi
Debrecen
Cluj
ROMANIA
ODESSA
Perekop
Crimea
Kerch
Krasnodar
Kuban
Armavir
Stavropol
Braşov
Galati
Sibiu
Simferopol
Yalta
Sevastopol
Balaklava
Ploeşti
Constanta
Tuapse
Maykop
Turnu Severin
BUCHAREST
Craiova
Ruse
Danube
Pleven
Varna
BLACK SEA
Sukhumi
Batumi
Sinop
Trabzon
BULGARIA
SOFIA
Burgas
Plovdiv
Edirne
Bosporus
Zonguldak
Samsun
ISTANBUL
Üsküdar
Kavalla
S.of Marmara
Gelibolu
Bandirma
Bursa
Sakarya
ANKARA
Kizil
Sivas
Thessaloniki
Eskisehir
Dardanelles
Larisa
Lésvos
Manisa
Afyon
Khalkis
Khios
Izmir
Aydin
Isparta
Konya
Antalya
G.of Antalya
ATHENS
Patras
Piraeus
Kalamai
Aegean Sea
Rhodes
Kárpathos
Iraklion
Crete
CYPRUS
Nicosia
Anatolia
TURKEY
Kayseri
Malatya
Euphrates
Adana
Gaziantep
Iskenderun
HALAB (ALEPPO)
Hama
Homs
SYRIA
BEIRUT
LEBANON
DAMASCUS
Haifa
ISR.
Dayr az Zawr
Al Hadithah
Mosul
Erbil
Kirkuk
IRAQ
BAGHDAD
An Najaf
Euphrates
Al Basrah
Tigris
Diyarbakir
Erzurum
L.Van
Van
Rizaiyeh (Urmia)
L.Urmia
Tabriz
GEORGIA
Kutaisi
TBILISI
Nukha
ARMENIA
Leninakan
YEREVAN
Nakhichevan
Araks
AZERBAIJAN
Kirovabad
BAKU
Lenkoran
Rasht
TEHRAN
Kuma
Pyatigorsk
Groznyy
Ordzhonikidze
Daghestan
Makhachkala
CASPIAN SEA
Astrakhan
Stepnoy
VOLGOGRAD
SARATOV
Uralsk
Ural
Guryev
Emba
Aktyubinsk
Chelkar
Aralsk
Novo Kazalinsk
Syr
Aral Sea
UZBEKISTAN
Kungrad
Amu
Nukus
Tashauz
Urgench
Fort Shevchenko
Kara Bogaz Gol
Krasnovodsk
TURKMENISTAN
Darvaza
Kizyl Arvat
Ashkhabad
Bojnurd
Bandar e Shāh
Shāhrūd
Arak
Kāshān
Kermānshāh
Esfahan
Yazd
Dezfūl
Ahvāz
Bandar e Shahpur
Abadan
Persian Gulf
Shirāz
IRAN
KAZAN
Izhevsk
Sarapul
Kirov
PERM
UFA
Sterlitamak
Abdulino
Ulyanovsk
KUYBYSHEV
Orenburg
Orsk
Magnitogorsk
CHELYABINSK
SVERDLOVSK
Kurgan
Tyumen
Tobolsk
Tavda
Ivdel
Irtysh
Ishim
Tara
OMSK
Petropavlovsk
Kustanay
Krasivoe
Tselinograd
L.Tengiz
Turgay
Baikonur
KAZAKHSTAN
R. S. F. S. R.
U. S. S. R.
RUSSIAN S. F. S. R.
REPUBLIC
MEDITERRANEAN SEA
Copyright:Vallardi Ind.Graf.

ATLANTIC OCEAN
ARCTIC OCEAN
North Pole
Greenland (Den.)
Reykjavik
ICELAND
Jan Mayen (Nor.)
Faeroe Is. (Den.)
Outer Hebrides
Shetland Is.
Norwegian Sea
West Spitsbergen
North East Land
Svalbard (Nor.)
Bear I.
Franz Josef Land
Barents Sea
Komsomolets I.
Oct. Revolution I.
Bolshevik I.
C.Chelyuskin
Severnaya Zemlya
Novaya Zemlya
Kara Sea
Taimyr Pen.
Kotelny I.
New Siberian
De Long Is.
IRELAND
DUBLIN
GLASGOW
EDINBURGH
Belfast
LIVERPOOL
BIRMINGHAM
LONDON
Plymouth
UNITED KINGDOM
North Sea
NORWAY
SWEDEN
FINLAND
DENMARK
COPENHAGEN
OSLO
STOCKHOLM
HELSINKI
Bergen
Trondheim
Narvik
Göteborg
Tampere
Oulu
Luleå
Murmansk
Kola Pen.
White Sea
Kolguyev I.
Vaigach I.
Arkhangelsk
Mezen
Pechora
Vorkuta
Gulf of Ob
Salekhard
Dikson
Tazovskoye
Dudinka
Turukhansk
Nordvik
Khatanga
Olenek
Bulun
Tiksi
Arctic Circle
Vilyuisk
PORTUGAL
LISBON
Oporto
Coruna
Bilbao
SPAIN
MADRID
VALENCIA
BARCELONA
Cadiz
Gibraltar (Br.)
Balearic Is.
MOROCCO
Ouahran
ALGERIA
EL DJEZAIR
FRANCE
PARIS
Le Havre
Nantes
Bordeaux
LYONS
Toulouse
MARSEILLES
Corsica
Sardinia
BRUSSELS
BELGIUM
NETHERL.
LUX.
W.GERM.
Rhine
Bonn
E.GERM.
HAMB.
BERLIN
Leipzig
SWITZ.
Bern
TURIN
GENOA
MILAN
MUNICH
AUSTRIA
VIENNA
PRAGUE
CZECHOSLOVAKIA
POLAND
WARSAW
KRAKOW
LODZ
Gdansk
Kaliningrad
Baltic Sea
Lithuania
Latvia
Estonia
RIGA
Vilnius
Tallin
MINSK
White Russia
ITALY
Florence
ROME
NAPLES
Adriatic Sea
Tyrrhenian Sea
Cagliari
Sicily
PALERMO
MALTA
Ionian Sea
Mediterranean Sea
TUNIS
TUNISIA
Tripoli
Benghazi
LIBYA
YUGOSLAVIA
ZAGREB
BELGRADE
Sarajevo
Tirana
ALBANIA
BUDAPEST
HUNGARY
ROMANIA
BUCHAREST
Cluj
BULGARIA
SOFIA
Constanta
GREECE
ATHENS
Thessaloniki
Aegean Sea
Crete
ISTANBUL
Bursa
Izmir
Eskisehir
TURKEY
ANKARA
Konya
Kayseri
Adana
Antalya
Samsun
Zonguldak
Erzurum
Black Sea
Sevastopol
ODESSA
Nikolayev
Kishinev
Moldavia
Ukraine
KIEV
LVOV
KHARKOV
DNEPROPETROVSK
VORONEZH
Orel
Gomel
LENINGRAD
L.Ladoga
Petrozavodsk
L.Onega
MOSCOW
YAROSLAVL
GORKI
Kotlas
Syktyvkar
UNION OF SOVIET SOCIALIST REPUBLICS
Kirov
Volga
KAZAN
PERM
SVERDLOVSK
Serov
Tobolsk
Tyumen
UFA
KUYBYSHEV
Penza
SARATOV
CHELYABINSK
Magnitogorsk
Petropavlovsk
Orenburg
Kustanay
Uralsk
Aktyubinsk
OMSK
NOVOSIBIRSK
Tomsk
Kemerovo
Novokuznetsk
Barnaul
Biysk
Rubtsovsk
Semipalatinsk
Pavlodar
Tselinograd
L.Tengiz
KARAGANDA
Kazakhstan
Balkhash
L. Balkhash
ALMA ATA
Siberia
Surgut
Yeniseysk
KRASNOYARSK
Angara
Abakan
Irkutsk
Cheremkhovo
Ust-Kut
Kirensk
Lena
Stony Tunguska
Lower Tunguska
Tuva
Kyzyl
Ulan Kom
ROSTOV
VOLGOGRAD
Krasnodar
Grozny
Astrakhan
Guryev
Caspian Sea
Batumi
Georgia
TBILISI
Armenia
YEREVAN
Azerbaijan
BAKU
Aral Sea
Syr Darya
Amu Darya
Kzyl Orda
Uzbekistan
Khiva
Bukhara
Samarkand
TASHKENT
Dzhambul
Frunze
Kirgizia
Andizhan
Tadzhikistan
Dushanbe
Turkmenistan
Krasnovodsk
Ashkhabad
Mashhad
CYPRUS
Nicosia
ALEXANDRIA
CAIRO
Suez
Asyut
Aswan
Nile
ARAB REPUBLIC OF EGYPT
Libyan Desert
BEIRUT
LEBANON
Tel Aviv
ISRAEL
Jerusalem
Amman
JORDAN
Damascus
SYRIA
Homs
HALAB
Mosul
Kirkuk
IRAQ
BAGHDAD
Al Basrah
Van
Tabriz
Rasht
Hamadan
TEHRAN
Kermanshah
Esfahan
Ahvaz
Abadan
Yazd
Kerman
Shiraz
Bushehr
Bandar Abbas
Zahedan
IRAN
KUWAIT
Kuwait
NEUTRAL ZONE
Nafud
Hail
Tabuk
Al Jawf
Hejaz
SAUDI ARABIA
Buraida
Riyadh
Medina
Jidda
Mecca
Nejd
Al Hufuf
BAHRAIN
Al Manamah
QATAR
Dawhah
U.A.E.
Persian Gulf
Tropic of Cancer
Rub' al Khali
Asir
Al Qunfidha
Red Sea
YEMEN
San'a
Hodeida
Ta'iz
SOUTH YEMEN
Aden
Mukalla
Saihut
G. of Aden
OMAN
Muscat
Sur
Al Masira I.
Kuria Muria Is.
Socotra (S.Yemen)
Ras Hafun
Arabian Sea
Herat
Mazar-i-Sharif
Maimana
AFGHANISTAN
Kabul
Kandahar
Helmand
Peshawar
Rawalpindi
Islamabad
Srinagar
Kashmir
Gilgit
Jammu
Amritsar
LAHORE
Quetta
Multan
PAKISTAN
Indus
Sukkur
Hyderabad
KARACHI
Gwadar
Bikaner
Jodhpur
Jaipur
DELHI
Saharanpur
Bareilly
AGRA
KANPUR
LUCKNOW
Allahabad
VARANASI (BANARAS)
Patna
Udaipur
Indore
AHMADABAD
Jamnagar
Baroda
Diu
Bhopal
Jabalpur
NAGPUR
Raipur
Ranchi
HOWRAH
CALCUTTA
Cuttack
BOMBAY
Pune
Sholapur
HYDERABAD
Godavari
Vizagapatam
Vijayawada
Krishna
Panjim (Goa)
Hubli
Mangalore
BANGALORE
MADRAS
Mysore
Pondicherry
Calicut
Deccan
Laccadive Is. (Ind.)
INDIA
MADURA
Tiruchirapalli
Trivandrum
Jaffna
Kandy
COLOMBO
SRI LANKA
MALDIVE IS.
Male
Equator
INDIAN OCEAN
Bay of Bengal
Andaman Is. (Ind.)
Port Blair
Nicobar Is. (Ind.)
Bandar Atjeh (Kutaradja)
MEDAN
Sibolga
Nias
Mergui Archip.
NEPAL
Katmandu
SIKKIM
BHUTAN
Punakha
Brahmaputra
Shillong
Imphal
BANGLADESH
DHAKA (DACCA)
Chittagong
BURMA
Mandalay
Lashio
Myitkyina
Akyab
Irrawaddy
Prome
Bassein
RANGOON
Moulmein
BANGKOK
Tibet
Lhasa
Shigatse
Gartok
Sinkiang-Uigur
Tahcheng
Karamai
Kuldja
Urumchi
Turfan
Aksu
Kashgar
Tarim
Lop Nor
Cherchen
Khotan
Ansi
Kiuchuan
Koko Nor
SINING
LANCHOW
CHINA
Chamdo
Yangtze Kiang
Mekong
Kantse
CHENGTU
Sadiya
MONGOLIA
Dzhargalantu
Dalan Dzadagad
Ulan Bator
Yinchwan
UGANDA
Kampala
L. Turkana
Lake Victoria
KENYA
Nairobi
Arusha
Mombasa
TANZANIA
Tabora
Zanzibar
Dar-es-Salaam
L. Malawi (Nyasa)
ZAMBIA
MALAWI
MOZ.
SOMALI REP.
Mogadishu
Kismayu
Obbia
Shibeli
ETHIOPIA
Addis Ababa
Harer
Hargeisa
Berbera
DJIBOUTI REP.
Djibouti
Eritrea
Asmera
Mesewa
Gonder
Blue Nile
White Nile
Khartoum
Omdurman
Kassala
El Obeid
SUDAN
Wadi Halfa
Port Sudan
Atbara
CHAD
Scale 1:34 000 000
0 250 500 750 1000 1250 1500 Kms.
0 250 500 750 1000 St.mls.
East from 80 Greenwich

MAP 5

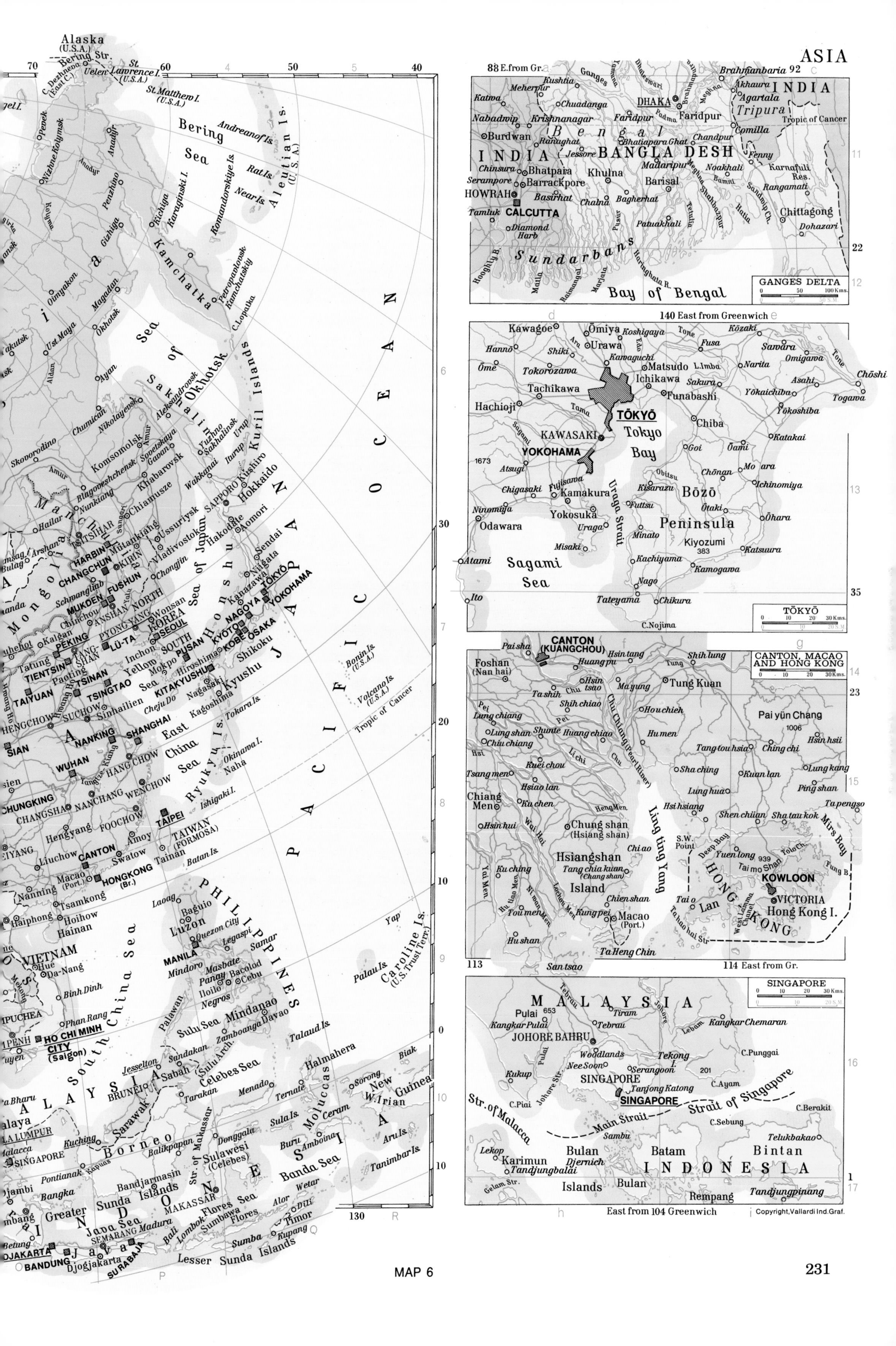

Alaska (U.S.A.)
Bering Str.
Bering Sea
Aleutian Is. (U.S.A.)
Sea of Okhotsk
Kamchatka
Sakhalin
Kuril Islands
Hokkaido
Honshu
Shikoku
Kyushu
Sea of Japan
Yellow Sea
East China Sea
South China Sea
Ryukyu Is.
TAIWAN (FORMOSA)
PHILIPPINES
Luzon
MANILA
Mindanao
Celebes Sea
Sulu Sea
Borneo
Sulawesi (Celebes)
Moluccas
Banda Sea
Java Sea
Flores Sea
Lesser Sunda Islands
Greater Sunda Islands
New Guinea
PACIFIC OCEAN
Tropic of Cancer
Caroline Is. (U.S. Trust Terr.)
Bonin Is. (U.S.A.)
Volcano Is. (U.S.A.)
TŌKYŌ
YOKOHAMA
NAGOYA
KYOTO
KOBE
OSAKA
SEOUL
PYONG YANG
PEKING
TIENTSIN
SHANGHAI
NANKING
WUHAN
CHUNGKING
CANTON
HONGKONG (Br.)
Macao (Port.)
TAIPEI
HO CHI MINH CITY (Saigon)
VIETNAM
Hainan
JAKARTA
SURABAJA
BANDUNG
MAKASSAR
GANGES DELTA
INDIA
BANGLA DESH
DHAKA
CALCUTTA
HOWRAH
Chittagong
Bay of Bengal
Sundarbans
Tropic of Cancer
TŌKYŌ
Tokyo Bay
KAWASAKI
YOKOHAMA
Bōzō Peninsula
Sagami Sea
Uraga Strait
140 East from Greenwich
CANTON, MACAO AND HONG KONG
CANTON (KUANGCHOU)
Ling ting Yang
Hsiangshan Island
KOWLOON
VICTORIA
Hong Kong I.
Mirs Bay
114 East from Gr.
SINGAPORE
MALAYSIA
JOHORE BAHRU
SINGAPORE
Strait of Singapore
Str. of Malacca
Main Strait
INDONESIA
Bintan
Batam
Bulan
East from 104 Greenwich
Copyright, Vallardi Ind. Graf.

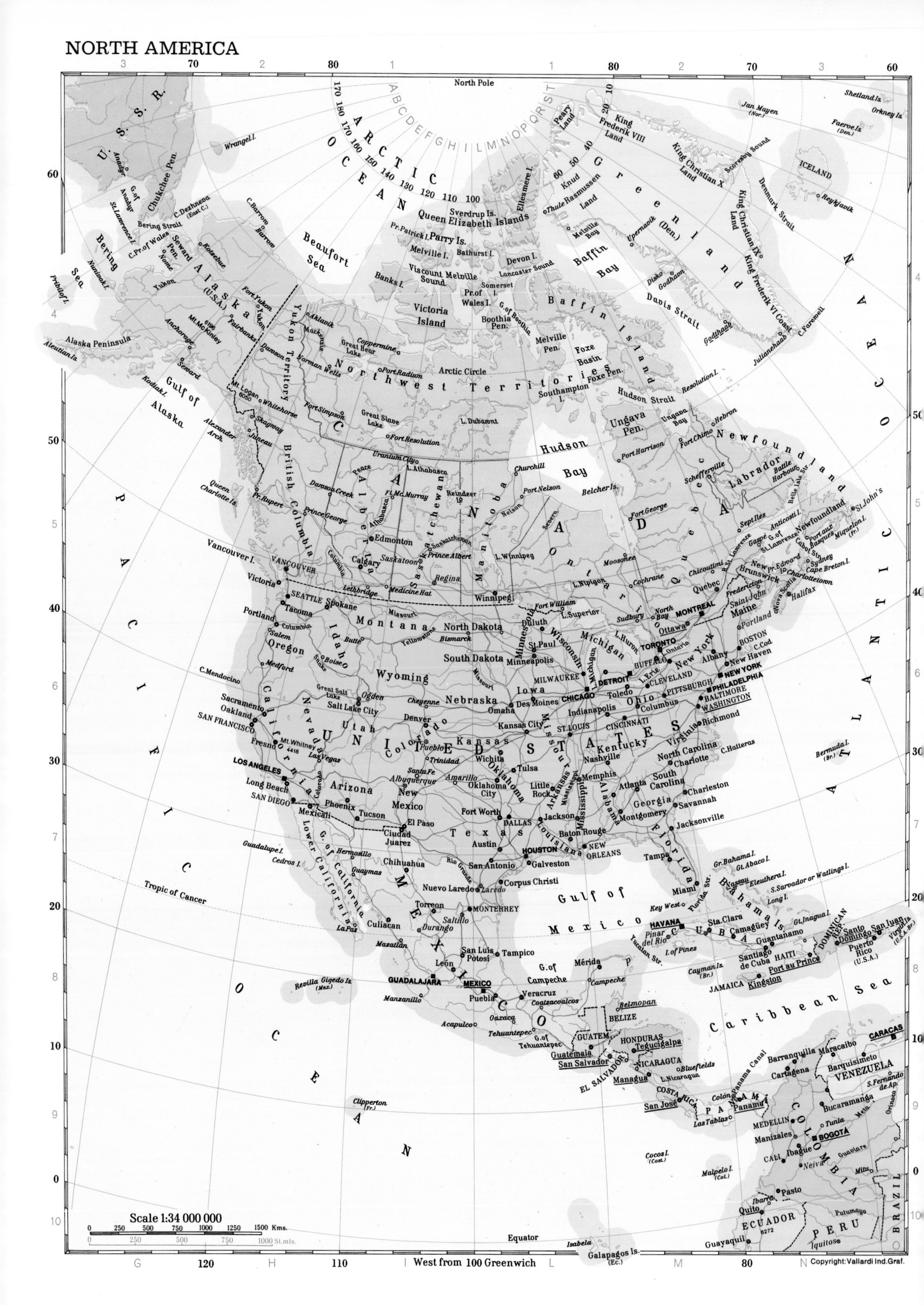
North Pole
ARCTIC OCEAN
U.S.S.R.
Greenland (Den.)
ICELAND
Reykjavik
Baffin Bay
Baffin Island
Davis Strait
Beaufort Sea
Bering Sea
Bering Strait
Alaska (U.S.A.)
Gulf of Alaska
Queen Elizabeth Islands
Victoria Island
Banks I.
Northwest Territories
Yukon Territory
Hudson Bay
Hudson Strait
Ungava Pen.
Labrador
Newfoundland
CANADA
British Columbia
Alberta
Saskatchewan
Manitoba
Ontario
Quebec
VANCOUVER
Victoria
Edmonton
Calgary
Winnipeg
MONTREAL
Ottawa
TORONTO
Quebec
Halifax
UNITED STATES
SEATTLE
Portland
SAN FRANCISCO
LOS ANGELES
SAN DIEGO
CHICAGO
DETROIT
NEW YORK
PHILADELPHIA
BALTIMORE
WASHINGTON
BOSTON
HOUSTON
DALLAS
NEW ORLEANS
Montana
North Dakota
South Dakota
Wyoming
Nebraska
Kansas
Iowa
Colorado
Utah
Nevada
California
Oregon
Idaho
Arizona
New Mexico
Texas
Oklahoma
Kentucky
North Carolina
South Carolina
Georgia
Florida
Gulf of Mexico
MEXICO
GUADALAJARA
Monterrey
Caribbean Sea
HAVANA
CUBA
Bahama Is.
JAMAICA
Kingston
HAITI
Port au Prince
DOMINICAN REP.
Santo Domingo
Puerto Rico (U.S.A.)
BELIZE
GUATEM.
HONDURAS
Tegucigalpa
EL SALVADOR
San Salvador
NICARAGUA
Managua
COSTA RICA
San José
PANAMA
CARACAS
VENEZUELA
COLOMBIA
BOGOTÁ
ECUADOR
Quito
Guayaquil
PERU
BRAZIL
PACIFIC OCEAN
ATLANTIC OCEAN
Tropic of Cancer
Arctic Circle
Equator
Galapagos Is. (Ec.)
Scale 1:34 000 000
West from 100 Greenwich
Copyright: Vallardi Ind.Graf.

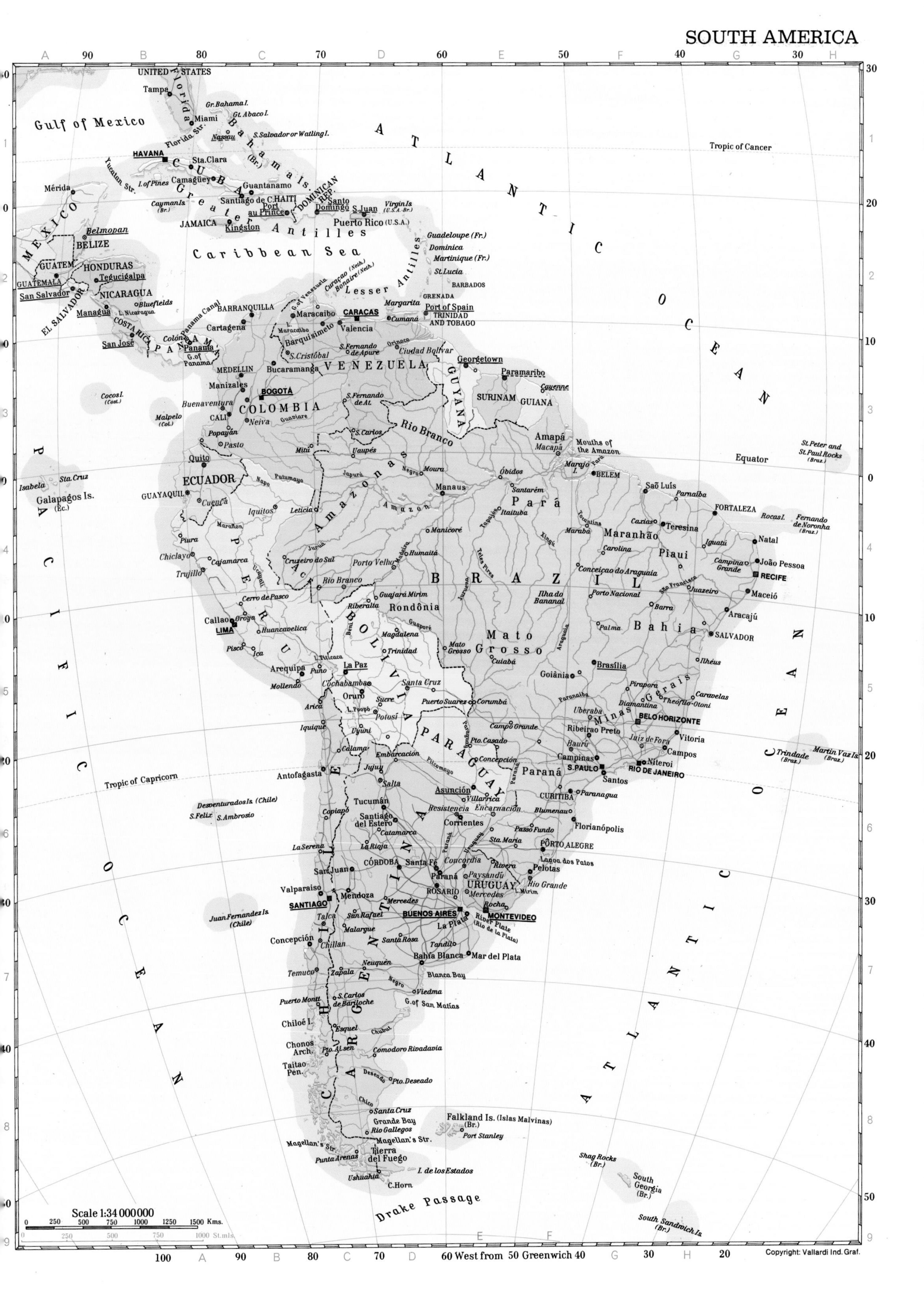
UNITED STATES
Tampa
Florida
Miami
Florida Str.
Gulf of Mexico
Gr.Bahama I.
Gt.Abaco I.
Nassau
S.Salvador or Watling I.
Bahamas Is. (Br.)
HAVANA
Sta.Clara
CUBA
Yucatan Str.
I.of Pines
Camagüey
Guantanamo
Mérida
Cayman Is. (Br.)
Santiago de C.
HAITI
DOMINICAN REP.
Santo Domingo
Port au Prince
S.Juan
Virgin Is. (U.S.A.-Br.)
Puerto Rico (U.S.A.)
JAMAICA
Kingston
Greater Antilles
MEXICO
Belmopan
BELIZE
GUATEM.
GUATEMALA
HONDURAS
Teguçigalpa
San Salvador
EL SALVADOR
NICARAGUA
Bluefields
Managua
L.Nicaragua
COSTA RICA
San José
Colón
Panama
PANAMA
Panama Canal
G.of Panama
Caribbean Sea
Guadeloupe (Fr.)
Dominica
Martinique (Fr.)
St.Lucia
BARBADOS
GRENADA
Lesser Antilles
Curaçao (Neth.)
Bonaire (Neth.)
G.of Venezuela
Margarita
Port of Spain
TRINIDAD AND TOBAGO
BARRANQUILLA
Cartagena
Maracaibo
L. Maracaibo
CARACAS
Cumaná
Valencia
Barquisimeto
S.Cristóbal
S.Fernando de Apure
Orinoco
Ciudad Bolívar
VENEZUELA
Georgetown
GUYANA
Paramaribo
SURINAM
Cayenne
GUIANA
MEDELLIN
Bucaramanga
Manizales
BOGOTÁ
Buenaventura
CALI
COLOMBIA
Neiva
Guaviare
Popayán
Pasto
S.Fernando de At.
S.Carlos
Rio Branco
Mitú
Vaupés
Cocos I. (Cost.)
Malpelo (Col.)
Quito
ECUADOR
Napo
Putumayo
Isabela
Sta.Cruz
Galapagos Is. (Ec.)
GUAYAQUIL
Cuenca
Iquitos
Leticia
Japurá
Negro
Moura
Manaus
Amazonas
Amazon
Óbidos
Santarém
Pará
Amapá
Macapá
Mouths of the Amazon
Marajó I.
BELEM
Equator
St.Peter and St.Paul Rocks (Braz.)
São Luís
Parnaíba
FORTALEZA
Rocas I.
Fernando de Noronha (Braz.)
Caxias
Teresina
Maranhão
Iguatú
Natal
Piauí
Campina Grande
João Pessoa
RECIFE
Maceió
Aracajú
SALVADOR
Bahia
Ilhéus
Marañon
Piura
Chiclayo
Cajamarca
Trujillo
PERU
Ucayali
Juruá
Cruzeiro do Sul
Acre
Porto Velho
Madeira
Humaitá
Manicoré
Tapajós
Itaituba
Xingú
Tocantins
Marabá
Carolina
Conceiçao do Araguaia
Rio Branco
BRAZIL
Cerro de Pasco
Guajará Mirim
Riberalta
Rondônia
Ilha do Bananal
Porto Nacional
São Francisco
Juazeiro
Barra
Callao
Oroya
LIMA
Huancavelica
Beni
Magdalena
Guaporé
BOLIVIA
Trinidad
Mato Grosso
Cuiabá
Araguaia
Palma
Pisco
Ica
Arequipa
Puno
L.Titicaca
La Paz
Brasília
Goiânia
Mollendo
Cochabamba
Santa Cruz
Oruro
Sucre
Potosí
Puerto Suarez
Corumbá
Pirapora
Minas Gerais
Caravelas
Teófilo-Otoni
Arica
L.Poopó
Uyuni
Iquique
Paranaíba
Uberaba
BELO HORIZONTE
Campo Grande
Ribeirao Preto
Juiz de Fora
Vitoria
Campos
Trindade (Braz.)
Martin Vaz Is. (Braz.)
Calama
Pto.Casado
Bauru
Campinas
S.PAULO
Niteroi
RIO DE JANEIRO
Santos
PARAGUAY
Embarcación
Pilcomayo
Concepción
Paraná
Antofagasta
Jujuy
Salta
Tropic of Capricorn
Asunción
Villarrica
CURITIBA
Paranagua
Desventurados Is. (Chile)
S.Felix
S.Ambrosio
Tucumán
Copiapó
Resistencia
Encarnación
Blumenau
Santiago del Estero
Corrientes
Florianópolis
Catamarca
Passo Fundo
Sta.Maria
La Serena
La Rioja
PÔRTO ALEGRE
Lagoa dos Patos
Uruguay
CÓRDOBA
Santa Fe
Concordia
Rivera
Pelotas
San Juan
Paraná
Paysandú
Rio Grande
URUGUAY
Valparaiso
Mendoza
ROSARIO
Mercedes
L.Mirim
SANTIAGO
Juan Fernandez Is. (Chile)
Talca
San Rafael
BUENOS AIRES
Rocha
MONTEVIDEO
La Plata
River Plate (Rio de la Plata)
Malargue
Concepción
Chillan
Santa Rosa
Tandil
Bahía Blanca
Mar del Plata
Neuquén
Temuco
Zapala
Negro
Blanca Bay
Viedma
Puerto Montt
S.Carlos de Bariloche
G.of San Matías
Chiloé I.
Esquel
Chubut
Chonos Arch.
Pto.Aisen
Comodoro Rivadavia
Taitao Pen.
Deseado
Pto.Deseado
CHILE
ARGENTINA
Chico
Santa Cruz
Grande Bay
Rio Gallegos
Falkland Is. (Islas Malvinas) (Br.)
Port Stanley
Magellan's Str.
Punta Arenas
Tierra del Fuego
I. de los Estados
Ushuaia
C.Horn
Shag Rocks (Br.)
South Georgia (Br.)
Drake Passage
South Sandwich Is. (Br.)
ATLANTIC OCEAN
PACIFIC OCEAN
Tropic of Cancer
Scale 1:34 000 000
0 250 500 750 1000 1250 1500 Kms.
0 250 500 750 1000 St.mls.
West from Greenwich
Copyright: Vallardi Ind. Graf.

ASIA
China
Yangtze-Kiang
Si-kiang
Hong Kong
Formosa Str.
East China Sea
Shikoku
Kyūshū
Tanega
Tokara Is.
Amami Is.
Ryukyu Is.
Okinawa
Miyako
Ishigaki
Taiwan (Formosa)
Ryukyu Trench
Daito Is.
Nanpo Shotō
Ramapo Trench (Izu Tr.)
Bonin Is.
Volcano Is.
Minami Tori Shima (Marcus)
Hainan
Batan Is.
Babuyan Is.
Philippine Basin
Parece Vela
Farallon de Pajaros
Paracel Island
Luzon
South China Sea
Asuncion
Agrihan
Pagan
Mariana Is.
Anatahan
Saipan
Tinian
Rota
Guam
Marianas Trench
Wake I.
PACIFIC
Taongi Atoll
Marshall Islands
Bikini A.
Rongelap A.
Ratak Chain
Ralik Chain
Eniwetok A.
Kwajalein Atoll
Maleolap A.
Arno A.
Ailinglapalap A.
Jaluit A.
Mili A.
Mindoro
Samar
Philippine Trench
Panay
Palawan
Negros
Leyte
Philippines
Sulu Sea
Mindanao
Apo
Balabac Str.
Kinabalu
Sulu Arch.
Yap Is.
Ulithi A.
Caroline Islands
Gaferut
MICRONESIA
Palau Is.
Woleai A.
Eauripik A.
Pulap A.
Hall Is.
Truk Is.
Ponape
Senyavin Is.
Kusaie
Mortlock Is.
Brunei
Bunguran Is.
Iran Ra.
Borneo
Celebes Sea
Talaud Is.
Pulo Anna
Tobi
Nukuoro A.
Makin
Kiribati Rep.
Tarawa Atoll
Kapuas
Schwaner Ra.
Mahakam
Barito
Str. of Makasar
G. of Tomini
Molucca Sea
Moluccas
Morotai
Halmahera
Mapia Is.
MELANESIA
Kapingamarangi Atoll
Nauru
Ocean I.
Nonouti
Waigeo
Dampier Str.
Kwoka
Biak
Japen
C. d'Urville
Ninigo Group
Admiralty Is.
St. Matthias Group
Sula Is.
Obi Is.
Sarera B.
Sukarnapura
New Hanover I.
New Ireland
Celebes
Rantemario
Ceram
Buru
Sukarno Peak
New Guinea
Central Range
Finisterre Ra.
Bismarck Archipelago
Solomon Is.
Bougainville
Nakumanu Is.
Billiton
Bandjarmasin
Java Sea
Makassar
Banda Sea
Kai Is.
Aru Is.
Choiseul
New Britain
Nanumea
Tuvalu
Semarang
Madura
Flores Sea
Java
Lesser Sunda Is.
Wetar
Tanimbar Is.
Babar Is.
Kolepom I.
Digul
Fly
G. of Papua
Port Moresby
Stanley Ra.
Trobriand or Kiriwina Is.
New Georgia
Sta. Isabel
Malaita
D'Entrecasteaux Is.
Duff Is.
Semeru
Bali
Lombok
Sumbawa
Flores
Timor
Arafura Sea
Torres Str.
C. York
Guadalcanal
S. Cristobal
Rennell
Santa Cruz Is.
Mitre I.
Java Trench
Sumba
Roti
Timor Sea
Melville I.
Cobourg Pen.
Louisiade Arch.
Darwin
Arnhem Land
G. of Groote Eylandt Carpentaria
Cape York Pen.
C. Melville
Coral Sea
Jos. Bonaparte Gulf
Roper
Victoria
Cooktown
Great Barrier Reef
Banks Is.
Espiritu Santo Is.
Aurora
Vanuatu Rep.
Malekula
Efate
Eromanga
Viti Levu
King Sd.
Kimberley
Barkly Tableland
Wellesley I.
Mitchell
Mt. Bartle Frere
Flinders
Mellish Reef
Chesterfield Iles
New Hebrides Tr.
Loyalty Is.
Uvea
Mt. Panié
New Caledonia
Maré
I. des Pins
Matthew
Fitzroy
Selwyn Ra.
Great Dividing Range
Bellona Reefs
Great Sandy Desert
De Grey
Barrow I.
Exmouth G.
N.W. Cape
Hamersley Ra.
L. Mackay
Macdonnell Ranges
L. Disappointment
Alice Springs
Diamantina
Barcoo
Rockhampton
Cato
South Fiji Basin
Ashburton
Gibson Desert
L. Amadeus
Mt. Augustus
AUSTRALIA
Musgrave Ranges
Cooper's Cr.
Bulloo
Warrego
Brisbane
Tasman Sea
Norfolk I.
Shark B.
Dirk Hartog I.
Murchison
L. Austin
Great Victoria Desert
L. Eyre
Flinders Ra.
L. Frome
Bourke
New England Ra.
C. Byron
Middleton Reef
L. Barlee
Nullarbor Plain
L. Torrens
Darling
Macquarie
Liverpool Ra.
Lord Howe I.
L. Gairdner
Gawler Ra.
Pt. Augusta
Lachlan
Sydney
L. Moore
L. Cowan
Great Australian Bight
Eyre's Pen.
Murray
Murrumbidgee
Canberra
North C.
Darling Range
Perth
Spencer G.
Adelaide
Australian Alps
New Zealand
Geographe B.
Kangaroo I.
Encounter B.
Melbourne
Port Phillip B.
Wilson's Prom.
Bass Strait
Tasman Basin
Cook Str.
C. Leeuwin
South Australian Basin
King I.
Furneaux Group
INDIAN OCEAN
Tasmania
Legges Tor
Hobart
South East Cape
Mt. Cook
Southern Alps
Christchurch
South Island
West C.
Stewart I.
Mellis Seamount
Scale 1:33 000 000
0 250 500 750 1000 1250 1500 Kms.
0 250 500 750 1000 St. mls.
East from 120 Greenwich

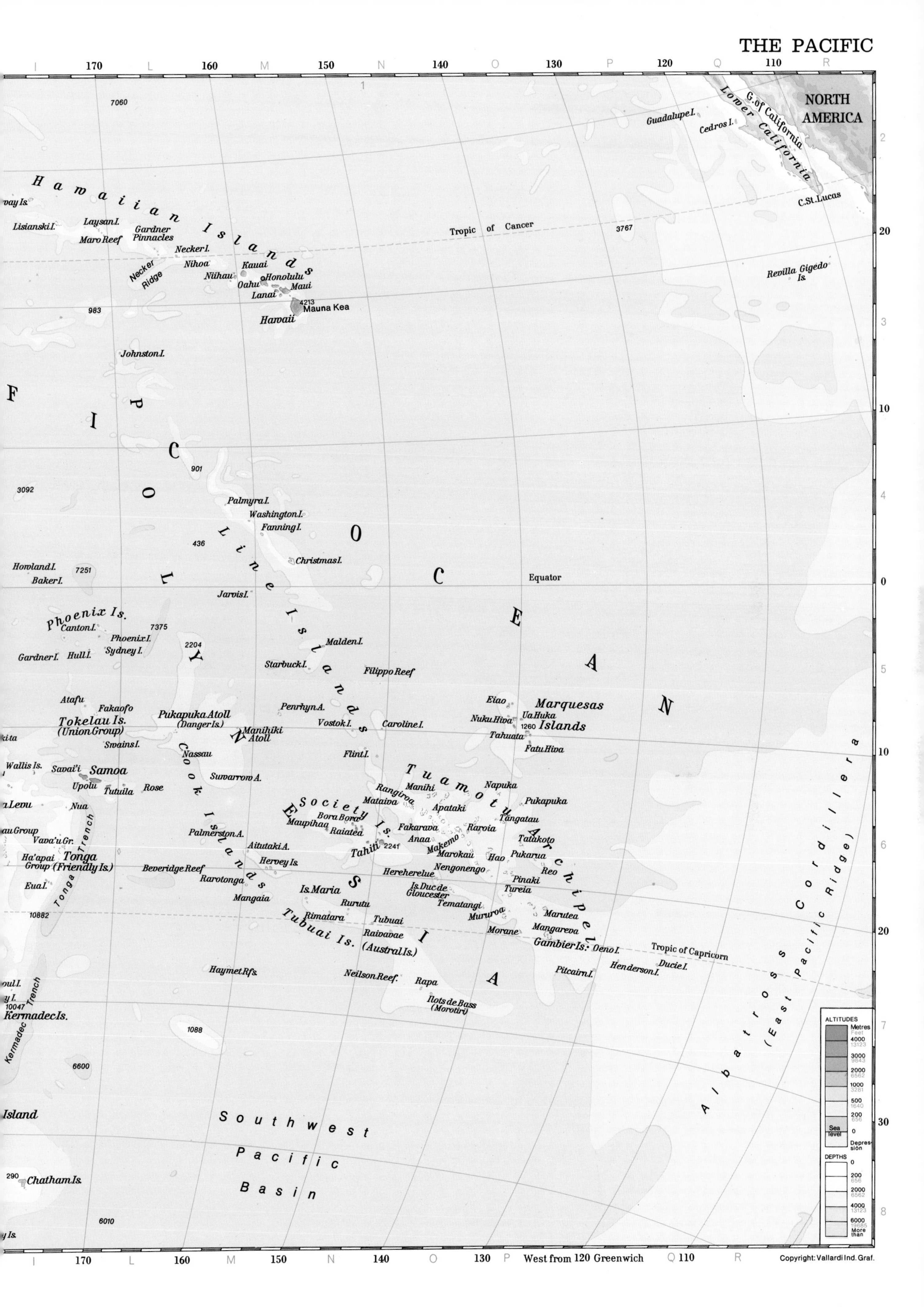

NORTH AMERICA
G. of California
Lower California
Guadalupe I.
Cedros I.
C. St. Lucas
Tropic of Cancer
3767
Revilla Gigedo Is.
Hawaiian Islands
Lisianski I.
Laysan I.
Maro Reef
Gardner Pinnacles
Necker I.
Necker Ridge
Nihoa
Niihau
Kauai
Oahu
Honolulu
Maui
Lanai
Hawaii
Mauna Kea
4213
983
7060
Johnston I.
PACIFIC OCEAN
POLYNESIA
901
3092
Palmyra I.
Washington I.
Fanning I.
436
Line Islands
Christmas I.
Howland I.
7251
Baker I.
Jarvis I.
Equator
Phoenix Is.
Canton I.
Phoenix I.
Sydney I.
7375
2204
Gardner I.
Hull I.
Malden I.
Starbuck I.
Filippo Reef
Atafu
Fakaofo
Tokelau Is. (Union Group)
Swains I.
Pukapuka Atoll (Danger Is.)
Manihiki Atoll
Penrhyn A.
Vostok I.
Caroline I.
Flint I.
Eiao
Marquesas Islands
Nuku Hiva
Ua Huka
1260
Tahuata
Fatu Hiva
Nassau
Suwarrow A.
Wallis Is.
Savai'i
Samoa
Upolu
Tutuila
Rose
Cook Islands
Nua
Vava'u Gr.
Ha'apai Group
Tonga (Friendly Is.)
Tonga Trench
Eua I.
10882
Palmerston A.
Aitutaki A.
Hervey Is.
Beveridge Reef
Rarotonga
Mangaia
Society Is.
Bora Bora
Maupihaa
Raiatea
Tahiti
2241
Tuamotu Archipelago
Rangiroa
Manihi
Mataiva
Apataki
Napuka
Pukapuka
Tangatau
Fakarava
Anaa
Raroia
Makemo
Marokau
Hao
Takakoto
Pukarua
Nengonengo
Hereherelue
Reo
Pinaki
Tureia
Is. Duc de Gloucester
Tematangi
Mururoa
Marutea
Morane
Mangareva
Gambier Is.
Oeno I.
Is. Maria
Rurutu
Rimatara
Tubuai
Raivavae
Tubuai Is. (Austral Is.)
Tropic of Capricorn
Ducie I.
Henderson I.
Pitcairn I.
Haymet Rfs.
Neilson Reef
Rapa
Ilots de Bass (Morotiri)
Albatross Cordillera (East Pacific Ridge)
Kermadec Is.
Kermadec Trench
10047
1088
6600
Southwest Pacific Basin
Chatham Is.
290
6010
ALTITUDES
Metres
Feet
4000
13123
3000
9843
2000
6562
1000
3281
500
1640
200
656
Sea level
0
Depression
DEPTHS
0
200
656
2000
6562
4000
13123
6000
19685
More than
West from 120 Greenwich

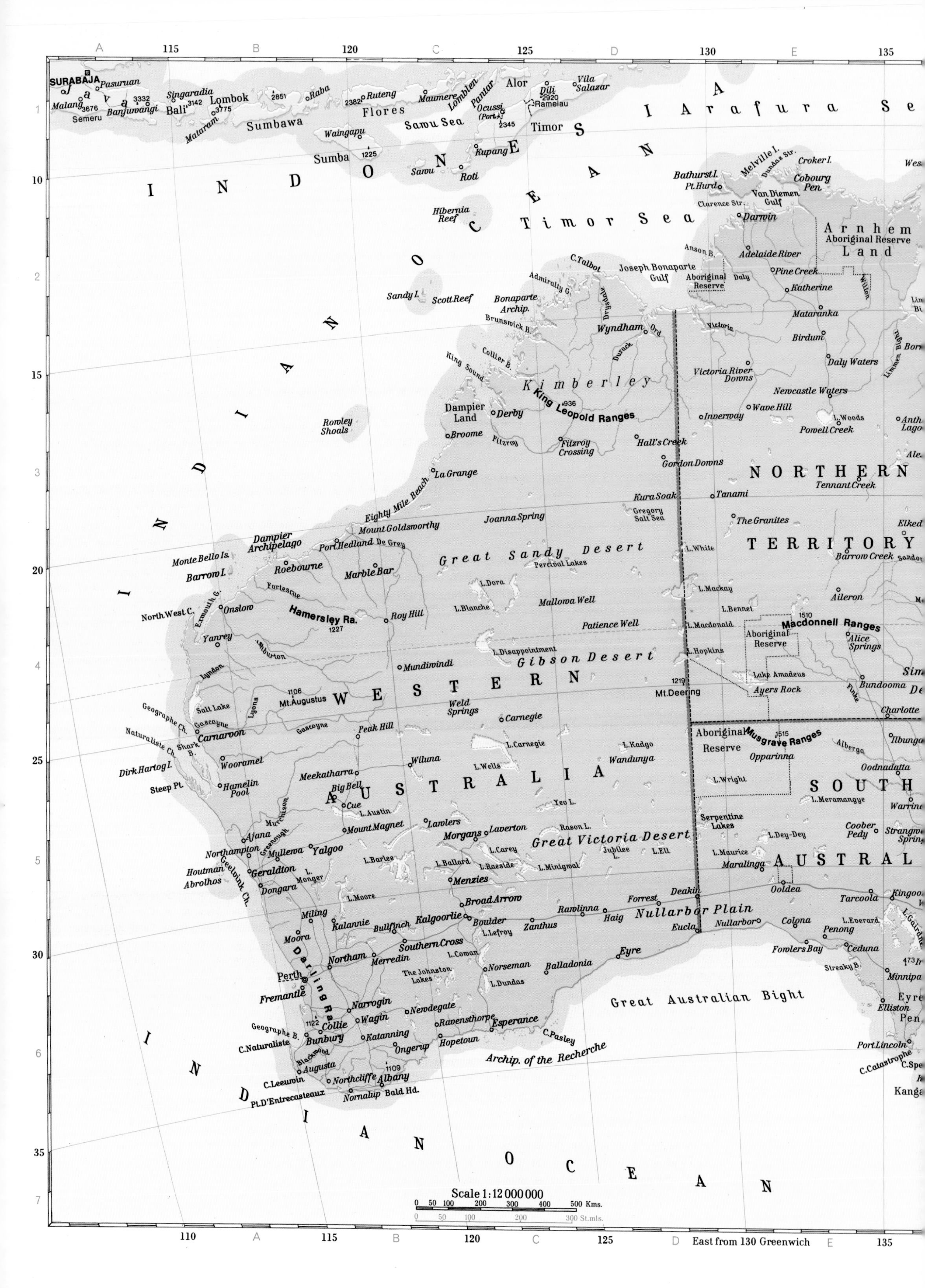
Timor Sea
Arafura Sea
Indian Ocean
Great Australian Bight
Western Australia
Northern Territory
South Australia
Kimberley
King Leopold Ranges
Great Sandy Desert
Gibson Desert
Great Victoria Desert
Nullarbor Plain
Hamersley Ra.
Macdonnell Ranges
Musgrave Ranges
Darling Ra.
Arnhem Land
Aboriginal Reserve
Perth
Fremantle
Darwin
Kalgoorlie
Geraldton
Albany
Esperance
Broome
Derby
Port Hedland
Carnarvon
Alice Springs
Ayers Rock
Sumbawa
Flores
Timor
Lombok
Bali
Sumba
Scale 1:12 000 000
East from 130 Greenwich

Coral Sea
Great Barrier Reef
QUEENSLAND
NEW SOUTH WALES
VICTORIA
Great Dividing Range
Gulf of Papua
Torres Strait
Papua New Guinea (Australia)
Cape York Peninsula
Tasmania
Bass Strait
INDIAN OCEAN
PACIFIC OCEAN
Tasman Sea
North Island
South Island
NEW ZEALAND
Tropic of Capricorn
BRISBANE
SYDNEY
MELBOURNE
ADELAIDE
Canberra
AUSTRALIAN CAPITAL TERRITORY
AUCKLAND
Wellington
Christchurch
Hobart
Scale 1:12 000 000
East from 170 Greenwich
Copyright:Vallardi Ind.Graf.

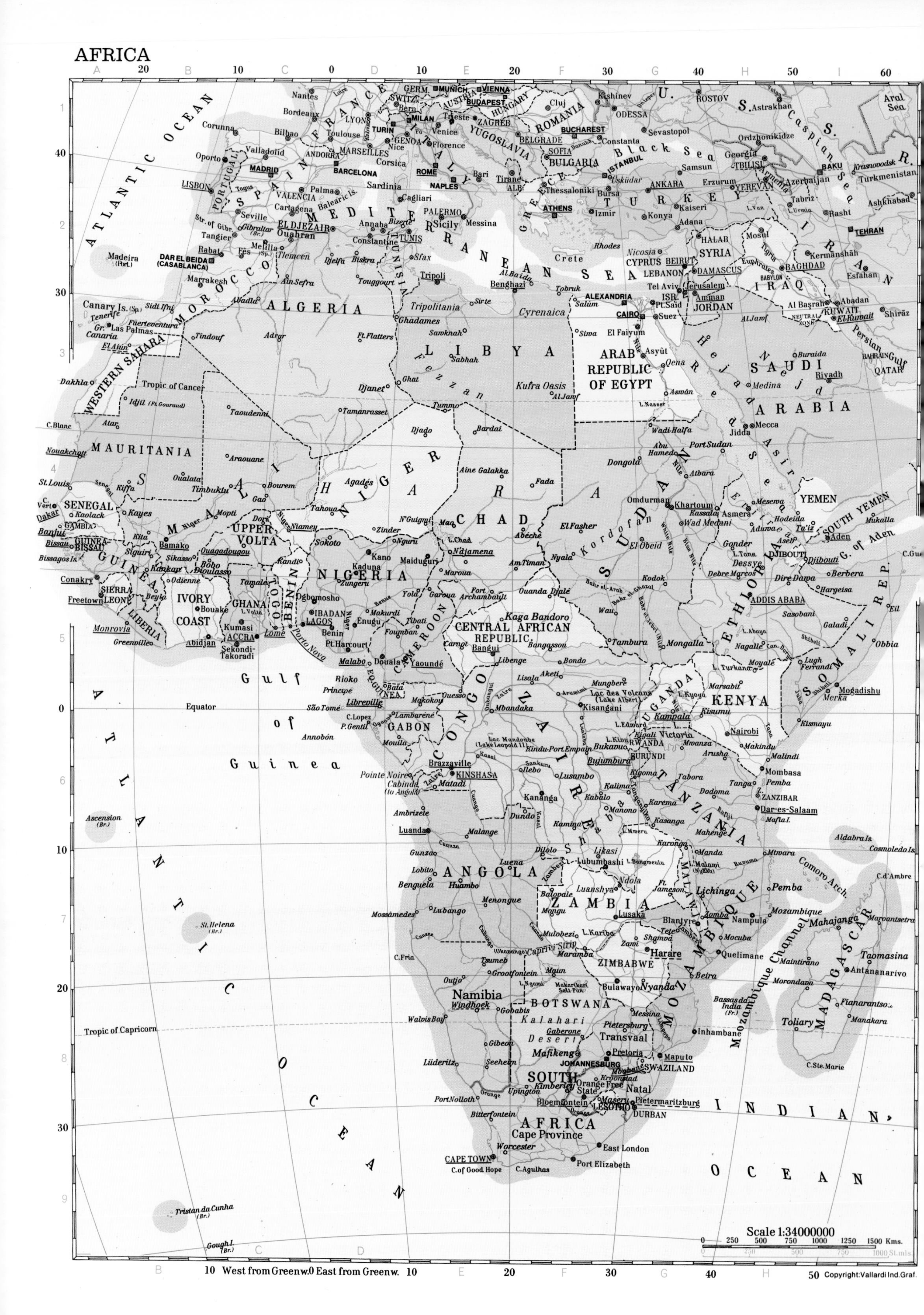
AFRICA
ATLANTIC OCEAN
MEDITERRANEAN SEA
Black Sea
Caspian Sea
Aral Sea
Gulf of Guinea
INDIAN OCEAN
Mozambique Channel
G. of Aden
Persian Gulf
Red Sea
Tropic of Cancer
Equator
Tropic of Capricorn
SPAIN
PORTUGAL
FRANCE
ITALY
YUGOSLAVIA
ROMANIA
BULGARIA
HUNGARY
AUSTRIA
GREECE
TURKEY
U.S.S.R.
SYRIA
LEBANON
CYPRUS
ISR.
JORDAN
IRAQ
IRAN
KUWAIT
SAUDI ARABIA
YEMEN
SOUTH YEMEN
MOROCCO
WESTERN SAHARA
ALGERIA
TUNISIA
LIBYA
ARAB REPUBLIC OF EGYPT
MAURITANIA
MALI
NIGER
CHAD
SUDAN
ETHIOPIA
DJIBOUTI
SOMALI REP.
SENEGAL
GAMBIA
GUINEA-BISSAU
GUINEA
SIERRA LEONE
LIBERIA
IVORY COAST
UPPER VOLTA
GHANA
TOGO
BENIN
NIGERIA
CAMEROON
CENTRAL AFRICAN REPUBLIC
EQU. GUINEA
GABON
CONGO
ZAIRE
UGANDA
KENYA
RWANDA
BURUNDI
TANZANIA
ANGOLA
ZAMBIA
MALAWI
MOZAMBIQUE
ZIMBABWE
Namibia
BOTSWANA
SOUTH AFRICA
SWAZILAND
LESOTHO
MADAGASCAR
LISBON
MADRID
BARCELONA
ROME
ATHENS
ISTANBUL
ANKARA
CAIRO
ALEXANDRIA
EL DJEZAIR
TUNIS
Tripoli
Benghazi
Khartoum
ADDIS ABABA
Nairobi
KINSHASA
Brazzaville
LAGOS
ACCRA
Dakar
Luanda
Lusaka
Harare
Windhoek
Pretoria
JOHANNESBURG
Maputo
CAPE TOWN
Durban
Dar-es-Salaam
Mogadishu
Antananarivo
Sahara
Kalahari Desert
Canary Is.
Madeira
Ascension (Br.)
St. Helena (Br.)
Tristan da Cunha (Br.)
Gough I. (Br.)
Scale 1:34000000
0 250 500 750 1000 1250 1500 Kms.
West from Greenw. 0 East from Greenw.
Copyright: Vallardi Ind. Graf.

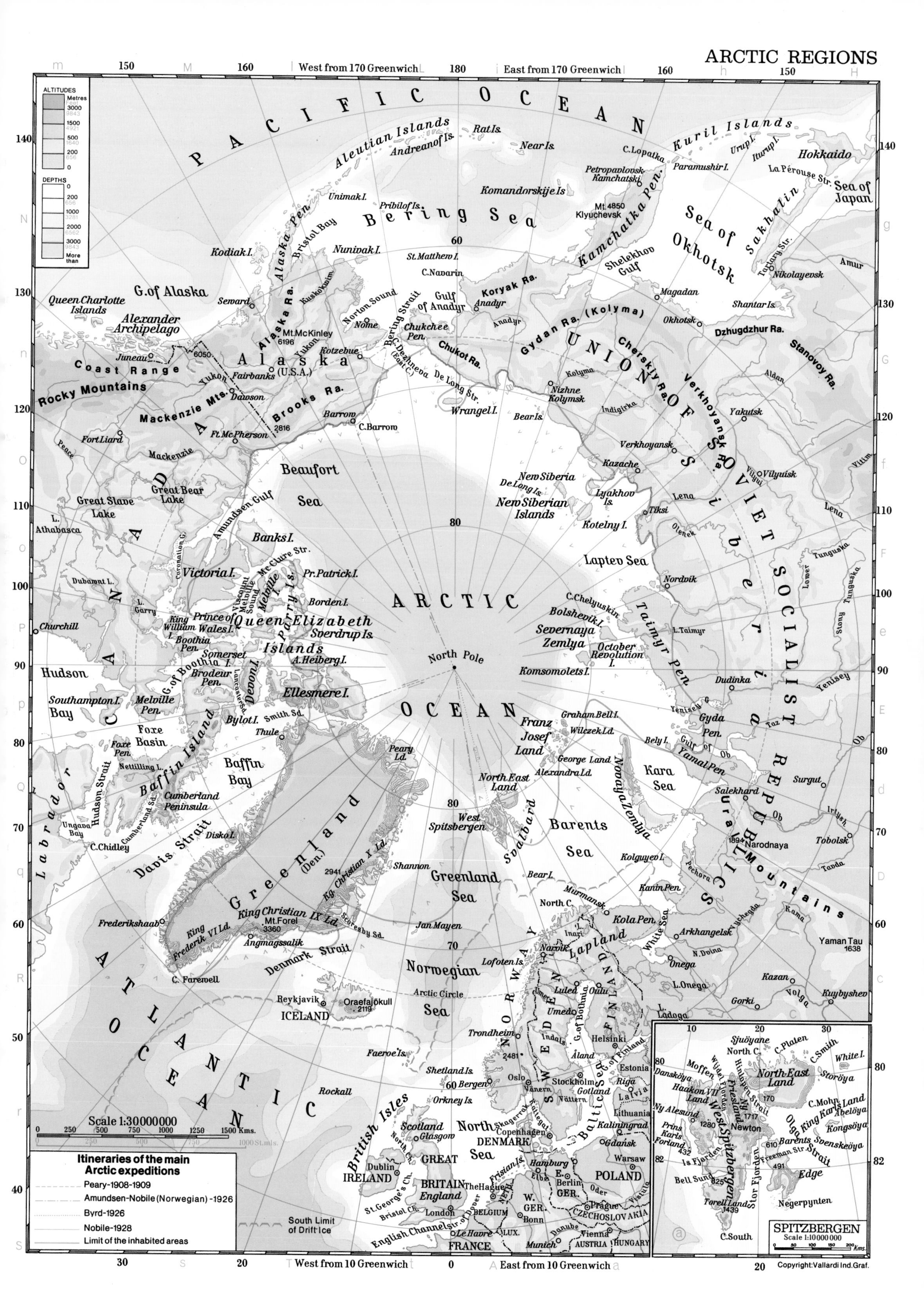
West from 170 Greenwich
East from 170 Greenwich
ALTITUDES
Metres
3000
1500
500
200
0
DEPTHS
0
200
1000
2000
3000
More than
PACIFIC OCEAN
Aleutian Islands
Andreanof Is.
Rat Is.
Near Is.
Kuril Islands
Urup I.
Iturup I.
Hokkaido
C. Lopatka
Paramushir I.
La Pérouse Str.
Sea of Japan
Petropavlovsk-Kamchatski
Komandorskije Is
Mt. 4850 Klyuchevsk
Kamchatka Pen.
Sakhalin
Sea of Okhotsk
Tartary Str.
Amur
Nikolayevsk
Unimak I.
Pribilof Is.
Bering Sea
Bristol Bay
Alaska Pen.
Kodiak I.
Nunivak I.
St. Matthew I.
C. Navarin
Shelekhov Gulf
Magadan
Shantar Is.
Koryak Ra.
Gulf of Anadyr
Anadyr
Okhotsk
Dzhugdzhur Ra.
Stanovoy Ra.
G. of Alaska
Queen Charlotte Islands
Alexander Archipelago
Seward
Alaska Ra.
Kuskokwim
Norton Sound
Nome
Bering Strait
Chukchee Pen.
Chukot Ra.
Gydan Ra. (Kolyma)
Mt. McKinley 6196
Kotzebue
C. Dezhneva (East C.)
De Long Str.
Juneau
6050
Alaska (U.S.A.)
Coast Range
Rocky Mountains
Yukon
Fairbanks
Dawson
Mackenzie Mts.
Brooks Ra.
Barrow
C. Barrow
2816
Wrangel I.
Bear Is.
Kolyma
Nizhne Kolymsk
Indigirka
Cherskiy Ra.
Verkhoyansk Ra.
Aldan
Yakutsk
Ft. Liard
Ft. McPherson
Peace
Mackenzie
Beaufort Sea
Verkhoyansk
Kazache
Vilyuisk
Vilyui
Vitim
Great Bear Lake
Great Slave Lake
L. Athabasca
Amundsen Gulf
New Siberia
De Long Is.
New Siberian Islands
Lyakhov Is.
Lena
Tiksi
Kotelny I.
Olenek
80
Banks I.
Coronation G.
Laptev Sea
Tunguska
Lower Tunguska
Stony Tunguska
Victoria I.
Mc Clure Str.
Pr. Patrick I.
Viscount Melville Sound
Melville I.
Parry Is.
Borden I.
Dubawnt L.
Nordvik
L. Garry
King William I.
Prince of Wales I.
Queen Elizabeth Islands
Sverdrup Is.
ARCTIC OCEAN
C. Chelyuskin
Bolshevik I.
Severnaya Zemlya
October Revolution I.
Taimyr Pen.
L. Taimyr
Churchill
Boothia Pen.
Somerset I.
G. of Boothia
A. Heiberg I.
North Pole
Komsomolets I.
Hudson Bay
Brodeur Pen.
Lancaster Sd.
Devon I.
Ellesmere I.
Dudinka
Yenisey
Southampton I.
Melville Pen.
Bylot I.
Smith Sd.
Franz Josef Land
Graham Bell I.
Wilczek Ld.
Gyda Pen.
Gulf of Ob
Taz
Ob
Foxe Basin
Thule
Bely I.
Yamal Pen.
Foxe Pen.
Baffin Island
Peary Ld.
George Land
Alexandra Ld.
Novaya Zemlya
Kara Sea
Netilling L.
Baffin Bay
North East Land
Surgut
Salekhard
Labrador
Hudson Strait
Cumberland Sd.
Cumberland Peninsula
Greenland (Den.)
West Spitsbergen
Svalbard
Barents Sea
Ural Mountains
Irtysh
Ungava Bay
Disko I.
Davis Strait
C. Chidley
Kg. Christian X Ld.
2941
Shannon
Kolguyev I.
Narodnaya 1894
Tobolsk
Tavda
Pechora
Greenland Sea
Bear I.
Murmansk
Kanin Pen.
Frederikshaab
King Frederik VI Ld.
King Christian IX Ld.
Mt. Forel 3360
Scoresby Sd.
Angmagssalik
North C.
Kola Pen.
White Sea
Kama
Arkhangelsk
Vychegda
Jan Mayen
Inari
Lapland
Yaman Tau 1638
70
Denmark Strait
Lofoten Is.
Narvik
N. Dvina
Onega
ATLANTIC OCEAN
C. Farewell
Norwegian Sea
Arctic Circle
NORWAY
SWEDEN
FINLAND
Luleå
Oulu
Kazan
L. Onega
Kuybyshev
Volga
Reykjavik
Oraefajökull 2119
ICELAND
Umeå
G. of Bothnia
Gorki
L. Ladoga
Trondheim
Indals
Helsinki
Åland
G. of Finland
Faeroe Is.
2481
Estonia
Shetland Is.
60 Bergen
Oslo
Stockholm
Riga
Vänern
Gotland
Latvia
Rockall
Orkney Is.
Vättern
Lithuania
Scale 1:30000000
Kms.
St. mls.
Scotland
Glasgow
North Sea
Skagerrak
Kattegat
Baltic Sea
Kaliningrad
British Isles
North Ch.
Copenhagen
DENMARK
Gdańsk
Itineraries of the main Arctic expeditions
Peary-1908-1909
Amundsen-Nobile (Norwegian) -1926
Byrd-1926
Nobile-1928
Limit of the inhabited areas
Dublin
IRELAND
GREAT BRITAIN
England
Frisian Is.
Hamburg
Elbe
E. GER.
Berlin
Warsaw
POLAND
Oder
Vistula
St. George's Ch.
Bristol Ch.
London
The Hague
NETH.
W. GER.
Bonn
Prague
CZECHOSLOVAKIA
South Limit of Drift Ice
English Channel
Str. of Dover
BELGIUM
LUX.
Le Havre
FRANCE
Munich
Danube
Vienna
AUSTRIA
HUNGARY
West from 10 Greenwich
East from 10 Greenwich
Sjuöyane
North C.
C. Platen
C. Smith
White I.
Moffen
Danskøya
Wijde Fjorden
Hinlopen Strait
North-East Land
Storöya
Haakon VII Land
Friesland
1717
170
C. Mohn
King Karls Land
Abelöya
Ny Alesund
1280
West Spitzbergen
Newton
Olga Strait
Kongsöya
Prins Karls Forland
432
Is Fjorden
610
Barents
Freeman Str.
Svenskeöya
491
Edge
Bell Sund
825
Stor Fjorden
Torell Land 1439
Negerpynten
C. South
SPITZBERGEN
Scale 1:10000000
Kms.

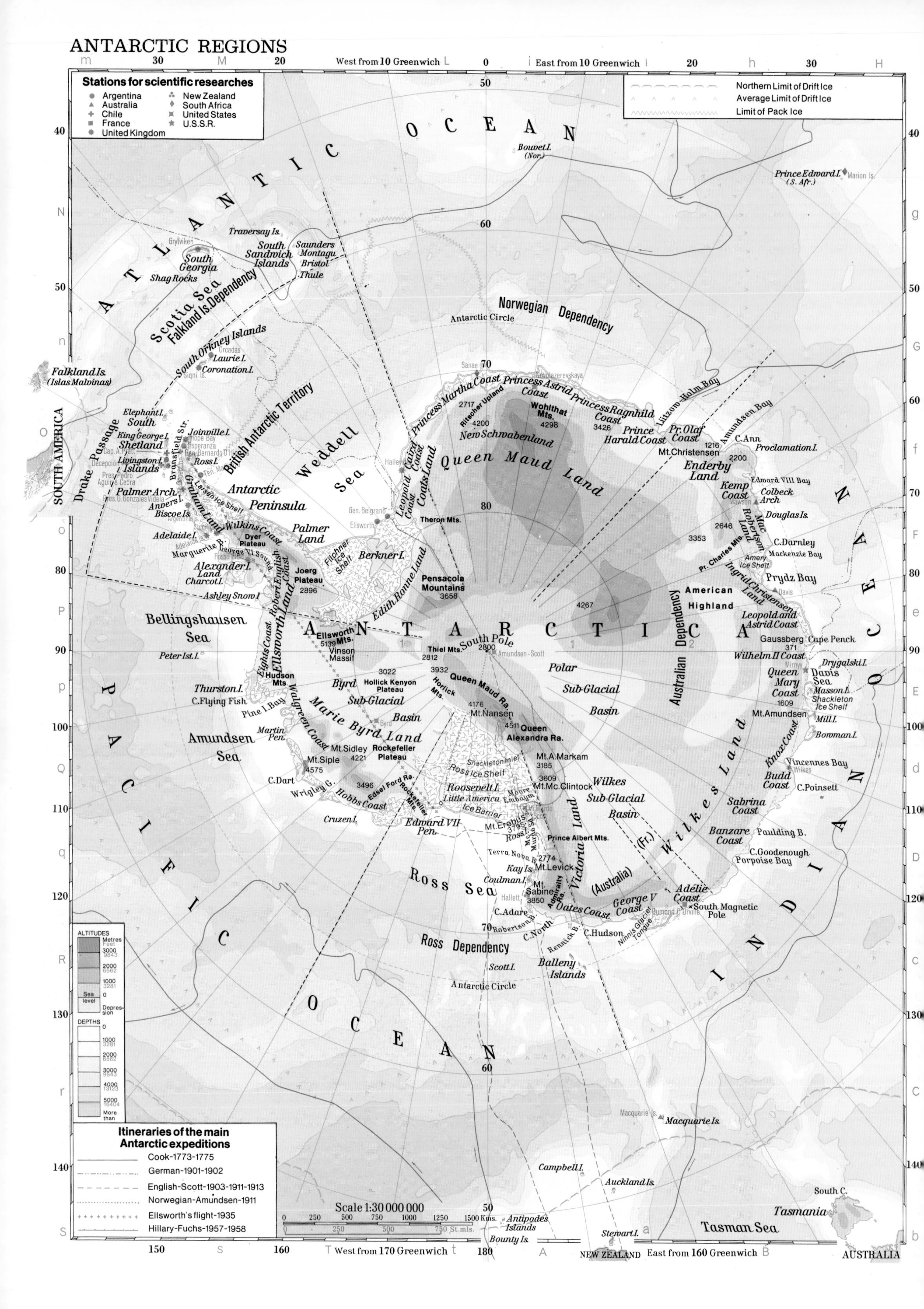

Stations for scientific researches
Argentina
Australia
Chile
France
United Kingdom
New Zealand
South Africa
United States
U.S.S.R.
Northern Limit of Drift Ice
Average Limit of Drift Ice
Limit of Pack Ice
West from 10 Greenwich
East from 10 Greenwich
ATLANTIC OCEAN
PACIFIC OCEAN
INDIAN OCEAN
ANTARCTICA
Bouvet I. (Nor.)
Prince Edward I. (S. Afr.)
Marion Is.
Traversay Is.
South Sandwich Islands
Saunders
Montagu
Bristol
Thule
Grytviken
South Georgia
Shag Rocks
Scotia Sea
Falkland Is. Dependency
South Orkney Islands
Orcadas
Laurie I.
Coronation I.
Signi Is.
Falkland Is. (Islas Malvinas)
SOUTH AMERICA
Drake Passage
Elephant I.
South Shetland Islands
King George I.
Livingston I.
Joinville I.
Hope Bay
Esperanza
Ross I.
Bransfield Str.
Palmer Arch.
Anvers I.
Biscoe Is.
Adelaide I.
Marguerite B.
George VI Sound
Alexander I. Land
Charcot I.
Ashley Snow I.
Graham Land
Antarctic Peninsula
British Antarctic Territory
Weddell Sea
Larsen Ice Shelf
Wilkins Coast
Dyer Plateau
Palmer Land
Filchner Ice Shelf
Berkner I.
Joerg Plateau
2896
English Coast
Robert English Coast
Eights Coast
Ellsworth Land
Edith Ronne Land
Gen. Belgrano
Ellsworth
Halley
Leopold Coast
Caird Coast
Coats Land
Theron Mts.
Pensacola Mountains
3658
Bellingshausen Sea
Peter Ist. I.
Thurston I.
C. Flying Fish
Pine I. Bay
Hudson Mts.
Ellsworth Mts.
5139
Vinson Massif
Thiel Mts.
2812
South Pole
2800
Amundsen-Scott
Byrd Sub-Glacial Basin
Hollick Kenyon Plateau
3022
3932
Horlick Mts.
Queen Maud Ra.
4176
Mt. Nansen
4511
Queen Alexandra Ra.
Marie Byrd Land
Walgreen Coast
Martin Pen.
Amundsen Sea
Mt. Sidley
4221
Rockefeller Plateau
Mt. Siple
4575
C. Dart
Wrigley G.
3496
Edsel Ford Ra.
Rockefeller Mts.
Hobbs Coast
Cruzen I.
Edward VII Pen.
Byrd
Shackleton Inlet
Ross Ice Shelf
Roosevelt I.
Little America
Moore Embaym.
Ice Barrier
Mt. Erebus
3795
Ross I.
Mt. A. Markam
3185
3609
Mt. Mc. Clintock
Prince Albert Mts.
Terra Nova B.
2774
Mt. Levick
Kay Is.
Coulman I.
Mt. Sabine
3850
Hallett
C. Adare
Admiralty Ra.
Victoria Land
Oates Coast
George V Coast
Robertson B.
C. North
Rennick B.
C. Hudson
Ninnis Glacier Tongue
Dumont d'Urville
Ross Sea
Ross Dependency
Scott I.
Balleny Islands
Antarctic Circle
Wilkes Sub-Glacial Basin
(Australia)
(Fr.)
Adélie Coast
South Magnetic Pole
Norwegian Dependency
Sanae
Novolazarevskaya
Princess Martha Coast
Princess Astrid Coast
Princess Ragnhild Coast
Ritscher Upland
2717
4200
Wohlthat Mts.
4298
New Schwabenland
Queen Maud Land
3426
Prince Harald Coast
Pr. Olaf Coast
Lützow-Holm Bay
Amundsen Bay
1216
C. Ann
Mt. Christensen
2200
Proclamation I.
Enderby Land
Edward VIII Bay
Kemp Coast
Mawson
Colbeck Arch
Mac Robertson Land
Douglas Is.
2646
3353
C. Darnley
Mackenzie Bay
Pr. Charles Mts.
Amery Ice Shelf
Ingrid Christensen Land
Prydz Bay
Davis
American Highland
4267
Australian Dependency
Leopold and Astrid Coast
Gaussberg
371
Cape Penck
Wilhelm II Coast
Drygalski I.
Mirnyy
Queen Mary Coast
Davis Sea
Masson I.
Shackleton Ice Shelf
1609
Mt. Amundsen
Mill I.
Bowman I.
Polar Sub-Glacial Basin
Knox Coast
Vincennes Bay
Wilkes
Budd Coast
C. Poinsett
Wilkes Land
Sabrina Coast
Banzare Coast
Paulding B.
C. Goodenough
Porpoise Bay
Macquarie Is.
Campbell I.
Auckland Is.
Antipodes Islands
Bounty Is.
Stewart I.
NEW ZEALAND
Tasman Sea
Tasmania
South C.
AUSTRALIA
ALTITUDES
Metres
Feet
3000
9843
2000
6562
1000
3281
Sea level
0
Depression
DEPTHS
0
1000
3281
2000
6562
3000
9843
4000
13123
5000
16404
More than
Scale 1:30 000 000
0
250
500
750
1000
1250
1500 Kms.
250
500
750 St. mls.
Itineraries of the main Antarctic expeditions
Cook-1773-1775
German-1901-1902
English-Scott-1903-1911-1913
Norwegian-Amundsen-1911
Ellsworth's flight-1935
Hillary-Fuchs-1957-1958
West from 170 Greenwich
East from 160 Greenwich

GAZETTEER

The first column of figures indicates the map number, and the second (letters and figures) the map reference.

A		
Aalborg, Denmark	3	H4
Aarhus, Denmark	3	I4
Abadan, Iran	5	G6
Abadla, Algeria	13	C2
Abakan, U.S.S.R.	5	N4
Abdulino, U.S.S.R.	4	S5
Abeche, Chad	13	F4
Aberdeen, U.K.	3	F4
Abidjan, Ivory Coast	13	C5
Abrolhos I., Australia	11	A5
Abu Hamed, Sudan	13	G4
Acapulco, Mexico	7	L8
Accra, Ghana	13	C5
Aconcagua Mt., Argentina	1	
Acre State, Brazil	8	C4
Adamawa Highlands, Nigeria/Cameroon	2	
Adana, Turkey	5	F6
Adare Cape, Antarctica	15	A2
Adavale, Australia	12	G5
Ad Dam, Saudi Arabia	5	G7
Ad Dawhah, Qatar	5	H7
Addis Ababa, Ethiopia	13	G5
Adelaide, Australia	12	F6
Adelaide I., Antarctica	15	O3
Adelaide River, Australia	11	E2
Adelie Coast, Antarctica	15	bC3
Aden, Gulf of, Africa	13	H4
Aden, Yemen	5	G8
Admiralty Gulf, Australia	11	D2
Admiralty Is., Pacific Ocean	9	E5
Admiralty Range, Antarctica	15	a2
Adriatic Sea, Europe	3	IL7
Adwa, Ethiopia	13	G4
Aegean Sea, Greece	4	N8
Afghanistan, Asia	5	I6
Agades, Niger	13	D4
Agra, India	5	L7
Agrigento, Sicily	3	I8
Agrihan I., Pacific Ocean	9	E3
Agulhas C., S. Africa	13	F9
Ahmadabad, India	5	L7
Ahvaz, Iran	5	G6
Aileron, Australia	11	E4
Ailinglapalap At., Pacific Ocean	9	G4
Ain Galakko, Chad	13	E4
Ain Sefra, Algeria	13	C2
Aitutaki At., Pacific Ocean	10	M6
Ajaccio, Corsica	3	H7
Ajana, Australia	11	A5
Aketi, Zaire	13	F5
Aklavik, Canada	7	F3
Aksu, China	5	M5
Aktyubinsk, U.S.S.R.	4	T5
Akureyri, Iceland	3	C2
Akyab, Burma	5	N7
Alabama State, U.S.A.	7	M6
Aland Is., Sweden	3	L3
Alaska, U.S.A.	7	CDE3
Alaska G. of, Alaska	7	E4
Alaska Pen., Alaska	7	CD4
Alaska Range, Alaska	14	mM2
Albania, Europe	3	LM7
Albacete, Spain	3	F8
Albany, Australia	11	B6
Albany, U.S.A.	7	N5
Al Basrah, Iraq	5	G6
Albatross Cordillera, Pacific Ocean	10	Q67
Al Baida, Libya	13	F2
Alberga R., Australia	11	E5
Albert L., Uganda/Zaire	13	G5
Alberta, Canada	7	H4
Albuquerque, U.S.A.	7	I6
Albury, Australia	12	H7
Aldabra Is., Indian Ocean	13	H6
Aldan R., U.S.S.R.	6	R4
Aleksandrovsk, U.S.S.R.	6	S4
Alesund, Norway	3	H3
Aleutian Is., Alaska	7	C4
Alexander Arch., Alaska	7	F4
Alexander I., Antarctica	15	o2
Alexandria, Australia	11	F3
Alexandria, A.R.E.	13	F2
Algeria, Africa	13	CD23
Algiers, Algeria	13	D2
Al Hufuf, Saudi Arabia	5	G7
Alice Springs, Australia	11	E4
Al Jawf, Libya	13	F3
Al Jawf, Saudi Arabia	5	F7
Allahabad, India	5	M7
Alma Ata, U.S.S.R.	5	L5
Al Manamah, Saudi Arabia	5	H7
Al Masira I., Muscat & Oman	5	H7
Almeria, Spain	3	F8
Alps, Europe	2	
Al Qunfidha Saudi Arabia	5	G8
Amadeus L., Australia	11	E4
Amapa, Brazil	8	E3
Amarillo, U.S.A.	7	I6
Amazon R., Brazil	8	D4
Amazonas region, Brazil	8	D4
Amboina, Indonesia	6	Q10
Ambrizete, Angola	13	E6
American Highland, Antarctica	15	F2
Amery Ice Shelf, Antarctica	15	F3
Amiens, France	3	G6
Amman, Jordan	5	F6
Amoy, China	6	P7
Amritsar, India	5	L6
Amsterdam, Netherlands	3	G5
Am Timan, Chad	13	F4
Amu Darya R., U.S.S.R.	5	I5
Amundsen Bay, Antarctica	15	g3
Amundsen Gulf, Canada	14	nO1
Amundsen Mt., Antarctica	15	d3
Amundsen Sea, Antarctica	15	Qq23
Amur R., U.S.S.R./China	6	R4
Anaa I., Pacific Ocean	10	N6
Anadyr Gulf of, U.S.S.R.	14	L3
Anadyr R., U.S.S.R.	14	Li3
Anatahan I., Pacific Ocean	9	E3
Anatolia, Turkey	2	
Anchorage, Alaska	7	E3
Ancona, Italy	3	I7
Andalsnes, Norway	3	H3
Andaman Is., India	5	N8
Andes Mts., S. America	1	
Andizhan, U.S.S.R.	5	L5
Andorra, Europe	3	G7
Andreanof Is., Asia	6	Z4
Andros Is., Bahamas	7	N7
Angara R., U.S.S.R.	5	NO4
Angers, France	3	F6
Angola, Africa	13	EF67
Angouleme, France	3	G6
Ankara, Turkey	5	F6
Ann C., Antarctica	15	G3
Annaba, Algeria	13	D2
Annobon I., Gulf of Guinea	13	D6
Annuello, Australia	12	G6
Anshan, China	6	Q5
Ansi, China	5	N5
Anson Bay, Australia	11	E2
Antalya, Turkey	5	F6
Antananarivo, Madagascar	13	H7
Antarctic Peninsula, Antarctica	15	On3
Anthony Lagoon, Australia	11	F3
Anticosti I., Canada	7	O5
Antofagasta, Chile	8	C6
Antwerp, Belgium	3	G5
Anvers I., Antarctica	15	O3
Aomori, Japan	6	S5
Apataki I., Pacific Ocean	10	N6
Apennines Mts., Italy	2	
Appalachian Mts., U.S.A.	1	
Arabia, Asia	5	G78
Arabian Desert, A.R.E.	2	
Arabian Sea, Asia	5	HI8
Aracaju, Brazil	8	G5
Arad, Romania	4	M6
Arafura Sea, Indonesia	9	D5
Araguaia R., Brazil	8	E5
Araks R., U.S.S.R.	4	R8
Aral Sea, U.S.S.R.	5	HI5
Aralsk, U.S.S.R.	4	U6
Aramac, Australia	12	H4
Araouane, Mali	13	C4
Ararat, Australia	12	G7
Ararat Mt., Turkey	2	
Aravalli Range, India	36	L7
Archer R., Australia	12	G2
Arel Heiberg I., Canada	14	Pp12
Arequipa, Peru	8	C5
Argentina, S. America	8	
Argentine I., Antarctica	15	O3
Arica, Chile	8	C5
Arizona State, U.S.A.	7	H6
Arkansas R., U.S.A.	1	
Arkansas State, U.S.A.	7	L6
Arkhangelsk, U.S.S.R.	4	Q3
Armavir, U.S.S.R.	4	Q7
Armenia, U.S.S.R.	4	QR78
Armidale, Australia	12	I6
Arnhem C., Australia	12	F2
Arnhem Land, Australia	11	EF2
Arno At., Pacific Ocean	9	H4
Arno R., Italy	3	I7
Arorae I., Pacific Ocean	9	H5
Arshan, China	6	P5
Arta, Greece	4	M8
Aru Is., New Guinea	6	R10
Arusha, Tanzania	13	G6
Aruwimi R., Zaire	13	F5
Arzamas, U.S.S.R.	4	Q4
Ascension Is., Atlantic Ocean	13	B6
Aseb, Ethiopia	13	H4
Ashburton R., Australia	11	B4
Ashley Snow I., Antarctica	15	o2
Ash-Shaab, South Yemen	5	G8
Askhabad, U.S.S.R.	5	H6
Asmera, Ethiopia	13	G4
Astrakhan, U.S.S.R.	4	R6
Asuncion I., Pacific Ocean	9	E3
Asuncion, Paraguay	8	E6
Aswan, A.R.E.	13	G3
Asyut, A.R.E.	13	G3
Atafu I., Pacific Ocean	10	I5
Atar, Mauritania	13	B3
Atbara, Sudan	13	G4
Atbara R., Sudan	13	G4
Athabasca L., Canada	7	I4
Athabasca R., Canada	7	H4
Athens, Greece	4	M8
Atlanta, U.S.A.	7	M6
Atlas Mts., Morocco	2	
Auckland, New Zealand	12	g13
Augsburg, Germany	3	I6
Augusta, Australia	11	B6
Augustus Mt., Australia	11	B4
Aurekun, Australia	12	G2
Aurora I., Pacific Ocean	9	G6
Austin, U.S.A.	7	L6
Austin Lake, Australia	11	B5
Austral Downs, Australia	12	F4
Australia	11/12	
Australian Alps, Australia	12	H7
Australian Capital Territory, Australia	12	H7
Australian Dependency, Antarctica	15	dEeF2
Austria, Europe	3	IL6
Avignon, France	3	G7
Awanui, New Zealand	12	g13
Ayan, U.S.S.R.	6	R4
Ayers Rock, Australia	11	E5
Ayr, Australia	12	H3
Ayutthaya, Thailand	5	O8
Azerbaijan, U.S.S.R.	4	R7
Azov, Sea of, U.S.S.R.	4	P6
B		
Babuyan Is., Philippines	9	C3
Babylon	13	H2
Badajoz, Spain	3	E8
Baffin Bay, Canada	7	NO2
Baffin I., Canada	7	MNO23
Baghdad, Iraq	5	G6
Baguio, Philippines	6	Q8
Bahama Is., Atlantic Ocean	7	N7
Bahia, Brazil	8	F5
Bahia Blanca, Argentina	8	D7
Bahrain, Asia	5	H7
Bahr-el-Arib R., Sudan	13	F5
Bahr-el-Jebel R., Sudan	13	G5
Bairnsdale, Australia	12	H7
Baker I., Pacific Ocean	10	I4
Baku, U.S.S.R.	4	R7
Balabac Str., Indonesia	9	B4
Balaklava, U.S.S.R.	4	O7
Bald Head, Australia	11	B7
Balearic Is., Mediterranean Sea	3	G8
Bali I., Indonesia	6	P10
Balikpapan, Indonesia	6	P10
Balkan Peninsula	2	
Balkhash, U.S.S.R.	5	L5
Balkhash L., U.S.S.R.	5	L5
Balladonia, Australia	11	C6
Ballarat, Australia	12	G7
Ballard L., Australia	11	C5
Balleney Is., Antarctica	15	a3
Balonne R., Australia	12	H5
Balovale, Zambia	13	F7
Balranald, Australia	12	G6
Baltic Sea, Europe	3	L4
Baltimore, U.S.A.	7	N6
Bamako, Mali	13	C4
Banaras, India	5	M7
Banda Sea, Indonesia	6	Q10
Bandar Abbas, Iran	5	H7
Bandar Atjeh, Indonesia	5	N9
Bandirma, Turkey	4	N7
Bandjarmasin, Indonesia	6	P10
Bandung, Indonesia	6	O10
Bangalore, India	5	L8
Bangassou, Cen. Africa	13	F5
Bangka I., Indonesia	6	O10
Bangkok, Thailand	5	O8
Bangladesh, Asia	5	MN7
Bangui, Cen. Africa	13	E5
Bangwelu L., Zambia	13	FG7
Banja Luka, Yugoslavia	3	L7
Banjul, Gambia	13	B4
Banks Is., Pacific Ocean	9	G6
Banks I., Australia	12	G12
Banks I., Canada	7	GH2
Banks Pen., New Zealand	12	g14
Banks Str., Australia	12	b9
Banzare Coast, Antarctica	15	c3
Baranovichi U.S.S.R.	4	N5
Barbados Is., Caribbean Sea	8	E2
Barcaldine, Australia	12	H4
Barcelona, Spain	3	G7
Barcoo R., Australia	12	G4
Bardai, Chad	13	E3
Bareilly, India	5	L7
Barents Sea	5	FG2
Bari, Italy	3	L7
Barlee Lake, Australia	11	B5
Barnaul, U.S.S.R.	5	M4
Baroda, India	5	L7
Barquisimeto, Venezuela	8	C3
Barra, Brazil	8	F5
Barraba, Australia	12	I6
Barranquilla, Colombia	8	C2

Index

Picture acknowledgements: Alan Hutchison 118; Novosti Press Agency 97; Syndication International 12 (above); John Topham 6 (far left), 39, 57 (above), 122 (below), 130 (above), 133 (above), 171; Zefa 9 (top), 100 (above), 133 (below). *Front cover:* Syndication International.